D1546238

TECHNIQUES OF CHEMISTRY

ARNOLD WEISSBERGER, *Editor*

VOLUME XVI

TECHNIQUES OF CHEMISTRY

VOLUME XVI

SEPARATIONS BY CENTRIFUGAL PHENOMENA

HSIEN-WEN HSU

The University of Tennessee

EDITOR:

EDMOND S. PERRY

Research Laboratories
Eastman Kodak Company
Rochester, New York

A WILEY-INTERSCIENCE PUBLICATION

JOHN WILEY & SONS

New York • Chichester • Brisbane • Toronto

Library of Congress Cataloging in Publication Data:

Hsu, Hsien-Wen, 1928-
  Separations by centrifugal phenomena.

  (Techniques of chemistry; v. 16)
  "A Wiley-Interscience publication."
  Includes index.
  1. Centrifugation. 2. Separation (Technology)
I. Perry, Edmond S. II. Title. III. Series.
QD61.T4 vol. 16 [QD54.C4] 542s [543'.083] 81-4991
ISBN 0-471-05564-6   AACR2

Printed in the United States of America

10 9 8 7 6 5 4 3 2 1

*To My Parents
and My Teachers,
Especially RBB and OAH*

INTRODUCTION TO THE SERIES

Techniques of Chemistry is the successor to the Technique of Organic Chemistry Series and its companion—Technique of Inorganic Chemistry. Because many of the methods are employed in all branches of chemical science, the division into techniques for organic and inorganic chemistry has become increasingly artificial. Accordingly, the new series reflects the wider application of techniques, and the component volumes for the most part provide complete treatments of the methods covered. Volumes in which limited areas of application are discussed can be easily recognized by their titles.

Like its predecessors, the series is devoted to a comprehensive presentation of the respective techniques. The authors give the theoretical background for an understanding of the various methods and operations and describe the techniques and tools, their modifications, their merits and limitations, and their handling. It is hoped that the series will contribute to a better understanding and a more rational and effective application of the respective techniques.

Authors and editors hope that readers will find the volumes in this series useful and will communicate to them any criticisms and suggestions for improvements.

The editor of the volume and I gratefully acknowledge the assistance of Professor Howard Brenner of the University of Rochester in the editorial work.

ARNOLD WEISSBERGER

Research Laboratories
Eastman Kodak Company
Rochester, New York

PREFACE

Separations using centrifugal phenomena have been widely used in laboratories and industries for many years. Such methods grew from nineteenth century farm-oriented milk separation to a fundamental technique for determining molecular properties of either natural or synthetic polymers in laboratories and to a basic unit operation component in modern industries today.

Over the last twenty-five years, important contributions to the general understanding of centrifugal separation and ultracentrifugal analysis have appeared in the literature. These contributions are both theoretical and experimental, covering widely diversified fields. Very little effort seems to have been made toward the aims of organizing all this new information in a coherent fashion. The scientific information is scattered in a variety of technical publications; keeping track of it is a difficult task, which is not made simpler by the fact that even the nomenclature has not been standardized.

The objective of writing this book has been to organize this technical information in as coherent a way as possible while emphasizing uniqueness in methodology and pertinent theory for various kinds of centrifugations. In this attempt a framework of transport phenomena to bring together some of the principles of centrifugal separations has been used.

The topics in the book are arranged according to gas, liquid, and solid phases, in the order of increasing phase densities. Much space has been devoted to liquid centrifugation because it is now one of the most basic and valuable techniques in chemical and biological laboratories. The proper application of many centrifugation techniques requires considerable expertise. It was realized that many of the pertinent details scattered in the literature for the laboratory techniques had to be drawn together first, so that the laboratory researchers as well as the novice would benefit from these modern technologies. In addition, many separational and characterizational examples are illustrated in details. It is hoped that this volume complements those in other volumes in the *Techniques of Chemistry* series.

Many of my former colleagues of the Molecular Anatomy (MAN) Program at Oak Ridge National Laboratory have contributed either directly or indirectly to the present treatise. I acknowledge gratefully the contributions of the following: Dr. J. W. Holleman, who read the entire draft and made numerous corrections and suggestions and whose friendship provided valuable encouragement; Messrs. J. N. Brantley, R. E. Canning, L. H. Elrod, W. W. Harris, L. C. Patrick, C. T. Rankin, Jr., and D. D. Wills, for their assistance and advice; the sharing of their extensive knowledge has been invaluable to me in many respects; Dr. N. G. Anderson, under whose leadership zonal centrifuges have been developed and who provided a scholarly atmosphere when I worked at the MAN Program on a part-time basis from October 1967 to January 1973.

I also wish to acknowledge the generous support and encouragement of the Department of Chemical, Metallurgical and Polymer Engineering of the University of Tennessee–Knoxville in providing the clerical staff and supplies for the preparation of the manuscript; Dr. H. F. Johnson, who gave constant encouragement and support in the preparation of the manuscript; Dr. E. E. Stansbury, who gave many constructive suggestions; Mrs. Kay Davis, Mrs. Sally O'Connor, and Mrs. Janet Deurlein, who typed the original and revised manuscripts; and Mr. T. H. DeHart, who helped in many photographic problems.

In writing a book such as this, it was necessary to use some data from the manuals of manufacturers and materials from many publications. I thank the following for giving permission to use their publications: Academic Press, Inc.; American Chemical Society; Americal Elsevier Publishing Co.; American Institute of Chemical Engineers; American Instrument Co.; American Society for Microbiology; Beckman Instruments Co., Spinco Division; British Biochemical Society; Butterworth Publishers, Inc.; Cambridge University Press; Department of Energy, Electro Nucleonics, Inc.; Instrumentation Specialities Co.; John Wiley & Sons, Inc.; Marcel Dekker, Inc.; McGraw-Hill Book Co.; Oak Ridge National Laboratory; The Rockefeller University Press; Union Carbide Corp.; W. B. Saunders Company; and The Wistar Institute of Anatomy and Biology.

I also thank Dr. R. K. Genung, Oak Ridge National Laboratory, who on many occasions has assisted me in the preparation of manuscripts; and finally, my wife Cecilia, who provided a comfortable atmosphere and also shared with me all the burdens that devolve upon an author's wife.

HSIEN-WEN HSU

Knoxville, Tennessee
June 1981

CONTENTS

Chapter III

CHAPTER IV

CHAPTER V

CHAPTER VI

CHAPTER VII

APPENDIX A

APPENDIX B

APPENDIX C

APPENDIX D

TECHNIQUES OF CHEMISTRY

ARNOLD WEISSBERGER, *Editor*

VOLUME XVI

Chapter I

INTRODUCTION

Separation by centrifugal phenomena is based on the transfer of materials from one phase to another by mechanical means utilizing differences in particle density and size in mixtures under an applied centrifugal force field.

Separation of immiscible liquids and insoluble particles has been occurring in nature since the beginning of the universe. The conscious application of centrifugal force to aid separation, however, is recent. Centrifugal separators began finding applications in the process industries as late as about 100 years ago. The earliest uses were in sugar manufacture and in separating cream from milk. The first continuous centrifugal separator was probably invented in 1877 by a Swedish engineer, Dr. Carl Gustaf Patrik DeLaval, in order to separate cream from milk. This hand-cranked farm tool was disclosed in an English patent in 1878 and a corresponding U.S. patent in 1881 [1]. The applications in the sugar and the dairy industries still account for a substantial number of the new separators made each year, but the number of other applications is constantly increasing. Types of centrifugal separators also proliferated from the original two to a wide diversity of machines.

Industrial centrifuges underwent rapid development during World War II, notably in the field of isotope separation or enrichment by gas centrifuges. A pilot plant-scale ultracentrifuge to separate ^{235}U by centrifugal force has been developed [2]. Centrifuges operated by remote control have been used in the isolation of plutonium [2]. The preparation of blood plasma, the concentration of rubber latex, and the separation of penicillin solvents are only a few of many other new uses of the centrifugal method.

The recent population explosion has presented us with various new problems, for example, water pollution, industrial wastes, shortage of resources, and so forth. Centrifuge separators have also been successfully applied to these new problems. For instance, centrifuges are used to purify and recover oils, coolants, and hydraulic fluids in factories, thus saving industry hundreds of thousands of dollars and helping conserve our diminishing resource of energy-providing petroleum. Using centrifuges in recycling operations minimizes handling, hauling, and disposal costs for used lubricants. In the oil industry, centrifuge separators recover otherwise wasted "slop" oil. Centrifuges are used on board ship and on oil platforms to remove salt water and contaminants from fuels and lubricating oils.

In the food industries, centrifuges are used in many operations, for example, in the centrifugal extraction of edible proteins from both animal and vegetable products. In wineries and breweries, centrifugal separators clarify the liquids by removing yeast and other solid matter. Centrifuges, many with programmable controls, are increasingly used to process pharmaceutical preparations. Impurities in coffee, tea, and fruit juices, for example, citrus juices, are removed with the aid of centrifuges. In pulp and paper mills, centrifuges help in producing tall oil, which in turn is used in the production of coatings and soaps and in many chemical processes. To help agricultural producers in their efforts to keep pace with the world's food requirements, centrifuges dewater organic liquids that are used in the production of fertilizers.

Industrial and municipal wastes are treated with the help of centrifuges to dewater sludge, thus aiding in the preservation of water resources. In mining operations, centrifuges recover minerals, such as uranium, with the aid of a new liquid ion-exchange method. In steel mills, centrifuges reclaim and recycle lubricating oils and thus play an important role in a manufacturing process that is basic to nearly every industry. Centrifuges are also used in the production of television tubes to recover rare earth phosphors that might be lost as waste.

In cheese plants, centrifuges separate small curd fines from whey to recover cheese that would otherwise be lost. Silver halide coatings used in photographic film and paper are recovered from wash water and put back into production with the aid of centrifuges. Thus, centrifuges are tools with a thousand uses.

Common industrial centrifugal separators handle tons of material at a moderate cost. The current operating conditions of these separators are as follows: temperatures from -62 to $230°C$; absolute pressures from 5 mm Hg to about 1000 kPa; particles sizes from to 1 μ in diameter or less up to 0.65 cm.

Centrifuges are not, of course, the answer to all separation problems. In liquid–liquid separations, the specific gravities of the phases must normally differ by a minimum of 3%. In removing solids, absolute clarity of the effluent liquor can rarely be expected. Often, 1 to 5% of the solids fed to the machine remain in the discharged liquid. Continuous and automatic filtering centrifuges are limited to relatively coarse, free-filtering solids.

Advances in centrifuge technology are not limited to large-scale industrial centrifuges. Small centrifuges are also used extensively in laboratories for research and development. Since the late Professor T. Svedberg began to apply centrifugal force for the study of colloid systems in the 1920s, progress in laboratory centrifugations has continued in two directions. The first line of development has produced the analytical ultracentrifuge, which incorporates an optical system for the analysis of a small amount of material which is being centrifuged through a homogeneous medium (the solvent) under ideal conditions. The analytical centrifuge is now widely used not only to characterize molecular weight but also to give insight into the size, shape, density, and the base composition and activity of biopolymers and other macromolecules. The second line of development has produced the preparative centrifuge, which uses centrifugal force to sediment the solid phase of materials. The preparative centrifuge is a simpler instrument and does not have an optical system. However, recent rapid technical progress and development permit the preparative centrifuge not only to completely separate several or all components in a mixture but also to perform analytical measurements. This versatile technique is known as density-gradient centrifugation.

The density gradient method involves a supporting column of fluid whose density increases from top to bottom of the centrifugal bottle. The extreme usefulness of centrifugal systems with swing-out buckets in density-gradient centrifugal separation has been well proved. However, the relatively small volumes that can be handled in high centrifugal fields, together with the mechanical handling procedures, impose certain limitations on the technique and the resolution. To eliminate these limitations, N. G. Anderson and his co-workers at Oak Ridge National Laboratory developed a series of density-gradient centrifuge rotors for the mass separation of subcellular particles including viruses. These centrifuges are known as "zonal centrifuges" [3].

1 CLASSIFICATION AND APPLICATION

Industrial centrifuges may be classified in several ways, but almost all of them fall into one of two classes. In one class are the machines that separate by sedimentation and depend on the difference in density between phases. The other class contains filtration machines, in which liquid is forced through a filter medium by centrifugal action. The former is generally classified as a *centrifugal settling machine*, the latter as a *centrifugal filter*. A few machines perform both functions in a single unit. Each class may be subdivided into various types, depending on the configuration of the machines.

The centrifugal settling machine may be further subclassified into three types, each with a broad field of application. These are (1) tubular type, (2) disk type, and (3) decanter. The centrifugal filter may be subclassified into (1) basket type, (2) push type, (3) screw-conveyer type, and (4) self-discharge type. In each type are machines that discharge solids intermittently or continuously.

Centrifugal Settling Machines

This class contains high-speed, high-force separators with manual removal of accumulated solids; high-speed separators with continuous or periodic discharge of a slurry or sludge; moderate-speed units with continuous discharge of sludge; and slow-speed, large-diameter separators with intermittent solids removal.

High-speed sedimentation centrifuges with manual solids discharge include tubular and disk machines. Their uses include liquid–liquid separations, clarification of liquids by the removal of small amounts of solids, concentration of emulsions, classification of fine solids, and the partial separation of gases of differing molecular weight.

Industrial tubular settling centrifuges measure approximately 10 to 15 cm (4 to 6 in.) in diameter, rotate at speeds up to 15,000 rpm, and generate centrifugal forces as high as 16,000 g. They are simple in construction, easy to clean, and have the best operating characteristics of any centrifuge on viscous liquids. Typical applications include purification of lubricating and fuel oils; harvesting of bacteria; clarification of nitrocellulose dope and molten chicle; dewaxing of petroleum residual stocks in hydrocarbon solution; and recovery of finely divided metal particles such as silver from film scrap, platinum from spent catalyst, and germanium from forming and sawing operations. Throughput normally ranges from 4 to 40 liters (1 to 10 gallons) of liquid per minute. A modified tubular settling centrifuge containing several concentric annular chambers connected in series is used for relatively simple clarification problems. Its chief application is in clarifying brewer's wort at rates up to about 120 liters (30 gallons) per minute. A sketch of a tubular settling centrifuge is presented in Fig. 1.1a.

The solid-wall disk centrifuge was first developed as a milk separator and still finds wide application in this service. It is also useful for concentrating other emulsions, such as natural and synthetic rubber latexes. It finds a wide variety of applications that overlap the applications of tubular machines to some extent. Typical uses include the purification of lubricating and fuel oils; separation of wash water from fats in refining vegetable oils, fish oils, and whale oil; separation of acid sludge from the acid treatment of petroleum stocks; separation of solvent extracts from fermentation broths; and removal of water from jet fuel. Industrial disk settling centrifuge separators range from 15 to 76 cm (6 to 30 in.) in inside diameter, with liquid throughputs as great as several hundred gallons per minute on easy separations. Some disk centrifuges include openings in the bowl wall for periodic discharge of accumulated solids. These openings may be individually valved ports that are self-actuated by the accumulation of solids or externally actuated from a hydraulic circuit, or they may be peripheral slots which are automatically uncovered at intervals by hydraulic action. Self-actuating valve bowls are used in recovering wool grease from scouring liquors. Externally actuated valve bowls find application when the flow rate required for satisfactory clarification is low and the density of the separated sludge is not much greater than that of the liquid. This type of settling machine is used to remove excess pulp from pineapple and orange juices.

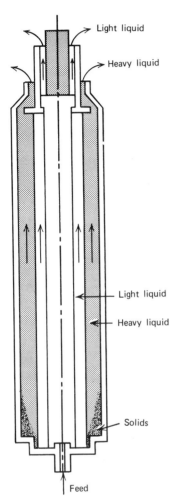

Light liquid

Heavy liquid

Light liquid

Heavy liquid

Solids

Feed

Fig. 1.1 Schematic diagrams of settling centrifuge machines (*a*) Tubular type (*left*). (*b*) Disk type. (*c*) Decanter type (*see p. 6*).

Slot-discharge centrifuges, with their large openings, can handle a wide variety of relatively coarse solids. They have been used in clarifying fruit and vegetable juices, concentrating precipitated proteins, and removing leafy particles from carnauba wax.

Disk settling centrifuges may be equipped with peripheral nozzles measuring 0.01 to 0.25 cm in diameter which discharge a concentrated slurry and may simultaneously separate one liquid phase from another. Strainers are usually included on the feed and recycle systems to remove solid particles that are larger than half the diameter of the nozzles. The uses of the nozzle-type settling centrifuge include the concentration and washing of cornstarch and

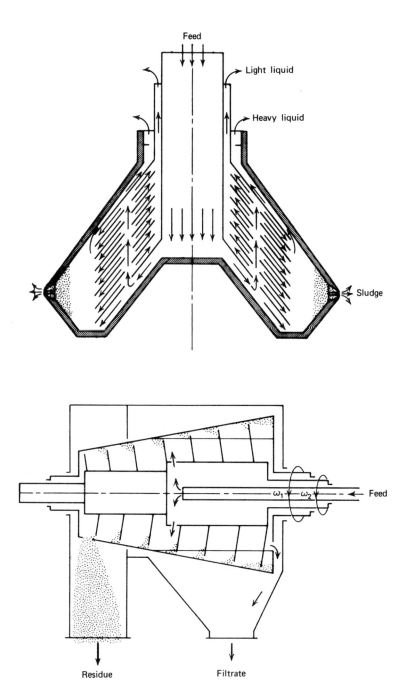

Feed

Light liquid

Heavy liquid

Sludge

ω_1 ω_2 Feed

Residue

Filtrate

Fig. 1.1 (*b*) and (*c*).

gluten, clarification of wet-process phosphoric acid, purification of heavy fuel oils, recovery of citrus oils from peel wash liquid, and a concentration of yeast from fermenter broths. A simple disk settling centrifuge is depicted in Fig. 1.1b.

Solid-wall settling centrifuges are often provided with internal conveyors that move the settled solids along the bowl wall, out of the pond of liquid, and out through slots at one end of the bowl. Liquid escapes through overflow ports in the cover plate at the other end of the bowl. The bowls may have a horizontal or vertical axis of rotation and may be conical, cylindrical, or part cylindrical and part conical. The conveyor rotates more slowly than the bowl, with a speed differential of $\frac{1}{20}$ to $\frac{1}{160}$ of the bowl speed. Typical applications include liquid removal from poly vinyl chloride and polyolefins, classification of dispersed clay to remove particles larger than 3 to 4 μ in diameter, recovery of supported metal catalysts from organic reactions, removal of solids from sewage and industrial wastes, dewatering of coarse crystals of potassium chloride and sodium chloride, and the dewatering of coal. Slow differential speeds are needed with easily disintegrated solids such as in separating meat tissue from melted animal fat. The moisture content of the discharged solids depends on the particle size and particle size range. A representative decanter settling centrifuge of this type is schematically shown in Fig. 1.1c.

Filtering Centrifuges

In contrast to the settling centrifugal machines, the centrifugal filter exists in a wide variety of forms for fully continuous, automatic intermittent, and batch operation. The centrifugal filter is normally used to remove a particulate solid phase from a fluid slurry. This is frequently followed by a displacement wash that is applied to the solid phase for reducing the impurity contents of the mother liquor. During operation, the solid phase is supported on a screen or some other permeable membrane, such as metal cloth, perforated metal plates, or fabrics of various materials, through which liquid passes. A cake 2 to 20 cm (1 to 7 in.) thick is deposited. The cake may be washed and spun dry, after which the basket is slowed down and the solids unloaded with a knife or, with some materials, by hand. If the solid phase is the desired product and the liquid is water, a centrifugal filter may be properly called a centrifugal dehydrator.

The difference between the centrifugal settling machine and the filtering centrifuge is that the former has a solid wall, whereas the latter is built with a perforated wall. The schematics of four basic types of filtering machines are depicted in Fig. 1.2.

2 LABORATORY CENTRIFUGES

Small centrifuges used in laboratories for research and development and in pharmaceutical industries for separation and purification may be classified in-

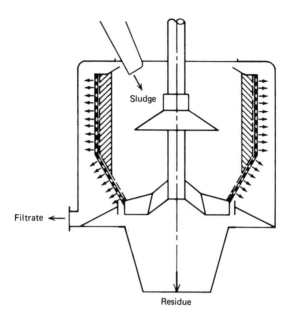

Fig. 1.2   Schematic diagrams of centrifugal filters. (*a*) Basket type (*above*). (*b*) Multistage push type (*right*). (*c*) Screw-conveyer type. (*d*) Self-discharge type (*see p. 10*).

to two categories, *analytical* and *preparative* centrifuges. An analytical centrifuge may contain only 1 ml or less solution, and the progress of sedimentation is followed by means of an optical system. The individual fractions are not normally separated at the end of the run, the object being simply to observe their behavior in the centrifugal field. A preparative centrifuge generally involves larger quantities of material, the individual fractions of which are required for other purposes. For convenience, centrifuges may be placed into one or the other category, although the techniques for which they are employed do not fall rigidly into this classification. That is, analytical investigations may be conducted with a preparative centrifuge, and it is possible, by using separation cells in the analytical ultracentrifuge, to prepare microquantities of material, say, at the nanogram level. The ultracentrifuge applies a high centrifugal force field to the separation and concentration of very small soluble or insoluble particles, down to those of molecular dimension.

The Analytical Ultracentrifuge

The analytical centrifuge is a very sophisticated piece of equipment and has an accurate temperature control and advanced optical systems which make it a very precise analytical instrument. Analytical rotors are used to study optically the sedimentation of samples during centrifugation. The centrifuges are com-

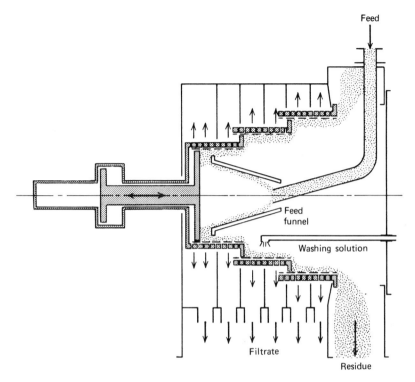

Fig. 1.2 (*b*).

posed of a rotor body, cell holders, quartz or sapphire windows, and sector-shaped compartments. The football-shaped rotors spin about the shorter L/D (length or diameter) axis. The center section of the rotor is similar to DeLaval high-strength turbine wheels where special contours permit higher-than-normal peripheral speeds with high stress loads. A typical analytical ultracentrifuge rotor is shown in Fig. 1.3*a*.

The preparative centrifuge is generally less complex and usually lacks an optical system. For this reason, it has not, until recently, been considered adequate for analytical work. Now, with the advent of tube-piercing devices and photoelectric scanning systems as well as advances in density-gradient centrifugation, preparative centrifuges are used also for analytical work. On the preparative scale, purity is often, to some extent, sacrificed for the sake of quantity.

Preparative centrifuges may be further classified according to the type of rotors used, namely, *swinging bucket rotors, fixed angle-head rotors,* and *zonal and continuous-flow rotors.* Swinging bucket and fixed angle-head rotors are used for handling samples in tubes.

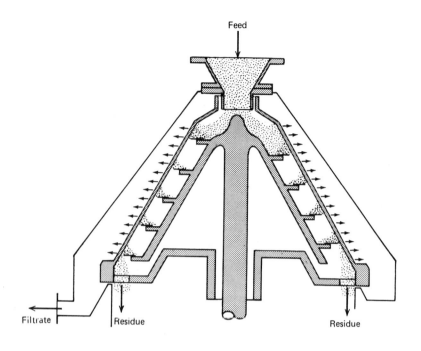

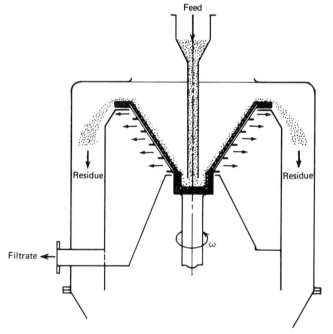

Fig. 1.2 (*c*) and (*d*).

Swinging Bucket Rotors

The swinging bucket centrifuges have preparative-size rotors with movable tubes. The tubes are loaded and unloaded in a vertical position. During centrifugation, the tubes swing 90° upward and seat firmly against the shoulder of the rotor in a horizontal position as the rotor accelerates. The rotor is similar to a spoked wheel with turning mounts on the spoke ends for supporting the buckets that hold the centrifuge tubes. This type of rotors is depicted in Fig. 1.3*b*.

Angle-Head Rotors

Fixed angle-head rotors have tubes which remain in one position at an angle during the run. The preparative angle-head rotors in common use are simple, rugged units (shown in Fig. 1.3*c*) suitable for differential or zonal isopycnic centrifugation. They are primarily used for pelleting. Angle-head centrifuges are composed of (a) rotor body, (b) top, (c) handle, and (d) tube with cap. Commercial models are obtainable with a variety of capacities, speed ranges, and angles of inclination. They are characterized by their solid construction and by their resemblance to the frustum of a cone. This gives rotational stability and permits high loads to be carried by the center section of the rotor. The tubes in which the fluid is centrifuged are inclined at different angles for high efficiency, for minimum stirring, or for other reasons, such as the amount to be centrifuged and the technique to be used.

Zonal and Continuous-Flow Rotors

These rotors are cylindrical, and the rotor volume therefore varies with the square of the radial distance from the center of rotation. The cylindrical cavity

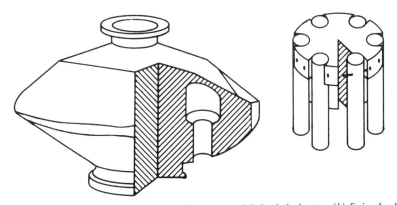

Fig. 1.3 Cross sections of laboratory centrifuge rotors: (*a*) Analytical rotor. (*b*) Swing-bucket rotor (*above*). (*c*) Angle-head rotor. (*d*) Tubular clarifier rotor (*see p. 12*). (*e*) Zonal centrifuge rotors (*see p. 13*). *Source:* Reproduced from Ref. 3, p. 79, Fig. 1, with the permission of the editor.

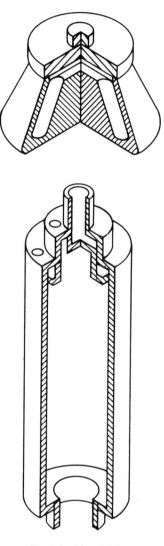

Fig. 1.3   (c) and (d).

of the bowl is divided into sector-shaped compartments by vanes attached to the core; the rotor is enclosed by a threaded lid. A rotating seal assembly, which may be fixed or removable, allows fluid to be pumped in and out of the cavity while the rotor is either spinning or at rest. Continuous-flow rotors of the clarifier type (Fig. 1.3d) have internal dams for separating different strata of a flowing stream. These rotors operate at high rotational speeds, high cen-

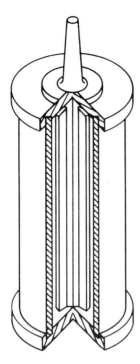

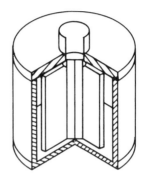

Fig. 1.3 (*e*).

trifugal fields, and large L/D ratios. The long rotor length provides adequate distance and time for a radial separation of different materials to be achieved. Zonal rotors can be made with large or small L/D ratios and still perform the same function (shown in Fig. 1.3*e*). With the same fluid volume, rotor material, and stress level, the large-L/D rotors produce high centrifugal fields with a sacrifice in radial sedimentation path. However, the small-L/D rotors operate with a large radial sedimentation path while producing smaller centrifugal fields. The separation desired dictates the rotor speed necessary and, indirectly, the L/D ratio used for a given volume. Use of different internal cores in the zonal centrifuge converts the rotor to one or more of the conventional centrifuges described above.

The selection of a rotor for a specific purpose is usually made on the basis of particle size and density, sample concentration, sample volume, and the type of separation to be performed. Physical specifications of Beckman swinging bucket and fixed angle-head rotors are presented in Tables 1.1 and 1.2, respectively. The first zonal centrifuge was built in 1954 by N. G. Anderson at Oak Ridge National Laboratory. Over 50 different zonal centrifuge rotor designs have been developed and evaluated. These have been grouped into a

Table 1.1 Physical Specifications

| | | Maximum Force and k | | | | | |
| | | 75,000 rpm | | 65,000 rpm | | 50,000 rpm | |
Type[a]	Rotor	Max. g	k	Max. g	k	Max. g	k
1. High-performance rotors	SW-65 Ti 65,000 rpm	420,000	46	420,000	46	249,000	78
	SW-60 Ti 60,000 rpm	485,000	45	485,000	45	337,000	65
	SW-50.1 50,000 rpm	300,000	59	300,000	59	300,000	59
2. Long, slender tubes	SW-41 Ti 41,000 rpm	286,500	125	286,500	125	286,500	125
	SW-40 Ti 40,000 rpm	284,000	137	284,000	137	284,000	137
	SW-27 1[c] 27,000 rpm	135,000	310	135,000	310	135,000	310
3. Big volume rotors	SW-27[c] 27,000 rpm	131,000	265	131,000	265	131,000	265
	SW-25.1 25,000 rpm	90,000	338	90,000	338	90,000	338
	SW-25.2 25,000 rpm	107,000	335	107,000	335	107,000	335

Source: Spinco DS-468, by O. M. Griffith (by permission of the Beckman Spinco Division).

[a] 1: High-performance rotors are designed to separate proteins, enzymes, and hormones from 1 to 20 S. Particles should have a difference of 2 to 4 S for effective separation. 2: Rotors with long tubes are designed to separate DNA, polysomes, ribosomes, and ribosomal subunits from 20 to 100 S. Particles should have a difference of 5 to 10 S for effective separation. 3: Large-volume rotors are designed to separate whole cells, subcellular organelles, and viruses of 80 S and above.

[b] A useful concept in rotor performance or selection is the k and k' factors. Those factors may be used to compare the efficiency of various rotors for the material that will be under investigation. The k factor provides an estimate of the time t (in hours) re-

of Beckman Swinging Bucket Rotors

Factor[b]					
40,000 rpm		k' Factors	Number Tubes,	Rotor Capacity	
Max. g	k	($\rho=1.1 \sim 1.9$)	Volume/Size	(ml)	Comments
159,000	122	261–109	3 × 5.0 ml $\frac{1}{2}$ × 2 in. 1.3 × 5.1 cm	15.0	Highest-speed swinging bucket rotor; 3 samples
215,000	102	288–123	6 × 4.4 ml $\frac{7}{16}$ × $2\frac{3}{8}$ in. 1.1 × 6.03 cm	26.4	Highest-force swinging bucket rotor to date; 6 samples
192,000	93	342–140	6 × 5.0 ml $\frac{1}{2}$ × 2 in. 1.3 × 5.1 cm	30.0	High performance and force for 50,000 rpm instruments
272,700	132	689–289	6 × 13.2 ml $\frac{9}{16}$ × $3\frac{1}{2}$ in. 1.4 × 8.9 cm	79.2	Highest forces for long, slender tubes
284,000	137	754–318	6 × 14.0 ml $\frac{9}{16}$ × $3\frac{3}{4}$ in. 1.4 × 9.5 cm	84.0	Slightly longer tubes
135,000	310	1709–724	6 × 17.0 ml $\frac{5}{8}$ × 4 in. 1.6 × 10.2 cm	102.0	Longest small-diameter tubes
131,000	265	1492–622	6 × 38.5 ml 1 × $3\frac{1}{2}$ in. 2.5 × 8.9 cm	231.0	Biggest-volume swinging bucket rotor; large samples
90,000	338	1881–791	3 × 34.0 ml 1 × 3 in. 2.5 × 7.6 cm	102.0	Early rotor; compare with Types SW-27, SW 27.1, SW-40 Ti, and SW-41 Ti
107,000	335	1862–783	3 × 60.0 ml $1\frac{1}{4}$ × $3\frac{1}{2}$ in. 3.2 × 8.9 cm	180.0	Early rotor; compare with Type SW-27

quired to pellet a particle of known sedimentation coefficient S (in Svedberg units) at the maximum speed of the rotor. $t = k/S_{20,w}$. By knowing the k factor for the rotor and the sedimentation coefficient of the particle, it is possible to estimate the pelleting time if the run is made at 20°C in aqueous solution. The k' factor is an estimate of the time required to move a zone of particles to the bottom of a centrifuge tube through a 5 to 20% w/w linear sucrose gradient (the most frequently used sucrose gradient) at the maximum rotor speed.
[c] Buckets are interchangeable; alternate bucket set available for either rotor.

Table 1.2 Physical Specifications

		Maximum Force and					
		75,000 rpm		*65,000 rpm*		*50,000 rpm*	
Type	*Rotor*	*Max.* g	k	*Max.* g	k	*Max.* g	k
High-performance rotors, small samples	75 Ti 75,000 rpm	503,500	35	378,225	46	223,802	96
	70 Ti 70,000 rpm	504,000	44	435,000	55	257,000	93
	65 65,000 rpm	368,800	45	368,800	45	218,200	76
	50.2 Ti 50,000 rpm	303,000	72	303,000	72	303,000	72
	50 Ti 50,000 rpm	226,600	77	226,600	77	226,600	77
High-performance rotors, big sample volume	60 Ti 60,000 rpm	326,600	63	326,600	63	251,800	90
	42.1 42,000 rpm	195,000	134	195,000	134	195,000	134
	35 35,000 rpm	142,800	225	142,800	225	142,800	225
Special applications	40.2 40,000 rpm	143,100	131	143,100	131	143,100	131
	40.3 40,000 rpm	143,200	81	143,200	81	143,200	81
	30.2 30,000 rpm	95,400	113	94,500	113	94,500	113
Initial processing and general-purpose rotors	50 50,000 rpm	198,700	66	198,700	66	198,700	66
	40 40,000 rpm	145,000	120	145,000	120	145,000	120
	30 30,000 rpm	105,700	209	105,700	209	105,700	209
	21 21,000 rpm	59,200	398	59,200	398	59,200	398
	19 19,000 rpm	53,700	776	53,700	776	53,700	776

Source: Spinco DS-468, by O. M. Griffith (by permission of the Beckman Spinco Division).

of Beckman Fixed Angle-Head Rotors

40,000 rpm Max. g	k	Number Tubes, Volume/Size	Rotor Capacity (ml)	Comments
143,234	122	8 × 13.5 ml ⁵⁄8 × 3 in. 1.6 × 7.6 cm	108	High-force rotor for small sample volume
164,000	146	8 × 38.5 ml 1 × 3¹⁄8 in. 2.5 × 8.6 cm	308	Highest-force fixed angle-head rotor, large volume
139,700	118	8 × 13.5 ml ⁵⁄8 × 3 in. 1.6 × 7.6 cm	108	High-force fixed angle rotor for 65,000 rpm instruments
193,000	113	12 × 38.5 ml 1 × 3¹⁄2 in. 2.5 × 8.9 cm	462	Highest-force rotor for 50,000-rpm instruments, large volume
145,000	120	12 × 13.5 ml ⁵⁄8 × 3 in. 1.6 × 7.6 cm	162	High-force rotor for 50,000-rpm instruments
161,100	141	8 × 38.5 ml 1 × 3¹⁄2 in. 2.5 × 8.9 cm	308	High force/volume for 65,000-rpm instruments
177,000	148	8 × 38.5 ml 1 × 3¹⁄2 in. 2.5 × 8.9 cm	308	High force/volume for 40,000-rpm instruments
142,800	225	6 × 94.0 ml 1¹⁄2 × 4 in. 3.8 × 10.2 cm	564	Largest-capacity high-performance rotor
143,100	131	12 × 6.5 ml ¹⁄2 × 2¹⁄2 in. 1.3 × 6.4 cm	78	Extremely shallow tube angle; for low concentrations
143,200	81	18 × 6.5 ml ¹⁄2 × 2¹⁄2 in. 1.3 × 6.4 cm	117	High forces for flotation runs
94,500	113	20 × 10.5 ml ¹⁄2 × 3¹⁄2 in. 1.3 × 8.9 cm	210	Steepest tube angle, small-diameter tubes; for flotation runs
127,100	103	10 × 10 ml ⁵⁄8 × 2¹⁄2 in. 1.6 × 6.4 cm	100	Early rotor; compare with Types 50 Ti, 65, and 75 Ti
145,000	120	12 × 13.5 ml ⁵⁄8 × 3 in. 1.6 × 7.6 cm	162	Identical to Type 50 Ti, except for lower speed and forces
105,700	209	12 × 38.5 ml 1 × 3¹⁄2 in. 2.5 × 8.9 cm	462	Early rotor; compare with Types 60 Ti and 42.1
59,200	398	10 × 94.0 ml 1¹⁄2 × 4 in. 3.8 × 10.2 cm	940	Early rotor; compare with Type 35
53,700	776	6 × 250-ml bottles	1500	Big volume for initial processing of cell particulates, other large particles

Table 1.3 Zonal Rotors Developed at Oak Ridge National Laboratory [4]

Description	Capacity (ml)	Speed (rpm)	Max. g	Construction Material
Series A				
I Sector-shaped Pyrex tubes for initial-rate zonal studies	68 (2 tubes)	3,000	2,500	Pyrex
II Gradient distributing rotor	250 (6 tubes)	2,200	1,100	Pyrex–brass–Al
III First zonal rotor. Spray injection and recovery	4000	2,000	939	Al
IV Demonstration of zonal centrifugation at higher speeds	625	18,000	23,500	Al
V Study of gradient behaviour at low speed. transparent rotor	1300	1,000	152	Lucite–Al
VI Large-scale nuclear isolation	3000	6,000	7,100	Lucite–Al
VII First reorienting gradient rotor	127	3,000	292	Lucite
VIII Reorienting gradient rotor	1450	6,000	2,817	Al
IX Transparent low-speed zonal rotor	1300	4,200	3,200	Lucite–Al
X Reduced-volume version of A-IX	300	4,200	3,200	Lucite–Al
XI Prototype low-speed analytical zonal centrifuge rotor	40 (2 tubes)	4,100	3,320	Steel–Lucite windows
XII[a,b] Redesign of A-IX to minimize mechanical deficiencies	1300	4,200	3,200	Lucite–Al
XIII Reduced-volume version of A-XII	350	4,200	3,200	Lucite–Al
XIV Continuous-flow version of A-XII	1300	4,200	3,200	Lucite–Al
XV 2nd continuous-flow version of A-XII	1300	4,200	3,200	Lucite–Al
XVI Reorienting-gradient rotor for nuclei	720	4,000	831	Lucite
XVII Rapid-batch nucleus isolation rotor	660	4,000	831	Lucite
XVIII[a,b] Redesign of A-XII to eliminate seals, using free-stream and scoop principle	1345	4,200	3,200	Lucite–Al
Series B				
I Prototype used in stability studies	1450	8,000	(Designed for 40,000)	Al

18

II	Initial high-speed studies; rotor unstable	1625	31,000	54,500	Al
III	Attempted development of high-speed reorienting rotor; unstable	1450	1,500	176	Al
IV[c]	First successful high-speed dynamically loaded zonal rotor	1693	40,000	90,900	Al
V	High-performance continuous-flow-with-pelleting rotor	128	40,000	90,900	Al
VI	Continuous-flow-with-banding rotor without septa; unsuccessful	1338	36,000	73,600	Al
VII	Continuous-flow-with-banding rotor without septa; unsuccessful	1012	36,000	73,600	Al
VIII	First successful continuous-sample-flow-with-banding rotor	932	36,000	73,600	Al
IX[c]	Improved version of B-VIII	754	40,000	90,900	Al
X	Prototype removable-seal zonal rotor	660	30,300	68,000	Al, steel
XI	Larger prototype removable-seal zonal rotors	1725	24,800	52,600	Al, steel
XII	Maraging steel B-IV	1693	Designed for 50,000; driven 46,000 by Spinco drive	142,240 at 50,000 rpm	Maraging steel
XIII	Skipped				
XIV[a,b,c]	Removable-seal zonal rotor	630	30,000 Al / 37,000 Ti	65,000 Al / 102,000 Ti	Al, Ti
XV[a,b,c]	Removable-seal zonal rotor	1665	24,000 Al / 28,000 Ti	57,000 Al / 77,000 Ti	Al, Ti
XVI[c]	B-IX with centrifuge valve	754	40,000	90,900	Al
XVII	Floater-trap version of B-XVI	754	40,000	90,900	Ti
XVIII	B-XV with windows				
XIX	Continuous-flow B-XIV	Not completed			
XX	Continuous-flow B-XV				
XXI	Gradient distributing rotor for 12-place angle head (30,000 rpm)	—	2,000	—	
XXII	Valve, trap, and other modifications	Not completed			
XXIII	Edge or center unloading B-XV	1450	24,000	57,000	Al
XXIV	B-XV for nuclei	1665	28,000	77,000	Ti
XXV	B-XV for DNA	1665	28,000	77,000	Ti
XXVI	B-IX reorienting gradient rotor	754	40,000	90,900	Al

19

Table 1.3 *(Continued)*

	Description		Capacity (ml)	Speed (rpm)	Max. g	Construction Material
XXVII	B-XIV for sedimenting particles to wall; supernatant removed by air pressure	Total Pellet	630 100	37,000	102,000	Ti
XXVIII	Sinker, swimmer, floater flu cleanup rotor		1700	5,000	—	Al–Ti
XXIX[a,b]	Detachable seal rotor for both center and edge unloading		1450	28,000	77,000	Ti
Series K						
I	Small-scale continuous-sample-flow-with-banding rotor		240	50,000	—	Ti
II[d]	Production scale continuous-sample-flow-with-banding for vaccine production		3600	35,000	83,637	Al
III	Titanium version of K-II		2340			Ti
IV[d]	Rate core for K-II		6980	35,000	83,637	Al
Series J						
I[d]	Small-scale continuous-sample-flow-with-banding rotor		825	100,000	994,000	Ti

Source: Q. Rev. Biophys. **1** (3), 217 (1968), with modifications.

[a] Available from MSE, London, England.

[b] Available from International Equipment Company, Needham Heights, Mass.

[c] Available from Spinco Division of Beckman Instruments, Palo Alto, Calif.

[d] Available from Electro-Nucleonics, 368 Passaic Avenue, Caldwell, N.J.

series of classes designated by letters, with each design in a given class designated by a Roman numeral, followed in some instances by an indication of the material of construction. The rotors fabricated in Oak Ridge are also identified by a number that indicates serially the number of the rotor of that type made. The following seven rotor series are of general interest:

A series.   Low-speed rotors for the separation of particles in the range of sizes visible in the light microscope (maximum speed circa 6000 rpm, although faster members of this series are under consideration).

B Series.   Intermediate-speed rotors (up to 45,000 rpm) for the separation of particles in the range of sizes visible in the electron microscope.

C Series.   High-speed preparative rotors (up to 150,000 rpm).

D Series.   Rotors for exploring the limits of available high centrifugal fields.

F Series.   Rotors for rapid centrifugal freezing.

K Series.   Rotors and rotor systems for commercial vaccine production and protein purification.

J Series.   Small-scale K series for laboratory use (up to 100,000 rpm).

The properties of a number of these rotors have been summarized by Anderson [4] and are given in Table 1.3. In a sense, K and J series rotors are a refined ultracentrifugal types of industrial and laboratory settling machines.

Rotors are generally made of an aluminum alloy whose surface can be protected by a hard anodic film to prevent corrosion. In recent years, there has been an increase in the use of titanium alloys, which are much more resistant to corrosion, especially by alkali salts, and will withstand much higher centrifugal forces before bursting. However, these rotors are more expensive and considerably heavier than the corresponding aluminum rotors.

References

1.   C. G. P. DeLaval, U.S. Pat. 247, 804 (1881).
2.   H. D. Smyth, *Atomic Energy for Military Purposes,* Princeton University Press, Princeton, N.J., (1945).
3.   N. G. Anderson, "The Development of Zonal Centrifuges and Ancillary System for Tissue Fractionation and Analysis," *National Cancer Institute Monograph 21,* 1966. U.S. Government Printing Office, Washington, D.C. (1966).
4.   N. G. Anderson, *Qt. Rev. Bioph.,1* (3), 217 (1968).

Chapter II

BASIC THEORY OF CENTRIFUGATION

In a gravitational field of force, a particle suspended in a less dense liquid medium tends to migrate through the fluid in a downward direction. For a given medium and fixed external conditions, the rate of this "sedimentation" may depend upon the size as well as the mass and shape of the particles, the difference in density between the particles and the local sustaining liquid density, and the liquid viscosity. Under some circumstances, no sedimentation may be observed at all; this is because any transport of solute particles due to sedimentation may be exactly counterbalanced by transport due to back diffusion. Back diffusion results from the concentration gradient created by the partial sedimentation of the solute particles concerned. However, if it is possible to place the solution in a field of force far stronger than gravity, the solute may sediment at a measurable rate. A force field that is far stronger than the gravitational field can be easily created by centrifugation.

If a particle of mass m is revolving at a radius r with an angular velocity ω, it is subjected to a centrifugal force $\mathbf{F}_c = m\omega^2 r/g_c$ in a radial direction and to a gravitational force $\mathbf{F}_g = mg/g_c$ in a vertical direction. In these equations, g is the acceleration due to gravity and g_c is the factor of proportionality or the unit conversion factor. The ratio of the centrifugal to the gravitational force is a measure of the separating power of the centrifuge; this ratio, Z, is known as the centrifuge effect, or relative centrifugal force:

$$Z = \frac{\mathbf{F}_c}{\mathbf{F}_g} = \frac{\omega^2 r}{g} \tag{2.1}$$

Separations can therefore be carried out very much more rapidly in a centrifuge than under the action of gravity. In addition, the degree of separation which is ultimately obtainable may be very much greater because the forces available are of a far higher order of magnitude. Thus, whereas relatively coarse solids can be separated from a fluid by gravitational sedimentation, very fine solids cannot be separated in this manner, because Brownian motion is sufficient to maintain an appreciable proportion of the particles in suspension. However, if the separating force is greatly increased by the use of a centrifuge, the forces giving rise to mixing become negligible in comparison, so that virtually complete separation is obtained.

A similar situation exists in the draining of mother liquor from a batch of crystals. Some of the liquid is held on the surfaces of the crystals by surface tension forces, and therefore only a limited degree of separation can be achieved by draining under gravity. If centrifugal separation is employed, a far higher proportion of the liquid can be removed and a substantially dry product obtained.

High-speed centrifuges are therefore used for the separation of colloidal suspensions and emulsions which would not separate under the action of the earths's gravitational field.

The basic theory of sedimentation in a centrifugal field has been the subject of many studies since the classical work of Svedberg and Pederson [1] in 1940. Many of the subsequent significant developments have been described recently in two excellent monographs by Fujita [2, 3]. Basic equations describing transport phenomena in centrifugation consist of the continuity equation and the equation of motion. The continuity equation for centrifugation in a cylindrical rotor rotating about its own axis at a constant angular velocity ω was originally derived by Lamm [4] for a closed system consisting of a solvent and a single homogeneous solute by relating the change of concentration with time to the divergence of the flow of solute due to sedimentation and diffusion. However, when one has to deal with more complicated systems or multicomponent systems, the rigorous approach of the thermodynamics of irreversible processes must be used. When more than two transport processes take place simultaneously, they may interfere and produce cross phenomena. Continuity equations for an isothermal, nonreacting multicomponent, and sedimenting system, using the irreversible thermodynamics approach, have been obtained by many investigators [5–11].

In this chapter, the derivation of basic centrifugal equations and of equations coupling diffusion and sedimentation are presented in a rigorous manner and in forms suitable for subsequent discussions.

1 CONSERVATION LAWS

An arbitrary extensive state function (mass, energy, entropy, etc.) $\phi(t)$ is considered. If the value of G valid for a volume element of a continuous system is divided by the volume of this space element, then one obtains an intensive quantity ϕ_v, the density of G, which varies generally in time and in position and does not depend on the system as a whole. It may be represented by the functional derivative

$$\phi_v(x, y, z, t) = \frac{\partial \phi(t)}{\partial V(x, y, z)} \qquad (2.1.1)$$

We may distinguish between two mechanisms for the change in time of $\phi(t)$:

$$\frac{\partial \phi}{\partial t} = P(\phi) + J(\phi) \qquad (2.1.2)$$

The term $P(\phi)$ corresponds to the production per unit time of the quantity ϕ within the volume V. It can be written as a volume integral:

$$P(\phi) = \int \sigma(\phi)\, dV \qquad (2.1.3)$$

where $\sigma(\phi)$ denotes the source of ϕ per unit time and unit volume. The second term in the right-hand side of (2.1.2), $J(\phi)$, represents the flow of the quantity ϕ through the boundary surface A. It can be written as a surface integral:

$$J(\phi) = \int j_n(\phi)\, dA \qquad (2.1.4)$$

The density of flow $\mathbf{j}(\phi)$ associated with ϕ projected along the inside normal to the surface is $j_n(\phi)$. In general, both the source and the flow terms in (2.1.2) may be positive or negative quantities, depending on the direction of flow, usually positive for inward flow and negative for outward flow.

Using (2.1.3) and (2.1.4), one obtains the so-called balance equation corresponding to the extensive variables ϕ as

$$\frac{\partial \phi}{\partial t} = \int \sigma(\phi)\, dV + \int j_n(\phi)\, dA \qquad (2.1.5)$$

In the case of a vectorial extensive variable $\phi(t)$, as, for example, the momentum of the system, the balance equation, (2.1.5), may still be used for each component ϕ_x, ϕ_y, and ϕ_z taken separately.

It is useful to write the balance equation, (2.1.2) or (2.1.5), in the symbolic form

$$d\phi = d_p\phi + d_i\phi \qquad (2.1.6)$$

where $d_p\phi$ denotes the source term and $d_i\phi$ is the flow term.

Equation 2.1.5 is valid for the whole volume space. Therefore, the application of Green's formula gives the balance equation directly in local form:

$$\frac{\partial \phi_v}{\partial t} = \sigma(\phi) - \text{div}\ \ \mathbf{j}(\phi) \qquad (2.1.7)$$

One of the advantages of this formulation is that all conservation equations can be expressed by the statement that the source term corresponding to a conserved quantity vanished.

Conservation of Mass

If there are s chemical reactions in the system, they may be symbolized as

$$\sum_{i=1}^{\nu} \alpha_{ik} A_i = 0 \qquad k = 1, 2, \ldots, s \qquad (2.1.8)$$

where A_i is the chemical symbol for component i and α_{ik} is the number of grams of component i produced per unit gram of reactant k. Since the mass of a reacting system is conserved while the number of moles in such a system is not, it is more convenient to use the gram basis rather than the mole basis in discussions of chemical reactions. The coefficient α_{ik} is negative if i is a reactant of the chemical reaction k and positive if i is a product of the reaction k. If component i does not participate in reaction k, then the corresponding α_{ik} is zero.

If the partial mass density of component i depends on the spatial position R and on the time t, the mass of component i in the fluid mixture at a time t is denoted by m_i and is given by

$$m_i(t) = \int_V \rho_i(R, t)\, dV \qquad i = 1.2, \ldots, \nu \qquad (2.1.9)$$

As a result of the chemical reactions that are taking place in the system, the mass of component i changes with time. If $d_k m_i$ is the change in mass of component i in a time interval dt due to reaction k, then the total change in mass of i during the interval dt is

$$dm_i = \sum_{k=1}^{s} d_k m_i \qquad (2.1.10)$$

The changes in masses of the components in the system as a result of the s chemical reactions are related by

$$\frac{d_k m_1}{\alpha_{1k}} = \frac{d_k m_2}{\alpha_{2k}} = \cdots = \frac{d_k m_\nu}{\alpha_{\nu k}} \qquad k = 1, 2, \ldots, s \qquad (2.1.11)$$

If one introduces the progress variable λ_k for the kth reaction and defines it as the grams of reaction occurred per gram of original reactants, one obtains

$$\frac{d_k m_i}{\alpha_{ik}} = m\, d\lambda_k \qquad \begin{array}{l} k = 1, 2, \ldots, s \\ i = 1, 2, \ldots, \nu \end{array} \qquad (2.1.12)$$

in which m is the total mass of the fluid system and is given as

$$m = \sum_{i=1}^{\nu} m_i(t) = \sum_{i=1}^{\nu} \int_V \rho_i(R, t)\, dV = \int_V \rho(R, t)\, dV \qquad (2.1.13)$$

One notes that because of (2.1.11), $d\lambda_k$ is independent of component i. From (2.1.10) and (2.1.11) one also finds that the change in mass of component i in the time interval dt due to all the chemical reactions is

$$dm_i = m \sum_{k=1}^{s} \alpha_{ik}\, d\lambda_k \qquad i = 1, 2, \ldots, \nu \qquad (2.1.14)$$

Equation 2.1.14 may be expressed in terms of mass density by dividing it by the total volume V of the system:

$$d\rho_i = \rho \sum_{k=1}^{s} \alpha_{ik}\, d\lambda_k \qquad i = 1, 2, \ldots, \nu \qquad (2.1.15)$$

If the macroscopic velocity of component i is $\mathbf{v}_i$, the current density of mass of component i is the mass of i flowing per unit area in unit time:

$$\mathbf{j}_i = \rho_i \mathbf{v}_i \qquad (2.1.16)$$

The internal source $\sigma(\phi)$ of mass of component i is the amount of i produced per unit volume and unit time by chemical reaction. From (2.1.15), one sees that

$$\sigma(\phi) = \frac{d\rho_i}{dt} = \rho \sum_{k=1}^{s} \alpha_{ik}\, \frac{d\lambda_k}{dt} \qquad (2.1.17)$$

The conservation of mass or the continuity equation for the component i is then obtained from (2.1.7) to give

$$\frac{\partial \rho_i}{\partial t} + \text{div}\,(\rho_i \mathbf{v}_i) = \rho \sum_{k=1}^{s} \alpha_{ik}\, \frac{d\lambda_k}{dt} \qquad (2.1.18)$$

The equation of continuity for the total mass of the system may be obtained by summing over all components:

$$\frac{\partial \rho}{\partial t} + \text{div}\,(\rho \mathbf{v}) = 0 \qquad (2.1.19)$$

in which the total density ρ and the mass-average velocity $\mathbf{v}$ are defined as

$$\rho = \sum_{i=1}^{\nu} \rho_i = \frac{1}{v} \qquad (2.1.20)$$

$$\mathbf{v} = \frac{\sum_{i=1}^{\nu} \rho_i \mathbf{v}_i}{\sum_{i=1}^{\nu} \rho_i} \quad \text{or} \quad \rho\mathbf{v} = \sum_{i=1}^{\nu} \rho_i \mathbf{v}_i \qquad (2.1.21)$$

The quantity v is the specific volume. The flow of component i with respect to the mass-average velocity (barycentric velocity) $\mathbf{v}$ is

$$\mathbf{J}_i = \rho_i(\mathbf{v}_i - \mathbf{v}) \qquad (2.1.22)$$

In terms of the substantial time derivative, $d/dt = \partial/\partial t + \Sigma_j \mathbf{v}_j \, \partial/\partial x_j$, and the flow of component i in (2.1.22), one can write the equation of continuity in the alternate form

$$\frac{d\rho_i}{dt} + \rho_i \operatorname{div} \mathbf{v} + \operatorname{div} \mathbf{J}_i = \rho \sum_{k=1}^{s} \alpha_{ik} \frac{d\lambda_k}{dt} \qquad (2.1.23)$$

From (2.1.21) and (2.1.22), it follows that

$$\sum_{i=1}^{\nu} \mathbf{J}_i = 0 \qquad (2.1.24)$$

If the fluid system is closed, the total mass is conserved:

$$dm = \sum_{i=1}^{\nu} dm_i = 0 \qquad (2.1.25)$$

Thus, from (2.1.14) and (2.1.25), it also follow that

$$\sum_{i=1}^{\nu} \alpha_{ik} = 0 \qquad k = 1, 2, \ldots, s \qquad (2.1.26)$$

Summing over all components i, (2.1.23) leads to

$$\frac{d\rho}{dt} = -\frac{1}{v^2} \frac{dv}{dt} = -\rho \operatorname{div} \mathbf{v} \qquad (2.1.27)$$

Conservation of Momentum

If $\mathbf{F}$ is the external force per unit mass acting on the fluid system, then the volume force due to external fields acting on V is $\int_V \rho\mathbf{F} \, dV$. The force $\mathbf{F}$ is the total force per unit mass acting on the volume element V and may be the result of several external fields. If these external forces should act selectively on the different components, then

$$\rho \mathbf{F} = \sum_{i=1}^{\nu} \rho_i \mathbf{F}_i \qquad (2.1.28)$$

where $\mathbf{F}_i$ is the force acting on component i per unit mass of component i.

The surface force acting on the surface of a volume element V of fluid may be regarded as the sum of the surface forces acting on each element of area dA. If $d\mathbf{f}$ is the force transmitted across an element of area dA, then

$$d\mathbf{f} = \bar{\bar{\sigma}} \cdot d\mathbf{A} \qquad (2.1.29)$$

where $d\mathbf{A}$ is a vector normal to the area $d\mathbf{A}$ and directed outward from V and has a magnitude $d\mathbf{A}$. The stress tensor $\bar{\bar{\sigma}}$ is equal to the negative of the pressure tensor and has nine components. The total surface force acting on V is obtained by integrating (2.1.29) over the surface A of V:

$$\mathbf{f} = \int_A \bar{\bar{\sigma}} \cdot d\mathbf{A} \qquad (2.1.30)$$

The total force acting on V is then the sum of the force acting on all components and the total surface force, the sum of (2.1.28) and (2.1.30).

The total momentum M in V is

$$M = \int_V \rho \mathbf{v}\, dV \qquad (2.1.31)$$

and its rate of change is

$$\frac{dM}{dt} = \int_V \frac{\partial}{\partial t} \rho \mathbf{v}\, dV \qquad (2.1.32)$$

Since the volume V is fixed with respect to the external coordinate axes, the order of integration and differentiation may be interchanged and the partial derivative notation used. The total momentum transferred out through the surface A per unit time is $\int_A \rho \mathbf{v}\mathbf{v} \cdot d\mathbf{A}$.

According to Newton's second law, the force acting on a body is equal to the rate of change of the momentum of the body. Equating the sum of the volume and surface forces to the net rate of change of momentum, one has

$$\int_V \frac{\partial}{\partial t}(\rho \mathbf{v})\, dV + \int_A \rho \mathbf{v}\mathbf{v} \cdot dA = \int_V \rho \mathbf{F}\, dV + \int_A \bar{\bar{\sigma}} \cdot d\mathbf{A} \qquad (2.1.33)$$

The surface integrals on each side of (2.1.33) may be converted into volume integrals by the use of Gauss' theorem. Then, (2.1.33) yields

$$\int_v \left\{ \frac{\partial}{\partial t} (\rho \mathbf{v}) + \nabla \cdot (\rho \mathbf{vv}) - \rho \mathbf{F} - \nabla \cdot \overline{\overline{\sigma}} \right\} dV = 0 \qquad (2.1.34)$$

Since the volume V is arbitrary, (2.1.34) is valid only for the integrand to vanish. Therefore, one obtains a form of the equation of motion:

$$\frac{\partial}{\partial t} (\rho \mathbf{v}) + \nabla \cdot (\rho \mathbf{vv}) = \rho \mathbf{F} + \nabla \cdot \overline{\overline{\sigma}} \qquad (2.1.35)$$

If one replaces the stress tensor $\overline{\overline{\sigma}}$ by the negative of the pressure tensor $\overline{\overline{\pi}}$ and uses the substantial derivative, (2.1.35) may be written as

$$\rho \frac{d\mathbf{v}}{dt} = -\nabla \cdot \overline{\overline{\pi}} + \sum_i^v \rho_i \mathbf{F}_i \qquad (2.1.36)$$

The pressure tensor appears in the form

$$\overline{\overline{\pi}} = \begin{bmatrix} p + P_{11} & P_{12} & P_{13} \\ P_{21} & p + P_{22} & P_{23} \\ P_{31} & P_{32} & p + P_{33} \end{bmatrix} \qquad (2.1.37)$$

in which p is the static pressure, which would be present without flow; and P_{ij} ($i, j = 1, 2, 3$) are viscous pressure, corresponding to the three additional normal pressures (P_{11}, P_{22}, P_{33}) and to the six tangential (shear) pressures ($P_{12}, P_{13}, P_{21}, P_{23}, P_{31}, P_{32}$) which act on each volume element of a streaming fluid.

The physical meaning will become much clearer if one writes (21.1.35) in the form of the balance equation (2.1.7):

$$\frac{\partial}{\partial t} (\rho \mathbf{v}) = \rho \mathbf{F} - \nabla \cdot (\rho \mathbf{vv} + \overline{\overline{\pi}}) \qquad (2.1.38)$$

By comparison with (2.1.7), one sees that the source associated to the ith component of the total momentum M is given by

$$\sigma(M_i) = \sum_{i=1}^v \rho_i \mathbf{F}_i \qquad (2.1.39)$$

Equation 2.1.39 shows that the total momentum is conserved in the absence of external forces. The flow of momentum contains both a convective current $\rho\mathbf{v}\mathbf{v}$ and a conduction current corresponding to the pressure tensor $\overline{\overline{\pi}}$. This tensor coincides with the total flow of momentum referring to the center of mass motion or barycentric velocity.

For many irreversible processes, one can distinguish between two stages during the course of the processes. In the first stage, the macroscopic motions die away because of viscous flow. In the second stage, slow processes such as diffusion, heat conduction, slow chemical reactions, and so on, are present, while all velocities and accelerations are very small. Here, local mechanical equilibrium, namely, the disappearance of the barycentric acceleration, is approximately satisfied, corresponding to the neglect of inertial forces. Simultaneously, the viscous pressures are very small. From (2.1.36), one obtains the condition

$$P_{ij} \approx 0 \quad (i, j = 1, 2, 3) \quad \sum_{i=1}^{\nu} \rho_i \mathbf{F}_i \approx \operatorname{grad} p \quad (2.1.40)$$

If these relations hold, one speaks of "creeping motion" in the fluid.

Conservation of Energy

According to the balance equation (2.1.7), the conservation of energy can be written as

$$\frac{\partial(\rho e)}{\partial t} = -\operatorname{div} \mathbf{J}_e \quad (2.1.41)$$

in which e is the specific total energy, the total energy per unit mass; and $\mathbf{J}_e$ is the total energy flux per unit surface per unit time. The specific total energy of a system is contributed by the specific internal energy u, the specific kinetic energy $\frac{1}{2}\mathbf{v}^2$, and the specific potential energy in external conservative force fields ψ_i; that is,

$$e = u + \frac{1}{2}\mathbf{v}^2 + \psi_i \quad (2.1.42)$$

Similarly, the total energy flux includes a convective term $p e \mathbf{v}$; an energy flux due to mechanical work performed on the system $\overline{\overline{\pi}} \cdot \mathbf{v}$; a potential energy flux due to the diffusion of the various components in the field of force $\sum_i \psi_i \mathbf{J}_i$; and heat flow $\mathbf{J}_Q$; that is,

$$\mathbf{J}_e = \rho e \mathbf{v} + \overline{\overline{\pi}} \cdot \mathbf{v} + \sum_i \psi_i \mathbf{J}_i + \mathbf{J}_Q \quad (2.1.43)$$

If one subtracts $[\partial\rho(\frac{1}{2}\mathbf{v}^2 + \psi_i)]/\partial t$ from (2.1.41), one obtains

$$\frac{\partial(\rho u)}{\partial t} = -\operatorname{div}(\rho u \mathbf{v} + \mathbf{J}_Q) - \bar{\bar{\pi}} : \operatorname{grad} \mathbf{v} + \sum_i \mathbf{J}_i \cdot \mathbf{F}_i \qquad (2.1.44)$$

With (2.1.27) and (2.1.37), (2.1.44) can be written in the form

$$\frac{du}{dt} = \frac{dq}{dt} - p\frac{dv}{dt} - vP_{ij} : \operatorname{grad} \mathbf{v} + v\sum_i \mathbf{J}_i \cdot \mathbf{F}_i \qquad (2.1.45)$$

in which

$$\mathbf{F}_i = -\operatorname{grad}\psi_i \qquad \frac{\partial\psi_i}{\partial t} = 0 \qquad (2.1.46)$$

and

$$\rho\frac{dq}{dt} = -\operatorname{div}J_Q \qquad (2.1.47)$$

have been used.

2 SECOND LAW OF THERMODYNAMICS

The entropy S has been postulated as a state function with the following properties:

1. The entropy is an extensive quantity. If a system consists of several parts, the total entropy is equal to the sum of the entropies of each part. The change of entropy dS can be divided into two parts: the entropy production d_pS due to changes inside the system, and the flow of entropy d_iS due to the interaction with the surroundings. Thus, in accordance with (2.1.6),

$$dS = d_pS + d_iS \qquad (2.2.1)$$

2. The entropy production d_pS due to changes inside the system is always positive:

$$d_pS \geq 0$$

For a closed system, which may only exchange heat with its surroundings, one has, according to the Carnot-Clausius theorem,

$$d_iS = \frac{dQ}{T}$$

where dQ is the heat supplied to the system by its surroundings, and T is the absolute temperature at which heat is received by the system.

The entropy concept introduces a clear distinction between two types of processes: reversible changes and irreversible changes (or processes). The entropy production d_pS vanishes when the system undergoes only reversible changes, and it is always positive if the system is subject to irreversible processes. Hence,

$$d_pS = 0 \quad \text{(reversible processes)} \quad (2.2.2)$$

$$d_pS > 0 \quad \text{(irreversible processes)} \quad (2.2.3)$$

For an isolated system, the entropy flow is by definition equal to zero, and (2.2.2) and (2.2.3) reduce to the classical formulation of the second law:

$$dS \geq 0 \quad \text{(isolated system)} \quad (2.2.4)$$

The inequality states that the entropy of an isolated system can never decrease. Following the definition of (2.1.3), the entropy production per unit time is

$$\frac{d_pS}{dt} = P[S] = \int_V \sigma(s)\, dV \geq 0 \quad (2.2.5)$$

where $\sigma(s)$ is the entropy source, the entropy production per unit time per unit volume. Since the inequality in (2.2.5) has to be valid for an arbitrary macroscopic volume, the inequality is also valid for the entropy source:

$$\sigma(s) \geq 0 \quad (2.2.6)$$

Likewise, the entropy flow $\mathbf{J}[s]$ is defined as

$$\frac{d_iS}{dt} = \mathbf{J}[s] = \int_A \mathbf{j}_s^T\, d\mathbf{A} \quad (2.2.7)$$

where $\mathbf{J}[s]$ is the component of entropy flow along the interior normal to the boundary surface A, and $\mathbf{j}_s^T$ is the total entropy flow density. By using the Gauss theorem and entropy per unit volume S_v one obtains the entropy balance equation from (2.1.7)

$$\frac{\partial S_v}{\partial t} + \nabla \cdot \mathbf{j}_s^T = \sigma(s) \quad (2.2.8)$$

3   THERMODYNAMICS OF IRREVERSIBLE PROCESSES

In the discussion of irreversible thermodynamics, the applicability of the Gibbs equilibrium thermodynamic equations is assumed, and hence processes that are too far removed from equilibrium are excluded [12]. Under these conditions, the rate of entropy production can be expressed as a bilinear form that contains two types of factors: the generalized forces or affinity F_i that drive the processes and the other to describe the responses to those forces, fluxes.

The extensive state function, entropy, depends functionally on extensive properties, X_1, X_2, ..., of the system. Now differentiating the entropy state function $S [X_1(t), X_2(t), ...]$ with respect to time for an isolated system, one has

$$\frac{d_pS}{dt} = \sum_i \frac{\partial S}{\partial X_i} \frac{dX_i}{dt} \geq 0 \qquad (2.3.1)$$

or

$$P[S] = \sum_i F_i \cdot \mathbf{J}_i \geq 0 \qquad (2.3.2)$$

in which

$$F_i = \frac{\partial S}{\partial X_i} = \text{affinity} \qquad (2.3.3)$$

$$\mathbf{J}_i = \frac{dX_i}{dt} = \text{flux} \qquad (2.3.4)$$

Thus, the rate of production of entropy is the sum of products of each flux with its associated driving force, affinity.

In adapting the equilibrium theory, a local entropy associated with the local extensive properties X_1, X_2, ..., X_i has the following relationship:

$$dS = \sum_i F_i \cdot dX_i \qquad (2.3.5)$$

Or, taking all quantities per unit volume,

$$dS_v = \sum_i F_i \cdot d(X_i)_v \qquad (2.3.6)$$

Equation 2.3.6 suggests a reasonable definition of the entropy flow density $\mathbf{j}_s{}^T$ if dS_v is d_iS_v:

$$\mathbf{j}_s{}^T = \sum_i F_i \cdot \mathbf{j}_i \qquad (2.3.7)$$

in which $\mathbf{j}_i$ is the current density of property i. By differentiation of (2.3.6) with respect to time in a space-fixed coordinate system, one obtains

$$\frac{\partial S_v}{\partial t} = \sum_i F_i \frac{\partial (X_i)_v}{\partial t} \qquad (2.3.8)$$

Substituting (2.3.7) and (2.3.8) into (2.2.8), one obtains the entropy source as

$$\sigma(s) = \frac{\partial S_v}{\partial t} + \nabla \cdot \mathbf{j}_s^T$$

$$= \sum_i F_i \frac{\partial (X_i)_v}{\partial t} + \nabla \cdot (\sum_i F_i \cdot \mathbf{j}_i)$$

$$= \sum_i \left[F_i \frac{\partial (X_i)_v}{\partial t} + \nabla F_i \cdot \mathbf{j}_i + F_i \nabla \cdot \mathbf{j}_i \right]$$

$$= \sum_i \left[F_i \left(\frac{\partial (X_i)_v}{\partial t} + \nabla \cdot \mathbf{j}_i \right) + \nabla F_i \cdot \mathbf{j}_i \right]$$

$$= \sum_i \nabla F_i \cdot \mathbf{j}_i \qquad (2.3.9)$$

In obtaining the last expression in (2.3.9) from the previous expression, the balance equation (2.1.7) was used together with the fact that the various extensive properties can be neither produced nor destroyed, so that the term in the parenthesis vanishes.

Finally, using (2.3.9), one obtains

$$\sigma(s) = \sum_i \nabla F_i \cdot \mathbf{j}_i \qquad or \qquad P[S] = \sum_i F_i \cdot \mathbf{J}_i \qquad (2.3.10)$$

The affinity is defined as the gradient of the entropy representation of intensive properties in continuous systems.

At thermodynamic equilibrium, the fluxes $\mathbf{j}_i$ and the corresponding driving forces or affinities F_i disappear. It is natural to describe the deviations from equilibrium and, thus, the irreversible processes relating that each local flux depends only upon the instantaneous local affinities and upon the local intensive properties:

$$\mathbf{J}_i = \mathbf{J}_i(F_j) \qquad or \qquad \mathbf{j}_i = \mathbf{j}_i(F_j) \qquad (2.3.11)$$

It should be noted that it is not assumed that each flux depends only on its own affinity but rather that each flux depends on all affinities. It is true that each flux tends to depend most strongly on its own associated affinity, but the dependence of a flux on other affinities as well is the source of the more interesting phenomena in the field of irreversible processes.

Each flux $\mathbf{J}_i$ or $\mathbf{j}_i$ is known to vanish as the affinities vanish, so we can expand $\mathbf{J}_i$ or $\mathbf{j}_i$ in powers of the affinity:

$$\mathbf{j}_i = \sum_j L_{ji} F_j + \frac{1}{2} \sum_k \sum_j L_{kji} F_k F_j + \cdots \qquad (2.3.12)$$

If the processes are not too far removed from equilibrium, the affinities are so small that all quadratic and higher-order terms in (2.3.12) can be neglected. A flux can be adequately described by the truncated approximate equation

$$\mathbf{j}_i = \sum_j L_{ji} F_j \qquad i = 1, 2, \ldots, \nu \qquad (2.3.13)$$

The above expansion is valid for either $\mathbf{J}_i$ or $\mathbf{j}_i$. The inference has been investigated by Prigogine [12] from a microscopic viewpoint. He has applied the Chapman–Enskag [13] kinetic theory of nonuniform gases to determine the domain of validity of using the Gibbs thermodynamic functions for nonequilibrium systems. In the Chapman–Enskag theory, the molecular distribution function is expanded into an infinite series:

$$f_i = f_i^{(0)}(1 + \phi_i^{(1)} + \phi_i^{(2)} + \cdots) \qquad (2.3.14)$$

where $f_i^{(0)}$ is the equilibrium molecular distribution function, $\phi_i^{(1)}$ is the first-order correction, and $\phi_i^{(2)}$, and so on, are subsequent corrections. For a system which is close to equilibrium, the successive corrections $\phi_i^{(1)}$, $\phi_i^{(2)}$, and so forth, become smaller and smaller. Prigogine found that the Gibbs thermodynamic functions are valid not only for an equilibrium system, characterized by the equilibrium distribution function $f_i^{(0)}$, but also for a nonequilibrium system which is characterized by a distribution function $f_i^{(0)}(1 + \phi_i^{(1)})$, that is, if a nonequilibrium system is sufficiently close to equilibrium that its distribution function is adequately represented by the first two terms on the right-hand side of (2.3.14). The first deviation from the equilibrium distribution is linear in the existing gradients, which is consistent with (2.3.13).

The equations represented by (2.3.13) are called *phenomenological equations*, and the coefficients L_{ji} are termed *phenomenological coefficients*, or *kinetic coefficients*. They are functions of temperature, pressure, compositions, and the frame of reference. They are not thermodynamic properties, but kinetic quantities. In spite of limitations imposed on (2.3.13), such as small departures from equilibrium or small values of F_j, it is perhaps surprising that so many physical processes of interests are linear and deviate only slightly from equilibrium. Substitution of (2.3.13) into (2.3.2) yields the entropy production

$$P[S] = \sum_i \sum_j L_{ij} F_i F_j \qquad (2.3.15)$$

In using (2.3.13) or (2.3.15), if the resulting phenomenological coefficients are well defined, then the matrix of these phenomenological coefficients is symmetrical, and the phenomenological coefficients defined in (2.3.13) satisfy the so-called Onsager's reciprocal relationship

$$L_{ij} = L_{ji} \qquad (2.3.16)$$

if, and only if, the fluxes $\mathbf{J}_i$ and the affinities F_i follow the Curie principle [17] that entities whose tensorial characters differ by an odd integer cannot interact in isotropic systems.

4   ENTROPY BALANCE

The Gibbs thermodynamic equation holds for each volume element without electrification and magnetization for sufficiently slow processes and is written as

$$T\,ds = du + p\,dv - \Sigma\,\mu_i\,dw_i \tag{2.4.1}$$

where T denotes the absolute temperature; s is the specific entropy; u is the specific internal energy; p is the equilibrium pressure; v is the specific volume; and μ_i and $w_i = \rho_i/\rho$ are the chemical potential and the mass fraction of species i, respectively.

Then, it is assumed that the total system is not in equilibrium, but within small mass elements a state of local instantaneous equilibrium exists, for which the local entropy s still obeys the equilibrium property. If we differentiate (2.4.1) with respect to time along the mass-average velocity, we have

$$T\frac{ds}{dt} = \frac{du}{dt} + p\frac{dv}{dt} - \sum_i^\nu \mu_i\frac{dw_i}{dt} \tag{2.4.2}$$

where the substantial derivative is $d/dt = \partial/\partial t + \mathbf{v}\cdot\nabla = \partial/\partial t + \Sigma_j\,\mathbf{v}_j$ $(\partial/\partial x_j)$, du/dt is given in (2.1.45), dv/dt can be obtained from (2.1.27), and dw_i/dt also can be obtained from (2.1.23).

The specific entropy s can be converted to the entropy per unit volume S_v by multiplying density ρ by the specific entropy s. Then, (2.2.8) may be rewritten into the form

$$\rho\frac{ds}{dt} = -\nabla\cdot\mathbf{j}_s + \sigma(s) \tag{2.4.3}$$

where the entropy flux $\mathbf{j}_s$ is the difference between the total entropy flux $\mathbf{j}_s^T$ and the convective term $\rho s\mathbf{v}$:

$$\mathbf{j}_s = \mathbf{j}_s^T - \rho s\mathbf{v} \tag{2.4.4}$$

Thus, the explicit form of the entropy balance equation (2.4.3) is obtained by rearranging (2.4.2) together with (2.1.45), (2.1.27), and (2.1.23), which gives

$$\rho \frac{ds}{dt} = -\frac{1}{T} \left\{ \text{div} \, \mathbf{J}_Q + \overline{\overline{\pi}} : \text{grad} \, \mathbf{v} - \sum_{i=1}^{\nu} \mathbf{J}_i \cdot \mathbf{F}_i \right.$$

$$\left. - \sum_{i=1}^{\nu} \mu_i \, \text{div} \, \mathbf{J}_i + \sum_{i=1}^{\nu} \sum_{k=1}^{s} \mathbf{J}_i \mu_i \alpha_{ik} \frac{d\lambda_k}{dt} \right\} \tag{2.4.5}$$

By rearranging (2.4.5) into the form of (2.4.3), one obtains

$$\rho \frac{ds}{dt} = -\text{div} \left(\frac{\mathbf{J}_Q - \sum_{i}^{\nu} \mu_i \mathbf{J}_i}{T} \right)$$

$$- \frac{1}{T^2} \cdot \mathbf{J}_Q \cdot \text{grad} \, T$$

$$- \frac{1}{T} \sum_{i=1}^{\nu} \cdot \mathbf{J}_i \cdot \left(T \, \text{grad} \, \frac{\mu_i}{T} - \mathbf{F}_i \right)$$

$$- \frac{1}{T} \overline{\overline{\pi}} : \text{grad} \, \mathbf{v}$$

$$- \frac{1}{T} \sum_{i=1}^{\nu} \sum_{k=1}^{s} \mathbf{J}_i \mu_i \alpha_{ik} \frac{d\lambda_k}{dt} \tag{2.4.6}$$

Comparison with (2.4.3) yields expressions for the entropy flux and the entropy source:

$$\mathbf{j}_s = \frac{1}{T} \left(\mathbf{J}_Q - \sum_{i}^{\nu} \mu_i \mathbf{J}_i \right) \tag{2.4.7}$$

$$\sigma(s) = -\frac{1}{T^2} \mathbf{J}_Q \cdot \text{grad} \, T - \frac{1}{T} \left\{ \sum_{i=1}^{\nu} \mathbf{J}_i \cdot \left(T \, \text{grad} \, \frac{\mu_i}{T} - \mathbf{F}_i \right) \right.$$

$$\left. + \overline{\overline{\pi}} : \text{grad} \, \mathbf{v} + \sum_{i=1}^{\nu} \sum_{k=1}^{s} \mathbf{J}_i \mu_i \alpha_{ik} \frac{d\lambda_k}{dt} \right\} \geq 0 \tag{2.4.8}$$

In rearranging (2.4.5) into (2.4.6) or to obtain (2.4.7) and (2.4.8), a constrained condition, the entropy source $\sigma(s)$ has to vanish if the thermodynamic equilibrium conditions are to be satisfied with the system. Equation 2.4.8 satisfies automatically this requirement when the driving forces or affinities become zero.

5 MECHANICAL EQUILIBRIUM

A fluid is said to be in mechanical equilibrium if the mass-average velocity $\mathbf{v}$ is independent of position R and of time t. For such a fluid, the equation of motion, (2.1.38), becomes

$$\rho \mathbf{F} - \nabla p = 0 \qquad (2.5.1)$$

The external forces are balanced by the pressure gradient. When there are no external forces acting on a system which is in mechanical equilibrium, the pressure of the system is uniform. Mechanical equilibrium is usually established in nonequilibrium systems much more rapidly than in thermodynamic equilibrium systems.

The Gibbs–Duhem equation written in terms of the gradients of temperature, pressure, and chemical potentials is

$$\nabla p = \sum_{i=1}^{\nu} \rho_i \, \nabla \mu_i + \rho s \, \nabla T \qquad (2.5.2)$$

In a mixture, the specific entropy is given by

$$\rho s = \sum_{i=1}^{\nu} \rho_i \overline{S}_i \qquad (2.5.3)$$

in which $\overline{S}_i$ is the partial specific entropy of component i. The chemical potential of component i may be written in terms of the partial specific quantities; therefore,

$$\mu_i = \overline{H}_i - T\overline{S}_i \qquad (2.5.4)$$

Substituting (2.5.3) and (2.5.4) into (2.5.2) and then substituting the resulting formulas into (2.5.1), one obtains

$$\nabla p - \rho \mathbf{F} = \sum_{i=1}^{\nu} \rho_i (\nabla \mu_i{}' + \overline{S}_i \, \nabla T) = 0 \qquad (2.5.5)$$

The quantity $\mu_i{}'$ is the total chemical potential of component i, including the external potentials, which are discussed later. For a system in mechanical equilibrium, the Gibbs–Duhem equation becomes

$$\sum_{i=1}^{\nu} \rho_i (\nabla \mu_i{}' + \overline{S}_i \, \nabla T) = 0 \qquad (2.5.6)$$

so that the affinities $\nabla \mu_i{}'$ and ∇T are not all independent.

6 MULTICOMPONENT DIFFUSION

The material currents in the fluid may be described in terms of the local average velocity $\mathbf{v}_i$ for each component i. The velocity $\mathbf{v}_i$ does not refer to the velocity of a particular molecule of i; rather, it refers to the average velocity with respect to the external coordinate system of the molecules of i in a microscopically large, macroscopically small reign of the fluid. Thus, local average velocity $\mathbf{v}_i$ is a field quantity and is a function of position and time, $\mathbf{v}_i(R, t)$. The local mass-average velocity $\mathbf{v}$ for the fluid is defined as

$$\mathbf{v} = \frac{\sum_i \rho_i \mathbf{v}_i}{\sum_i \rho_i} \tag{2.1.21}$$

This velocity $\mathbf{v}$, in general, is a function of position R and time t. A coordinate system fixed relative to the local mass average in one part of the fluid system will usually move with respect to a coordinate system fixed relative to the local mass average in another part of the fluid system.

The mass current density, or mass flux $\mathbf{j}_i$, of component i relative to the external coordinate fixed in space is as given by

$$\mathbf{j} = \rho_i \mathbf{v}_i \tag{2.1.16}$$

Diffusion current densities for each of the ν-chemical components in the fluid system may be defined in various ways; for example, the diffusion current density relative to:

1. The local mass average velocity [also defined in (2.1.22)]:

$$\mathbf{J}_i = \rho_i(\mathbf{v}_i - \mathbf{v}) \qquad \mathbf{v} = \frac{\sum_i \rho_i \mathbf{v}_i}{\sum_i \rho_i} \qquad \sum_i \mathbf{J}_i = 0 \tag{2.6.1}$$

2. The local molar average velocity:

$$\mathbf{J}_i^m = \rho_i(\mathbf{v}_i - \mathbf{v}^m) \qquad \mathbf{v}^m = \frac{\sum_i n_i \mathbf{v}_i}{\sum_i n_i} \tag{2.6.2}$$

where n_i is the number of moles of component i.

3. The local volume average velocity:

$$\mathbf{J}_i^v = \rho_i(\mathbf{v}_i - \mathbf{v}^v) \qquad \mathbf{v}^v = \sum_i \rho_i \overline{V_i} \mathbf{v}_i \qquad \sum_i \overline{V_i} \mathbf{J}_i^v = 0 \tag{2.6.3}$$

where $\overline{V}_i$ is the partial specific volume of component i. Since $\overline{V}_i$ is used in the definition of the diffusion flux, the assumption of local equilibrium is implied.
4. One of the components of the system:

$$\mathbf{J}_i^o = \rho_i(\mathbf{v}_i - \mathbf{v}_o) \qquad \mathbf{J}_o^o = 0 \qquad (2.6.4)$$

where $\mathbf{v}_o$ is the local mean velocity of component o. For solutions, $\mathbf{J}_i^o$ is usually defined relative to the solvent, that is, component o ($i = 1, 2, \ldots, \nu - 1$).

For a ν-component mixture in isothermal, isotropic, and nonreacting systems, Fick's law of diffusion is

$$\mathbf{J}_i = -\sum_{j=2}^{\nu} D_{ij} \operatorname{grad} \rho_i \qquad (i, j = 2, 3, \ldots, \nu) \qquad (2.6.5)$$

where $\mathbf{J}_i$ and ρ_i are considered dependent quantities. The $(\nu - 1)^2$ quantities of D_{ij} are the diffusion coefficients of the system. In conjunction with the thermodynamics of irreversible processes, one can identify the various affinities, the driving forces, from (2.4.8). Thus, the flux can be written in terms of various affinities from (2.3.13). For the case of Fick's diffusion in a ν-multicomponent system, one writes

$$\mathbf{J}_i = -\sum_{j=2}^{\nu} L_{ij} (\operatorname{grad} \mu_j)_{T,p} = -\sum_{j=2}^{\nu} \sum_{l=2}^{\nu} L_{ij}\mu_{jl} \operatorname{grad} \rho_l \qquad (i = 2, 3, \ldots, \nu)$$

$$(2.6.6)$$

in which

$$\mu_{jl} = \left(\frac{\partial \mu_j}{\partial \rho_l} \right)_{T,P,\rho_{j \neq l}} \qquad (2.6.7)$$

By comparing (2.6.5) with (2.6.6), it becomes obvious that the phenomenological flux equations (2.3.13) lead to the generalized form of Fick's diffusion flux equations. If one changes the reference coordinate system, the new flux $\mathbf{J}_i^\alpha$ can be obtained by use of coordinate transformation factors ϵ_{ij} in such a way that

$$\mathbf{J}_i^\alpha = -\sum_{j=2}^{\nu} \sum_{l=2}^{\nu} \epsilon_{ij} D_{jl} \operatorname{grad} \rho_l \qquad (i = 2, 3, \ldots, \nu) \qquad (2.6.8)$$

while the phenomenological flux equations (2.3.13) are invariant under coordinate transformations. It is advantageous to define fluxes in the form of (2.3.13), so that the transport coefficients immediately relate to the phenomenological coefficients which occur according to the choice of reference velocity and of concentration gradient.

7 EXTERNAL FORCE FIELDS

When a system is placed in a gravitational field, its thermodynamic proper-ties vary from point to point. The term gravitational field here implies that the field is either constant, such as the earth's gravitational field, or a centrifugal field. The force per unit mass acting on component i containing δ ionic species consists of the gravitational force $\mathbf{g}$, the electrical force $\mathbf{F}_e$, and the specific force $\mathbf{F}_c$ in a centrifugal field and is

$$\mathbf{F}_i = \mathbf{g} + \mathbf{F}_e + \mathbf{F}_c \qquad (2.7.1)$$

where $\mathbf{g}$ is a constant vector with the constant magnitude of 9.8 m/sec^2 and is the same for all components in the system. The electrical force $\mathbf{F}_e$ acting on ionic component i is

$$\mathbf{F}_{ei} = Z_i E = -Z_i \, \nabla \psi \qquad (2.7.2)$$

where Z_i is the specific charge of ionic species i which is related to the valence z_i and the molecular weight (grams per mole) M_i of the ion by

$$Z_i = \frac{z_i \mathfrak{F}}{M_i} \qquad (2.7.3)$$

in which $\mathfrak{F}$ is the charge of one mole of univalent ions (9.65 $\times$ 10^4 coulombs, or 1 faraday); E is the external electric field; and ψ is the electrical potential. The chemical potential $\mu_i{}'$ of ionic species i is

$$\mu_i{}' = \mu_i + Z_i \psi \qquad (2.7.4)$$

where μ_i is the chemical potential in the absence of any external fields. The electric neutrality condition of the system is

$$\sum_{i=1}^{\delta} Z_i \rho_i = 0 \qquad (2.7.5)$$

The specific force $\mathbf{F}_c$ acting on component i in a centrifugal field is

$$\mathbf{F}_c = -Z_i \, \nabla \psi + \omega^2 \mathbf{r} + (2\mathbf{v}_{ri} \times \boldsymbol{\omega}) \qquad (2.7.6)$$

where ω is the angular velocity of rotation in the laboratory-fixed coordinate system and $\mathbf{r}$ is the vector distance from the center of rotation. The term $2\mathbf{v}_{ri} \times \boldsymbol{\omega}$ is the Coriolis force. Since $|\mathbf{v}_{ri}| \ll \omega r$, the Coriolis force is small compared with $\omega^2 \mathbf{r}$, and one may usually neglect it without problems.

Thus, the total chemical potential $\mu_i{}'$ of ionic species i is

$$\mu_i{}' = \mu_i + Z_i \psi + \phi_c \qquad (2.7.7)$$

where $\phi_c = -\frac{1}{2}\omega^2 r^2$ is the centrifugal potential.

If the system is in mechanical equilibrium and in an isothermal condition, one obtains from (2.5.1), (2.5.5), and (2.7.7)

$$\nabla p = \rho \mathbf{F} = -\sum_{i=1}^{\nu} \rho_i Z_i \nabla \psi - \rho \nabla \phi_c = -\rho \nabla \phi_c \qquad (2.7.8)$$

where the electrical neutrality condition (2.7.5) has been used to obtain the last expression. In actual practice, most systems in a constant gravitational field are in mechanical equilibrium, and systems subjected to a centrifugal field attain mechanical equilibrium as soon as the angular velocity ω becomes constant.

The isothermal gradient of the total chemical potential in terms of independent variables may be obtained by substituting (2.7.8) into the gradient of (2.7.7):

$$(\nabla \mu_i')_T = (\nabla \mu_i)_{T,P} + Z_i \nabla \psi + (1 - \rho \overline{V_i}) \nabla \phi_c \qquad (2.7.9)$$

8   DIFFUSION AND SEDIMENTATION

If external forces are exerted on a mixture of chemical components, then the diffusion phenomena in such a system are influenced by these forces. In uniformly rotating systems, the various components are subjected to centrifugal as well as to Coriolis forces, which both arise as a result of the rotation of the system. If one considers isothermal, nonreacting ν-component mixtures in which viscous phenomena can be neglected, one has for the entropy source from (2.4.8)

$$\sigma(s) = \frac{1}{T} \sum_{i=1}^{\nu} \mathbf{J}_i \cdot [\mathbf{F}_i - (\text{grad } \mu_i)_T] \qquad (2.8.1)$$

The force $\mathbf{F}_i$ is denoted in (2.7.1). But here we consider only that

$$\mathbf{F}_i = \omega^2 \mathbf{r} + (2\mathbf{v}_{ri} \times \omega) \qquad (2.8.2)$$

In practice, one can neglect the Coriolis force $(2\mathbf{v}_{ri} \times \omega)$. Then, for an isotropic system, one may use the phenomenological equations (2.3.13) to write the flux expression from (2.8.1) identifying affinities in the system. For this case, one has

$$\mathbf{J}_i^\alpha = \frac{1}{T} \sum_{j=1}^{\nu-1} L_{ij}^\alpha \left[(1 - \rho \overline{V_j}) \omega^2 \mathbf{r} - \sum_{l=1}^{\nu-1} \mu_{jl} \, \text{grad } \rho_l \right] \qquad (2.8.3)$$

where the abbreviation (2.6.7) has been used and the superscript α denotes an arbitrary reference coordinate system. Equation 2.8.3 expresses the transport of component i due to centrifugal sedimentation and also due to back diffusion resulting from the concentration gradient created by the partial sedimentation of the component i.

One can substitute (2.8.3) into (2.1.23) with the $d\lambda_k/dt = 0$ to obtain the continuity equation for component i in a centrifugal force field. The phenomenological coefficients L_{ij}^α and $L_{ij}^\alpha \mu_{jl}$ have to be expressed in appropriate transport coefficients in accordance with the coordinate systems referred to. If one chooses to use another concentration unit, for instance, in terms of molarity, (2.8.3) may be rewritten as

$$\mathbf{J}_i^\alpha = \sum_{j=1}^{\nu-1} L_{ij}^\alpha \left[(1 - \rho\overline{V}_j)\omega^2\mathbf{r} - \sum_{l=1}^{\nu-1} \mu_{jl}\,\mathrm{grad}\,C_l \right] \qquad (2.8.4)$$

or in appropriate transport coefficients:

$$\mathbf{J}_i^\alpha = s_i\omega^2\mathbf{r}C_i - \sum_{l=1}^{\nu-1} D_{il}\frac{\partial C_l}{\partial r} \qquad (2.8.5)$$

in which

$$s_i = \frac{1}{C_i}\sum_{j=1}^{\nu-1} L_{ij}^\alpha(1 - \rho\overline{V}_j) \qquad (2.8.6a)$$

$$D_{il} = \sum_{l=1}^{\nu-1} L_{il}^\alpha \left(\frac{\partial\mu_i}{\partial c_l}\right)_{T.P.C_{m\neq l}} \qquad (2.8.6b)$$

It is seen that both s_i and D_{il} so defined are not only functions of temperature, pressure, and solute concentrations but also functions of the reference frame considered. The quantity s_i is the sedimentation coefficient of i.

9 EQUATION OF MOTION IN ROTATING SYSTEMS [18]

In the subsequent derivation, the term "particle" is understood to cover dissolved substances and particles of both microscopic and macroscopic dimensions, that is, everything except the suspending fluid.

The equation of motion of the particle is then given by Newton's second law:

$$m_i\frac{d^2\mathbf{r}_i}{dt^2} = \mathbf{F}_i \qquad (2.9.1)$$

where $\mathbf{F}_i$ is the sum of the force acting on the particle i and $d^2\mathbf{r}_i/dt^2$ is the acceleration of the particle relative to the mass-average velocity. Since the system is assumed to be rotating at constant angular velocity ω, the differential operator d/dt represents the changes relative to the rotating coordinate system, which can be written in terms of a differential operator $\partial/\partial t$ representing changes relative to the space-fixed coordinate system as follows:

$$\frac{d(G)}{dt} = \frac{\partial(G)}{\partial t} + \omega \times (G) \tag{2.9.2}$$

Applying this operator twice to the position vector r_i, one obtains from (2.9.1)

$$m_i \left[\frac{\partial \mathbf{v}_i}{\partial t} + 2\omega \times \mathbf{v}_i + \omega \times (\omega \times r_i) \right] = \mathbf{F}_i \tag{2.9.3}$$

where $\mathbf{v}_i = \partial r_i/\partial t$ is the particle velocity measured relative to the coordinate fixed in space and ω is time independent.

The force $\mathbf{F}_i$ consists of two parts: $\mathbf{F}_1$ from the drag on the particle, Stokes' resistance; and $\mathbf{F}_2$ from the buoyancy of the particle. The drag force as given by Stokes' resistance law for a spherical particle is

$$\mathbf{F}_1 = -3\mu\pi D_p \mathbf{v}_i \tag{2.9.4}$$

where μ is the local viscosity of the suspending medium and D_p is the particle diameter. Nonspherical particles can be handled in the usual fashion by the introduction of appropriate shape factors.

The buoyancy force on the particle is given by

$$\mathbf{F}_2 = -\int_A p\mathbf{n} \, dA \tag{2.9.5}$$

where p is the pressure in the fluid, $\mathbf{n}$ is the outward-directed normal at the particle surface, and the integral is to be evaluated over the surface of the particle. Green's theorem is used to convert the surface integral into an integral over the volume of the particle so that

$$\mathbf{F}_2 = -\int_A p\mathbf{n} \, dA = -\int_V (\operatorname{grad} p) \, dV \tag{2.9.6}$$

If the particle is sufficiently small so that $(\operatorname{grad} p)$ does not change in the region of space occupied by the particle, we finally obtain for the buoyant force

$$\mathbf{F}_2 = -V_i \operatorname{grad} p \tag{2.9.7}$$

where V_i is the particle volume.

If the suspending medium is assumed to be an incompressible fluid with density ρ_m, then the equation of motion for the fluid rotating as a rigid body gives the pressure gradient

$$-\operatorname{grad} p = \rho_m \omega \times (\omega \times \mathbf{r}_i) \tag{2.9.8}$$

Using (2.9.3), (2.9.4), (2.9.7), and (2.9.8) and assuming the particle to be a sphere, the volume of the particle and its mass being $\pi D_p^3/6$ and $\rho_i \pi D_p^3/6$, respectively, with ρ_i the particle density, one obtains the equation of motion for centrifugal sedimentation for a spherical particle of the ith component:

$$-\left(\frac{18\mu}{(\rho_i - \rho_m)D_p^2}\right) \mathbf{v}_i = \left(\frac{\rho_i}{\rho_i - \rho_m}\right) \frac{\partial \mathbf{v}_i}{\partial t}$$

$$+ 2\left(\frac{\rho_i}{\rho_i - \rho_m}\right) \omega \times \mathbf{v}_i + \omega \times (\omega \times \mathbf{r}_i)$$

$$\tag{2.9.9}$$

The terms on the right-hand side of (2.9.9) represent the particle acceleration, the Coriolis effect, and the effect of the centrifugal field. If ω is time dependent, a term $-\mathbf{r}_i \times (d/dt)\, \omega(t)$ must be added to the right-hand side of (2.9.9).

Equation 2.9.9 describes the behavior of particles or solutes in a centrifuge rotor subjecting to the constant angular velocity ω. In theory, the continuity equation (2.1.23) with an appropriate flux in a centrifugal force field given in (2.8.3) and the equation of motion in a steady rotational field (2.9.9) together with appropriate boundary conditions completely specify all the centrifugal phenomena. Upon obtaining the solutions to those simultaneous coupled differential equations, one should be able to obtain all the necessary information needed in the centrifugation phenomena. In practice, the solutions to those simultaneous coupled differential equations are not easy to obtain. Therefore, many studies have to rely on experimental correlations for finding ways to incorporate the results into the analytical procedures to elucidate the centrifugal phenomena.

SYMBOLS

A	Area
A_i	Chemical symbol for component i
C_i	Concentration of component i in molarity (number of moles of i in 1000 g solution)
d_p	Infinitesimal contribution from a source as defined in (2.1.6)
d_i	Infinitesimal contribution from a flow as defined in (2.1.6)

D_p	Diameter of particles
D_{ij}	Multicomponent diffusivity of i to j
e	Specific total energy
$\mathbf{f}$	Surface force
f_i	Distribution function of component i as defined in (2.3.14)
$\mathbf{F}_1$	Drag force on a particle as defined in (2.9.4)
$\mathbf{F}_2$	Buoyancy force on a particle as defined in (2.9.5)
$\mathbf{F}_i$	Total force acting on component i
F_i	Affinity of component i as defined in (2.3.3)
$\mathfrak{F}$	Charge of one mole of univalent ions (1 faraday, or 9.65×10^4 coulombs)
$\mathbf{g}$	Gravitational acceleration
g_c	Unit conversion factor
G	An arbitrary function
$\overline{H}_i$	Partial specific enthalpy of component i
$\mathbf{j}$	Flux with respect to the coordinate system fixed in space
$\mathbf{J}$	Flux with respect to the mass-average velocity of the system
$\mathbf{J}_i$	Flow of component i as defined in (2.3.4)
$\mathbf{J}(\phi)$	Flow of the quantity ϕ
$L_{ji}, L_{kji}, \ldots$	Kinetic coefficients, or phenomenological coefficients, for the flux of component i as defined in (2.3.13)
m	Mass of a molecule
m_i	Mass of component i
M	Total momentum as defined in (2.1.31)
M_i	Molecular weight of component i
n_i	Number of moles of component i
p	Static pressure
P_{ij}	Viscous pressure
$P(\phi)$	Production of quantity ϕ per unit time
q	Heat per unit volume
Q	Total heat
r	Radial distance in a cylindrical coordinate
$\mathbf{r}_i$	Radial position of particle i
$\mathbf{R}$	Position vector
s	Specific entropy
S	Total entropy
$\overline{S}_i$	Partial specific entropy of component i
t	Time

T	Temperature
u	Specific internal energy
v	Specific volume ($= 1/\rho$)
V	Total volume
$\mathbf{v}$	Mass-average velocity as defined in (2.1.21)
$\mathbf{v}_i$	Macroscopic velocity of component i
$\mathbf{v}_{ri}$	Radial velocity component of component i
w_i	Mass fraction of component i
x	Rectangular coordinate
y	Rectangular coordinate
z	Rectangular coordinate
z_i	Valence of ionic species i
Z	Centrifugal effect as defined in (2.1)
Z_i	Specific charge of ionic species i as defined in (2.7.3)

Greek Symbols

α_{ik}	The stoichiometric coefficient, the number of grams of i produced in the kth reaction
ϵ_{ij}	Coordinate transformation tensor
λ_k	Progress variable of the kth reaction as defined in (2.1.2)
μ_i	Chemical potential of component i
$\mu_i{}'$	Total chemical potential of component i including external potential
μ_{jl}	Quantity as defined in (2.6.7)
$\overset{=}{\pi}$	Pressure tensor
ρ	Mass density
ρ_i	Mass density of component i
$\sigma(\phi)$	Intensity of source of quantity ϕ, source of ϕ per unit time per unit volume
$\overset{=}{\sigma}$	Stress tensor
ϕ	An extensive quantity
ϕ_i	Correction factor for the distribution function of component i
ϕ_c	Centrifugal potential

Subscripts

c	Centrifugal force field
e	Electrical force

i	Component i
x, y, z	Differentiate with respect to x, y, and z
v	Differentiate with respect to v or per unit volume basis
Q	Total heat
s	Entropy
k	The kth reaction
m	Fluid medium

Superscripts

(0), (1), (2), ... (i)	Indices for the order of corrections
m	Relative to molar average quantity
v	Relative to volumetric average quantity
o	Relative to the oth component (solvent)
T	Total quantity
α	Relative to an arbitrary α-coordinate system

References

1. T. Svedberg and K. O. Pedersen, *The Ultracentrifuge*, Clarendon, Oxford, 1940.
2. H. Fujita, *Mathematical Theory of Sedimentation Analysis*, Academic, New York, 1962.
3. H. Fujita, *Foundations of Ultracentrifugal Analysis*, Wiley, New York, 1975.
4. O. Lamm, *Z. Phys. Chem.* (Leipzig), **A143**, 177 (1929); *Ark. Mat. Astron. Fys.*, **21B** (2), (1929).
5. S. R. deGroot, P. Mazur, and J. Th. G. Overbeck, *J. Chem. Phys.*, **20**, 1825 (1952).
6. S. R. deGroot and P. Mazur, *Non-Equilibrium Thermodynamics*, North-Holland, Amsterdam, 1962.
7. D. D. Fitts, *Non-Equilibrium Thermodynamics*, McGraw-Hill, New York, 1962.
8. R. Haase, *Thermodynamics of Irreversible Processes*, Addison-Wesley, Reading, Mass., 1969.
9. G. J. Hooyman, H. Holtan, Jr., P. Mazur, and S. R. deGroot, *Physica*, **19**, 1095 (1953).
10. G. J. Hooyman, "Thermodynamics of Diffusion and Sedimentation," in *Conference on the Ultracentrifuge*, J. W. Williams, Ed., Academic, New York, 1963.
11. J. W. Williams, K. E. van Holde, R. L. Baldwin, and H. Fujita, *Chem. Rev.*, **58**, 715 (1958).
12. I. Prigogine, *Physica*, **15**, 272 (1949).
13. D. Enskog, Thesis, Uppsala, 1917; S. Chapman and T. G. Cowling, *The*

Mathematical Theory of Non-Uniform Gases, Cambridge University Press, Cambridge, 1952.

14. L. Onsager, *Phys. Rev.,* **37**, 405; *ibid.,* **38**, 2265 (1931).
15. H. B. G. Casimir, *Rev. Mod. Phys.,* **17**, 343 (1945).
16. G. J. Hooyman and S. R. deGroot, *Physica,* **21**, 73 (1955).
17. P. Curie, *J. de Phys., 3rd Series,* **3**, 393 (1894).
18. A. S. Berman, "Theory of Centrifugation: Miscellaneous Studies," in *National Cancer Institute Monograph 21,* N. G. Anderson, Ed., U.S. Government Printing Office, Washington, D.C., June 1966.

Chapter III

GAS CENTRIFUGES

A gas centrifuge is a device that separates individual gases from a mixture by the action of a centrifugal force. If a mixture of gases is subjected to a high centrifugal force field, the heavier molecules tend to concentrate in the direction of the field. Thus, the method is suitable for the separation of isotopic gas mixtures.

Uranium isotopes were first centrifugally separated in 1940 by J. W. Beams at the University of Virginia. At that time, however, the technology of high-speed rotating machinery needed to produce enough enriched uranium 235 for the weapons program was inadequate, and the centrifuge method was abandoned in 1943 in favor of the gaseous diffusion process. Nevertheless, the basic theory of the gas centrifuge was developed during that period. Of the methods which have been used or tested for production of ^{235}U, as listed in Table 3.1, it now appears that enrichment by the gas centrifuge is technologically and economically favored. For example, the method requires less than 10% as much power and has a much higher separation or enrichment factor than the gaseous diffusion method. Therefore, it appears quite certain that the gas centrifuge will become a preferred method in the nuclear industries.

Due to the extremely small differences in density among gases in a mixture, the separation by gas centrifuges is different from most centrifugations in that the degree of separation effected by a single centrifuge unit is far less than the degree of separation desired between product and waste. Thus, it is necessary to connect centrifugal units in series, which is known as a cascade.

In this chapter, various terminologies used in cascade separation are defined, and a brief review of cascade theory developed by Benedict and Pigford [1] is presented. Then, the theory of three types of gas centrifugation developed by Cohen [2], that is, evaporative gas centrifugation, concurrent gas centrifugation, and countercurrent gas centrifugation, is outlined. Among these gas centrifugation methods, countercurrent gas centrifugation is the most promising for large-scale separational operations, because a high separation factor or enrichment factor can be achieved in a single unit. Following the first development of theory by Cohen, a few articles appeared [3-9]. Several aspects of Cohen's theory on the promising countercurrent gas centrifuge were analyzed in detail. Improvements in the analysis are focused on to find a more realistic radial concentration profile and also an axial flow velocity profile.

50

Table 3.1 Methods for Separation ^{235}U

Method	Developed by	Status
Gaseous diffusion of UF$_6$	U.S.A.: H. C. Urey, J. R. Dunning, Kellex Corp., Carbide and Carbon Chem. Corp.[a]	Three large plants in operation Total cost $3,600 million Total power consumption 5,400,000 kW
	England	Plant in operation
Thermal diffusion of UF$_6$	U.S.A.: P. H. Abelson[a]	Plant run for short time, then shut down
	Germany: W. Groth[b] U.S.S.R.: A. E. Brodskii[c]	Process tried—no separation
Electromagnetic	U.S.A.: E. O. Lawrence, Stone & Webster, Tennessee Eastman[a]	Plant run during World War II, then shut down
Centrifugation of UF$_6$ gas	U.S.A.: J. W. Beams, Standard Oil Devel. Co., Westinghouse[a]	Demonstrated in pilot plant
	Germany: W. Groth, P. Harteck et al.[d]	5% Enrichment obtained in laboratory machine

Source: M. Benedict and T. H. Pigford, *Nuclear Chemical Engineering*, McGraw-Hill, New York, 1957, p. 364 (by permission of McGraw-Hill).
[a]From H. D. Smyth, "Atomic Energy for Military Purposes," Princeton University Press, Princeton, N.J., 1945.
[b]From P. Hareck, private communication to Benedict and Pigford, October 1953.
[c]From A. E. Brodskii, *Acta Physicochim. U.R.S.S.*, *17*:224, 1942.
[d]From Beyerle et al., *Chem.-Ing.-Tech.*, *21*:331, 1949, Beiheft No. 59, pp. 1–72.

Several important contributions in these two areas are presented describing current researches.

To improve the separational performance of the countercurrent flow gas centrifuge, the effect of imposing either a thermal gradient [10–13] or an electric magnetic force field [14] on a centrifugal force field have been analyzed. However, these analyses are not included in the discussion here because they are still in a preliminary stage.

1 GAS CENTRIFUGES IN CASCADE SEPARATION [1]

In a mixture, one isotope is usually the more valuable and constitutes the desired product of a separation plant (e.g., ^{235}U). It is customary to express composition in a mixture of two isotopes in terms of the atom fraction of the desired isotope. In the discussions to follow, x will be used to represent *atom fractions*, with superscripts denoting the location of a stream in a separation unit and subscripts denoting the serial number of a unit or stage in a cascade. In a binary mixture, *the abundance ratio* ξ is defined as the ratio of the number of atoms of desired isotope to the number of atoms of the other isotope. In terms of atom fractions, the abundance ratio is given by

$$\xi = \frac{x}{1 - x} \tag{3.1.1}$$

The simplest type of separating unit or stage is one which receives one feed stream and delivers two product streams, one partially enriched in desired isotope, the other partially depleted. Figure 3.1 illustrates flow through such a unit or stage. Feed, with flow rate L and composition x, is separated into a heads fraction somewhat enriched in the desired isotope, with flow rate L' and composition x', and a tails fraction somewhat depleted in the desired isotope, with flow rate L'' and composition x''.

By definition,

$$x'' < x < x' \tag{3.1.2}$$

By material balance on both isotopes,

$$L = L' + L'' \tag{3.1.3}$$

and on a desired isotope,

$$Lx = L'x' + L''x'' \tag{3.1.4}$$

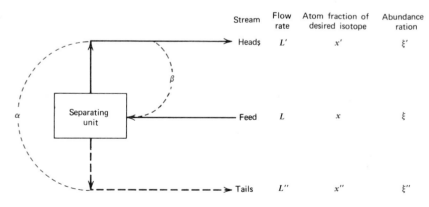

Fig. 3.1 Schematic diagram of stage flow rates and composition *Source:* Benedict and Pigford, *Nuclear Chemical Engineering*, Ref. 1, by permission of McGraw-Hill.

The ratio of head to feed is known as the *cut* θ and is given as

$$\theta = \frac{L'}{L} \tag{3.1.5}$$

so that (3.1.4) may be written

$$x = \theta x' + (1 - \theta)x'' \tag{3.1.6}$$

or

$$\theta = \frac{x - x''}{x' - x''} \tag{3.1.7}$$

Separation Factors

The degree of separation achieved by a single stage is known as the *stage separation factor*, or simply the *separation factor* α. This is defined as the abundance ratio in the heads stream to the abundance ratio in the tails:

$$\alpha = \frac{\xi'}{\xi''} = \frac{x'}{x''} \frac{(1 - x'')}{(1 - x')} \tag{3.1.8}$$

In an ideal stage between liquid and vapor, for example, the separation factor equals the relative volatility. The separation factor defined in this way is useful because it is independent of composition. The ratio x'/x'', on the other hand, varies strongly with composition.

The separation factor for a single centrifugal unit may be obtained from a mechanical equilibrium distribution of components. When a gas of mass density ρ is subjected to centrifugal acceleration of $\omega^2 r$, it experiences a force of $\rho\omega^2 r$ per unit volume. At mechanical equilibrium, from (2.5.1) one has

$$\frac{dp}{dr} = \rho\omega^2 r = \frac{Mp}{RT}\omega^2 r \tag{3.1.9}$$

in which $\omega^2 r$ is substituted for **F**. Equation 3.1.9 may be integrated to give the ratio of pressure between radius r' and radius r'' as

$$\frac{p''}{p'} = \exp\left\{\frac{M\omega^2[(r'')^2 - (r')^2]}{2RT}\right\} \tag{3.1.10}$$

This result is similar to the exponential law for the change of barometric pressure with altitude.

In a mixture of gases, similar results are obtained for the partial pressure of each component. In a binary mixture, at equilibrium:

Light component:

$$\frac{p''x''}{p'x'} = \exp\left\{\frac{M_1\omega^2[(r'')^2 - (r')^2]}{2RT}\right\} \qquad (3.1.11)$$

Heavy component:

$$\frac{p''(1 - x'')}{p'(1 - x')} = \exp\left\{\frac{M_2\omega^2[(r'')^2 - (r')^2]}{2RT}\right\} \qquad (3.1.12)$$

where M_1 and M_2 are molecular weights of light and heavy components and x' and x'' are the mole fractions of light component at radii r' and r'', respectively. The separation factor α for the single centrifugal unit is obtained from (3.1.8) as

$$\alpha = \frac{\xi'}{\xi''} = \exp\left\{\frac{(M_2 - M_1)\omega^2[(r'')^2 - (r')^2]}{2RT}\right\} \qquad (3.1.13)$$

It is important to note that α is a function of the difference in molecular weights rather than the ratio $\Delta M/M$, as was the case in the diffusion methods. This gives the centrifuge an important advantage for separating isotopes of heavy elements. Comparisons of separation factors for various isotope separation processes, as given by Benedict and Pigford, are presented in Table 3.2.

Another useful measure of the degree of separation effected by a stage is the *heads separation factor* β, defined by

$$\beta = \frac{\xi'}{\xi} = \frac{x'}{x}\frac{1 - x}{1 - x'} \qquad (3.1.14)$$

Clearly, one has

$$1 < \beta < \alpha \qquad (3.1.15)$$

From the definition of α and β, it follows that

$$x' = \frac{\beta x}{1 + x(\beta - 1)} = \frac{\alpha x''}{1 + x''(\alpha - 1)} \qquad (3.1.16)$$

$$x = \frac{x'}{x + \beta(1 - x')} = \frac{\alpha x''}{\alpha x'' + \beta(1 - x'')} \qquad (3.1.17)$$

$$x'' = \frac{x'}{x' + \alpha(1 - x')} = \frac{\beta x}{\beta x + \alpha(1 - x)} \qquad (3.1.18)$$

A relation between β, α, and θ may be obtained from (3.1.16), (3.1.18), and (3.1.6):

$$\beta - 1 = \frac{(\alpha - 1)(1 - \theta)}{1 + \theta(\alpha - 1)(1 - x')} \qquad (3.1.19)$$

Ideal Cascades

One type of separational design that is easy to treat theoretically, which leads to minimum total interstage flow and which is approximated by all isotope-separation plants designed for minimum cost and the most efficient cascade, is the so-called ideal cascade. An ideal cascade is one in which [15] (1) the heads separation factor β is constant, (2) the heads stream and tails stream fed to each stage have the same composition:

$$x''_{i+1} = x'_{i-1} = x_i \qquad (3.1.20)$$

Clearly, the remixing of materials of different mole fraction, to minimize the entropy change in the cascade operation, is to be avoided.

The above condition may also be written in terms of abundance ratios:

$$\xi''_{i+1} = \xi'_{i-1} = \xi_i \qquad (3.1.21)$$

From the definition of the heads separation factor,

$$\xi_i' = \beta \xi_i \qquad (3.1.22)$$

In an ideal cascade, because of (3.1.21), one has

$$\xi_i' = \beta \xi''_{i+1} \qquad (3.1.23)$$

Similarly,

$$\xi'_{i+1} = \beta \xi_{i+1} = \xi_i' \qquad (3.1.24)$$

By multiplying (3.1.22) and (3.1.23) together, one obtains

$$\xi'_{i+1} = \beta^2 \xi''_{i+1} \qquad (3.1.25)$$

But from the definition of the stage separation factor, (3.1.8),

$$\xi'_{i+1} = \alpha \xi''_{i+1} \qquad (3.1.26)$$

So that

$$\beta^2 = \alpha \quad \text{or} \quad \beta = \alpha^{1/2} \qquad (3.1.27)$$

Table 3.2 Comparison of Separation Factors

Separation Method	Isotopes					
	H–D		C^{12}–C^{13}		^{235}U–^{238}U	
	Working Substances	Separation Factor	Working Substances	Separation Factor	Working Substances	Separation Factor
Electrolysis	H_2–H_2O	7				1.000
Chemical exchange	H_2–H_2O	2.8	HCN–CN–CO	1.013		
Distillation	H_2	1.81		1.01		1.000
Gaseous diffusion (minimum power)	H_2	1.20	CH_4	1.030	UF_6	1.0042
Mass diffusion (minimum vapor)	H_2–H_2O	1.19	CH_4–H_2O	1.016	UF_6–C_7F_{16}	1.0022
Thermal diffusion (minimum power)	H_2	1.052	CH_4	1.0026	UF_6	?
Gas centrifuge[a] (minimum length)	H_2	1.009	CH_4	1.009	UF_6	1.026

Source: M. Benedict and T. H. Pigford, *Nuclear Chemical Engineering*, McGraw-Hill, New York, 1957 p. 515 (by permission of McGraw-Hill).

[a] Peripheral velocity = 25,100 cm/sec.

This relationship between the heads separation factor and the stage separation factor is the key property of an ideal cascade.

A relation between composition and stage number may be derived from (3.1.24) and (3.1.22) to yield

$$\xi_n{}' = \beta^{n-i}\xi_i{}' = \beta^{n-i+1}\xi_i = \beta^{n-i+1}\xi_{i+1}'' \tag{3.1.28}$$

Then, letting $i = 0$, one obtains

$$n = \frac{\ln \dfrac{x_p(1 - x_w)}{x_w(1 - x_p)}}{\ln \beta} - 1 = 2\,\frac{\ln \dfrac{x_p(1 - x_w)}{x_w(1 - x_p)}}{\ln \alpha} - 1 \tag{3.1.29}$$

Thus, the number of stages required for a given separation in an ideal cascade is just twice the minimum number needed at total reflux, Fenske's equation [16], minus 1. The quantities x_p and x_w are compositions of product enriched in the desired isotope and of waste depleted in the desired isotope.

By analogy with fractional distillation, the material balances in a cascade must satisfy

Overall balance:

$$F = P + W \tag{3.1.30}$$

Component balance:

$$Fx_F = Px_p + Wx_w \tag{3.1.31}$$

where F, P, and W are molar flow rate of feed, product, and waste, respectively. If L_i is the stream flow rate in the ith stage above the feed stage and the cascades are numbered from the product end down in the enriching section, the stage material balances are

Overall balance:

$$L_i{}' = L_{i+1}'' + P \tag{3.1.32}$$

Component balance:

$$L_i{}'x_i{}' = L_{i+1}''x_{i+1}'' + Px_p \tag{3.1.33}$$

And in the stripping section, where the direction of net flow is reversed, stage material-balances are

Overall balance:

$$L_j{}' = L_{j+1}'' - W \tag{3.1.34}$$

Component balance:

$$L_j{}'x_j{}' = L_{j+1}''x_{j+1}'' - Wx_w \tag{3.1.35}$$

A relation for the difference in composition between heads from one stage $x_i{}'$ and tails from the next higher stage x_{i+1}'' can be obtained from (3.1.32) and (3.1.33):

$$x_{i+1}'' - x_i{}' = \frac{x_i{}' - x_p}{L_{i+1}''/p} \qquad (3.1.36)$$

At total reflux $L_{i+1}''/p \to \infty$, x_{i+1}'' and $x_i{}'$ are equal.

The reflux ratio required to bring about conditions defining an ideal cascade can be found from (3.1.36):

$$\frac{L_{i+1}''}{P} = \frac{x_p - x_i{}'}{x_i{}' - x_{i+1}''} \qquad (3.1.37)$$

Since $x_{i+2}'' = x_i{}' = x_{i+1}$ for an ideal cascade and x_{i+1}' is given in terms of x_{i+1}'' by (3.1.16) with $\alpha = \beta^2$, then

$$\frac{L_{i+1}''}{P} = \frac{x_p - \beta^2 x_{i+1}''/[\beta^2 x_{i+1}'' + \beta(1 - x_{i+1}'')]}{\beta^2 x_{i+1}''/[\beta^2 x_{i+1}'' + \beta(1 - x_{i+1}'')] - x_{i+1}''}$$

$$= \frac{x_p[\beta^2 x_{i+1}'' + \beta(1 - x_{i+1}'')] - \beta^2 x_{i+1}''}{\beta^2 x_{i+1}'' - x_{i+1}''[\beta^2 x_{i+1}'' + \beta(1 - x_{i+1}'')]}$$

$$= \frac{x_p[\beta x_{i+1}'' + 1 - x_{i+1}''] - \beta x_{i+1}''}{(\beta - 1)x_{i+1}''(1 - x_{i+1}'')}$$

$$= \frac{1}{(\beta - 1)}\left[\frac{x_p}{x_{i+1}''} - \frac{\beta(1 - x_p)}{1 - x_{i+1}''}\right] \qquad (3.1.38)$$

In the stripping section, the reflux ratio corresponding to (3.1.38) is

$$\frac{L_j{}'}{W} = \frac{x_{j+1}'' - x_w}{x_j{}' - x_{j+1}''} \qquad (3.1.39)$$

The corresponding relation in the stripping section for an ideal cascade is $x_j'' = x_{j-1} = x_{j-2}'$, so that

$$\frac{L_j{}'}{W} = \frac{1}{\beta - 1}\left[\frac{1 - x_w}{1 - x_j{}'} - \frac{\beta x_w}{x_j{}'}\right] \qquad (3.1.40)$$

An equation for the reflux ratio in the enriching section as a function of stage number may be obtained from (3.1.28) together with the condition of an ideal cascade (3.1.20) for x_{i+1}'' as

$$x_{i+1}'' = x_i = \frac{x_p \beta^{i-1}}{\beta^n(1 - x_p) + x_p \beta^{i-1}} \tag{3.1.41}$$

Then, substituting (3.1.41) into (3.1.38),

$$\frac{L_{i+1}''}{P} = \frac{1}{\beta - 1} [x_p(1 - \beta^{i-n}) + (1 - x_p)\beta(\beta^{n-i} - 1)] \tag{3.1.42}$$

Similarly, in the stripping section, the reflux ratio in terms of stage number becomes

$$\frac{L_j'}{W} = \frac{1}{\beta - 1} [x_w \beta(\beta^j - 1) + (1 - x_w)(1 - \beta^{-j})] \tag{3.1.43}$$

Total Flow Rates

The total interstage flow rate of heads or tails is a measure of the size of the separation plant. An expression for the total flow rate may be derived by summing (3.1.42) and (3.1.43). The total heads flow rate in the stripping section is obtained from (3.1.43):

$$J_w' = \sum_{j=1}^{f-1} L_j' = \frac{W}{\beta - 1} [x_w \beta \sum_{j=1}^{f-1} (\beta^j - 1) + (1 - x_w) \sum_{j=1}^{f-1} (1 - \beta^{-j})] \tag{3.1.44}$$

The quantity f is the feed plate location in the cascade. Since

$$\sum_{j=1}^{f-1} \beta^i = \beta \frac{\beta^{f-1} - 1}{\beta - 1}$$

(3.1.44) becomes

$$J_w' = \frac{W}{\beta - 1} \left\{ \beta x_w \left[\beta \frac{\beta^{f-1} - 1}{\beta - 1} - (f - 1) \right] \right.$$

$$\left. + (1 - x_w) \left[(f - 1) - \frac{1 - \beta^{-(f-1)}}{\beta - 1} \right] \right\} \tag{3.1.45}$$

The term $f - 1$ may be obtained from (3.1.29) by replacing x_p by x_F, which gives

$$f - 1 = \frac{\ln \dfrac{x_F(1 - x_w)}{x_w(1 - x_F)}}{\ln \beta} - 1 \quad \text{or} \quad \beta^f = \frac{x_F(1 - x_w)}{x_w(1 - x_F)} \tag{3.1.46}$$

Then,

$$J_w' = \frac{W}{\beta - 1} \left\{ [1 - x_w(\beta + 1)] \frac{\ln \dfrac{x_F(1 - x_w)}{x_w(1 - x_F)}}{\ln \beta} \right.$$
$$\left. - \frac{\beta(x_F - x_w)(1 - 2x_F)}{(\beta - 1)x_F(1 - x_F)} \right\} \tag{3.1.47}$$

By a similar procedure, the total tails flow rate in the stripping section is

$$J_w'' = \sum_{j=1}^{f-1} L_j'' = \frac{W}{\beta - 1} \left\{ [\beta - x_w(\beta + 1)] \frac{\ln \dfrac{x_F(1 - x_w)}{x_w(1 - x_F)}}{\ln \beta} \right.$$
$$\left. - \frac{(x_F - x_w)[\beta^2 - (\beta^2 + 1)x_F]}{x_F(1 - x_F)(\beta - 1)} \right\} \tag{3.1.48}$$

The total tails flow rate in the enriching section is obtained from (3.1.42):

$$J_p'' = \sum_{i=f-1}^{n} L_i'' = \frac{P}{\beta - 1} \left\{ \frac{[x_p(\beta + 1) - \beta] \ln \dfrac{x_p(1 - x_F)}{x_F(1 - x_p)}}{\ln \beta} \right.$$
$$\left. + \frac{(x_p - x_F)}{x_F(1 - x_F)} \frac{[\beta^2 - (\beta^2 + 1)x_F]}{\beta - 1} \right\} \tag{3.1.49}$$

and the total heads flow rate in the enriching section is

$$J_p' = \sum_{i=f-1}^{n} L_i' = \frac{P}{\beta - 1} \left\{ \frac{[x_p(\beta + 1) - 1] \ln \dfrac{x_p(1 - x_F)}{x_F(1 - x_p)}}{\ln \beta} \right.$$
$$\left. + \beta \frac{(x_p - x_F)(1 - 2x_F)}{x_F(1 - x_F)(\beta - 1)} \right\} \tag{3.1.50}$$

The total flow in the entire cascade, J_T, is the sum of $J_p{}'$, $J_p{}''$, $J_w{}'$, and $J_w{}''$:

$$J_T = \frac{\beta + 1}{(\beta - 1) \ln \beta} [W(1 - 2x_w) \ln \frac{x_F(1 - x_w)}{x_w(1 - x_F)}$$

$$+ P(2x_p - 1) \ln \frac{x_p(1 - x_F)}{x_F(1 - x_p)} \qquad (3.1.51)$$

By use of (3.1.30) and (3.1.31), (3.1.51) can be written

$$J_T = \frac{\beta + 1}{(\beta - 1) \ln \beta} [W(2x_w - 1) \ln \xi_w$$

$$+ P(2x_p - 1) \ln \xi_p - F(2x_F - 1) \ln \xi_F] \qquad (3.1.52)$$

Equation 3.1.52 states that the total flow in the plant is the product of two factors, the first a function only of the heads separation factor and the second a function only of the flow rates and compositions of feed, product, and waste. The first factor is a measure of the relative ease or difficulty of the separation; it is large when β is close to unity and small when β differs markedly from unity. The second factor is a measure of the magnitude of the job of separation; it is proportional to the throughput and is large when product and waste differ substantially in composition from feed and is small when these compositions are nearly equal. The second factor has been termed the *separative duty*, which has the same dimensions as those used for the flow rates. It is a measure of the rate at which a cascade is performing separation.

The total internal flow rate in an ideal cascade, J_T, can be generalized into a convenient form:

$$J_T = \frac{\beta + 1}{(\beta - 1) \ln \beta} U \qquad (3.1.53)$$

where U is the separative duty of a cascade and is given by

$$U = \sum_i L_i \phi_i \qquad (3.1.54)$$

and

$$\phi_i = (2x_i - 1) \ln \xi_i \qquad (3.1.55)$$

in which L_i is any member of streams of molar flow rate, positive for a product and negative for a feed; and x_i and ξ_i are the composition and abundance ratio of that stream, respectively. The function ϕ_i defined by (3.1.55) is called the *separation potential*.

The importance of the separative duty in isotope separation lies in the fact that it is a measure of the magnitude of an isotope separation job. Many of the characteristics of the separation system that make important contributions to its cost are proportional to the separative duty.

The separative duty is analogous to the heat duty of an evaporator or to the number of transfer units in a countercurrent diffusional operation, such as a packed column used in gas absorption or humidification and dehumidification equipment. The separation potential is analogous to the enthalpy per mole of the streams entering or leaving an evaporator, or to the concentration, or to the vapor content per unit weight of dry gas of the streams entering or leaving a packed column.

The first factor in a total flow rate which indicates the degree of difficulty in a separation is analogous to the height of the transfer unit in a packed column operation. It is a characteristic of the system.

The Separative Power

A net change in the separative duty is called *the separative power*. With the simplest separating unit describing the cut and the material balance given in (3.1.3) through (3.1.6), the net variation of the separating duty for a desired component is

$$\delta U = \theta L \phi(x') + (1 - \theta) L \phi(x'') - L \phi(x) \qquad (3.1.58)$$

Expanding $\phi(x')$ and $\phi(x'')$ in Taylor's series about x, (3.1.58) becomes

$$\delta U = \phi(x)[\theta L + (1 - \theta)L - L] + \frac{d\phi}{dx}[\theta L(x' - x)$$

$$+ (1 - \theta)L(x'' - x)] + \frac{d^2\phi}{dx^2}\left[\theta L \frac{(x' - x)^2}{2} + (1 - \theta)L \frac{(x'' - x)^2}{2}\right]$$

$$+ \cdots \quad (3.1.59)$$

The coefficients of $\phi(x)$ and $d\phi/dx$ vanish by the conservation of mass given in (3.1.6), which can also be rearranged to give

$$\phi(x' - x) = -(1 - \theta)(x'' - x) \qquad (3.1.60)$$

Then, from (3.1.1) and (3.1.14), one obtains

$$x' - x = \frac{\xi'}{1 + \xi'} - \frac{\xi}{1 + \xi} = \frac{\xi' - \xi}{(1 + \xi')(1 + \xi)}$$

$$= (\beta - 1)x(1 - x') \quad (3.1.61)$$

Thus, (3.1.59) becomes

$$\delta U = \frac{\theta}{1 - \theta} \frac{L(\beta - 1)^2}{2} \frac{d^2\phi}{dx^2} [x(1 - x)]^2 \qquad (3.1.62)$$

Since the separation potential ϕ defined in (3.1.55) is a function only of composition and is dimensionless, one obtains

$$\frac{d^2\phi}{dx^2} = \frac{1}{x^2(1 - x^2)} \qquad (3.1.63)$$

Then, the separative power δU, (3.1.62), reduces to

$$\delta U = \frac{\theta}{1 - \theta} \frac{L(\beta - 1)^2}{2} \qquad (3.1.64)$$

2   BASIC DIFFERENTIAL EQUATION OF THE GAS CENTRIFUGE

The starting point for the derivation is the continuity equation (2.1.18) and the mass-average velocity of the mixture, (2.1.21). If $\mathbf{v}_i$ is the velocity of the component i, one can define a vector $\mathbf{u}_i$ as the velocity of the ith component in the mixture relative to the mass-average velocity $\mathbf{v}$:

$$\mathbf{u}_i = \mathbf{v}_i - \mathbf{v} \qquad (3.2.1)$$

With the use of this definition, the continuity equation (2.1.18) for a non-reacting system becomes

$$\frac{\partial \rho_i}{\partial t} + \nabla \cdot \rho_i (\mathbf{v} + \mathbf{u}_i) = 0 \qquad (3.2.2)$$

In the rigorous kinetic theory of ν-component gas mixtures, the diffusion velocity of the ith component is [17]

$$\mathbf{u}_i = \left(\frac{n^2}{n_i \rho}\right) \sum_{j=1}^{\nu} m_j D_{ij} d_j - \frac{D_i^T}{n_i m_i} \nabla \ln T \qquad (3.2.3)$$

where $n = \Sigma_i n_i$ and denotes the total number density of the mixture, n_i is the number density (number of molecules per unit volume), m_i is the mass of a molecule of the ith species, and $\rho = \Sigma_i n_i m_i$ and denotes the overall density of the gas. D_{ij} and D^T are the ordinary and thermal diffusion coefficients, and the symbol d_j includes the gradients of the mole fraction and pressure and also the effects of the external forces, $\mathbf{F}_K$, acting on the molecules:

$$d_j = \nabla \left(\frac{n_j}{n} \right) + \left(\frac{n_j}{n} - \frac{n_j m_j}{\rho} \right) \nabla \ln p$$

$$- \left(\frac{n_j m_j}{p\rho} \right) \left[\frac{\rho}{m_j} \mathbf{F}_j - \sum_{K=1}^{\nu} n_K \mathbf{F}_K \right] \quad (3.2.4)$$

For a binary mixture, one obtains the diffusion velocity of the desired component 1 (excluding all the external forces) from (3.2.3) and (3.2.4):

$$\mathbf{u}_1 = - \frac{M_2}{x_1 \langle M \rangle} D_{12} \left[\nabla x_1 + \frac{x_1 x_2 (M_2 - M_1)}{\langle M \rangle} \nabla \ln p \right]$$

$$- \frac{D^T}{x_1 M_1} \nabla \ln T \quad (3.2.5)$$

where x_i is the mole fraction of species i in the mixture and M_i is the molecular weight of species i; $\langle M \rangle$ is the molecular weight of the mixture defined by $\langle M \rangle = x_1 M_1 + x_2 M_2$.

If n is the molar density of the mixture, then the density of component 1 in the mixture is $\rho_1 = n x_1 M_1$. Assuming that the gas is ideal and the equation of state is $p/\rho = nRT$ and that the gas is isothermal, the continuity equation (3.2.2), together with the diffusion velocity (3.2.5), leads to

$$p \frac{\partial x_1}{\partial t} = - \nabla \cdot (p x_1 \mathbf{v})$$

$$+ \nabla \cdot \left\{ \frac{M_2}{\langle M \rangle} p D_{12} \left[\nabla x_1 + \frac{x_1 (1 - x_1)(M_2 - M_1)}{\langle M \rangle} \nabla \ln p \right] \right\} \quad (3.2.6)$$

Further assuming that the gas is rotating uniformly at a constant angular velocity ω, that a hydrodynamic steady state prevails, that there are no pressure gradients axially and azimuthally, that the contribution of viscosity in the equation of motion of fluid vanishes, and that there is a pseudoequilibrium in a radial direction, the continuity equation (3.2.6), written for the coordinate system of interest, namely, cylindrical coordinates, becomes

$$p \frac{\partial x_1}{\partial t} = \frac{1}{r} \frac{\partial}{\partial r} \left\{ \frac{M_2}{\langle M \rangle} p D_{12} r \left[\frac{\partial x_1}{\partial r} + \frac{x_1 (1 - x_1)(M_2 - M_1)}{RT} \omega^2 r \right] \right\}$$

$$+ \frac{\partial}{\partial z} \left\{ \frac{M_2}{\langle M \rangle} p D_{12} \frac{\partial x_1}{\partial z} \right\} - \frac{1}{r} \frac{\partial}{\partial r} (p x_1 r \mathbf{v}_r) - \frac{\partial}{\partial z} (p x_1 \mathbf{v}_z) \quad (3.2.7)$$

where v_r and v_z are the radial and axial component of the mass-average velocity v, respectively.

The binary diffusion coefficient D_{12} for a dilute gas may be expressed from the kinetic theory [17] to give

$$D_{12} = 2.628 \times 10^{-3} \frac{[T^3(M_1 + M_2)/2M_1M_2]^{1/2}}{p\sigma_{12}^2 \Omega_{12}^{(1,1)*}(T_{12}^*)} f_D \quad \text{(in cm}^2\text{/sec)}$$

$$(3.2.8)$$

where p is the pressure (in atmospheres); T is temperature (in °K); $\Omega_{12}^{(1,1)*}(T_{12}^*)$ is the collision integral for ordinary diffusion; $T_{12}^* = KT/\epsilon_{12}$ denotes the reduced temperature; σ_{12} and ϵ_{12} are parameters in the intermolecular potential function in Å and °K, respectively; and f_D is the correction factor for the second approximation, which is nearly unity. Since D_{12} is inversely proportional to p, at an isothermal condition, the quantity $D_{12}P$ is nearly constant.

Further simplifications in the basic centrifuge equation (3.2.7) are possible for the separation of uranium 235. If either $(M_2 - M_1)/M_2 \ll 1$, or $x_1 \ll 1$, or both are true, then $M_2/\langle M \rangle$ is approximately equal to unity. The mass-average velocity of the gas mixture in a cascade centrifuge may be given by the components

$$v_r = v_r(r,z) \quad (3.2.9a)$$

$$v_\theta = 0 \quad (3.2.9b)$$

$$v_z = v_z(r,z) \quad (3.2.9c)$$

3 SEPARATIONAL METHODS IN GAS CENTRIFUGES

In gas centrifuges, three types of separation methods have been attempted. The idea of the evaporative centrifuge was introduced in early 1920 by Mulliken [18]. In his method, a small amount of liquid is introduced into the centrifuge, forming a layer at the periphery. During the rotation of the rotor, vapor is removed slowly through a shaft along the axis. In this way, a simple differential distillation, or Rayleigh distillation, of the liquid is accomplished. An increased concentration change in the residual liquid is thus obtained. In 1938, Beams [19] successfully developed vacuum chamber centrifuges, which were free from vibration and thermally isolated to eliminate convective currents, to separate isotopes. Beams and Skarstrom [20,21] at the University of Virginia, using the evaporative centrifuge method on CCl_4, reported a 13% change in the ^{35}Cl to ^{37}Cl ratio. Shortly thereafter, Humphreys [22], using the same technique on ethyl bromide, altered the ^{79}Br to ^{81}Br abundance ratio by 11%.

The simple flowthrough centrifugation, as it was called the concurrent flow method, is a method flowing a gaseous mixture continuously through a cream separator. A single stream of gas enters one end of a rotor through a hollow shaft, and two streams are taken off the other end, one from the periphery and the other near the axis. This method produces a small change in concentration per machine.

The countercurrent flow type was designed to attain considerable separation in a single centrifuge, thus reducing the number of stages required and the amount of material circulated between stages. As originally proposed by Urey [23], circulation was to be established by continuous distillation of UF_6 from the bottom cap of the rotor. The vapor was to be condensed on the top cap. The heavy liquid would then be forced out to the periphery and would flow down the walls to the bottom cap, countercurrent to the vapor flow, and complete the cycle.

The theory of gas centrifugal separation methods is presented here following the theory developed by Cohen [2].

The Evaporative Centrifuge

The evaporative centrifuge method limits the amount of the end product, since only a definite amount of material can be placed initially in the centrifuge rotor. If one wishes to know the change in the abundance of the vapor either at the center or at the edge, it is necessary to know one member of this ratio apart from the other. Also, if vapor is removed from the center of the rotor at a finite rate, the composition of the mixture will never reach a steady value, so x_1 and x_2 will be functions of time as well as distance. As in any enrichment method, the rate of change of composition of the residue depends on the number of molecules removed:

$$\frac{\partial N_1(t)}{\partial t} \sim N'(t) \quad \text{or} \quad \frac{\partial N_1(t)/\partial t}{\partial N_2(t)/\partial t} = \frac{N_1'(t)}{N_2'(t)}$$

where $N_1(t)$ is the total number of molecule 1 at any time t throughout the rotor, $N(t)$ is the total number of all molecules present, and $N_1'(t)$ is the number of molecule 1 removed from the rotor. However, the composition of the vapor removed will be the same as that at the center, so

$$\frac{N_1'(t)}{N_2'(t)} = \frac{x_1(0,t)}{x_2(0,t)}$$

Except for the negligibly small number of molecules in the vapor state, the total number of molecule 1 in the rotor will be at the edge in the liquid state.

Let R be the distance from center to periphery, that is, $R \geq r \geq 0$. Then

$$N_1(t) \cong N_1(R,t)$$

Therefore,

$$\frac{\partial N_1(R,t)}{\partial N_2(R,t)} = \xi(0,t) = \alpha\xi(R,t) = \alpha\frac{N_1(R,t)}{N_2(R,t)} \tag{3.3.1}$$

After integration of (3.3.1), one has

$$\ln \frac{N_1(R,T)}{[N_2(R,t)]^\alpha} = \text{constant} \tag{3.3.2}$$

Before centrifuging at $t = t_0$, the following relationship exists:

$$N_1(R,t) = N_1(R,t_0) = N_1(t_0)$$

The above relationship implies that there is no significance to the variable r when $t = t_0$; therefore, one can also write

$$\frac{N_1(R,T)}{[N_2(R,t)]^\alpha} = \frac{N_1(t_0)}{[N_2(t_0)]^\alpha}$$

or

$$\frac{N_1(R,t)}{N_2(R,t)} = \frac{N_1(t_0)}{N_2(t_0)} \left[\frac{N_2(R,t)}{N_2(t_0)} \right]^{\alpha-1} \tag{3.3.3}$$

Since, $\alpha - 1 \ll 1$, the bracket term is almost unity. Then,

$$\frac{N_1(R,t)}{N_2(R,t)} \cong \frac{N_1(t_0)}{N_2(t_0)}$$

or

$$\frac{N(R,t)}{N_2(R,t)} \cong \frac{N(t_0)}{N_2(t_0)}$$

Therefore, by rearranging the above relation, one has

$$\frac{N_2(R,t)}{N_2(t_0)} = \frac{N(R,t)}{N(t_0)} = \frac{N(t_0) - N'(t)}{N(t_0)} = 1 - \frac{N'(t)}{N(t_0)} = 1 - \theta \tag{3.3.4}$$

in which $\theta = N'(t)/N(t_0)$, defined in (3.1.5) as the cut. Substituting (3.3.4) into (3.3.3) together with (3.3.1), one obtains

$$\xi(R,t) = \frac{\xi(0,t)}{\alpha} = \xi(t_0)[1 - \theta]^{\alpha-1} \tag{3.3.5}$$

Since the quantity $\xi(t_0)$ is the known initial abundance ratio and α can be calculated from (3.1.13), the altered abundance ratio at either the center or the edge can be calculated from an arbitrary cut θ.

In the centrifuge rotor, there are two opposing gradients set up, the concentration gradient and the pressure gradient. Each of these will cause the molecules to diffuse, the former causing a diffusion toward the edge, the latter causing a diffusion toward the center. If no material is removed from the rotor, the number of molecules diffusing per second past a given area due to one cause will equal the number due to the other cause, hence mechanical equilibrium is established as given by (2.6.4). However, as soon as vapor is removed at the center by pumping, this equilibrium is upset and never regained as long as vapor removal continues. The effect is to increase the number of molecules moving toward the center per second.

In applying the equation of the gas centrifuge to the evaporative centrifuge, x is independent of z and $\mathbf{v}_z = 0$. (The subscript 1 will be dropped, because no confusion arises.) Under the steady-state operation, the gas centrifuge equation (3.2.7) reduces to

$$\frac{d}{dr}\left\{pD_{12}\left[r\frac{dx}{dr} + \frac{x(1-x)(M_2 - M_1)}{RT}\omega^2 r^2\right]\right\} - \frac{d}{dr}(pxr\mathbf{v}_r) = 0 \tag{3.3.6}$$

The quantity $M_2/\langle M\rangle$, being approximately unity, is used in obtaining (3.3.6) from (3.2.7). Let vapor be removed from the axis at a molar flow rate L, which can be expressed as

$$L = -\rho\mathbf{v}_r A = -\frac{2\pi r Z p}{RT} \tag{3.3.7}$$

where Z is the length of centrifuge, A is the cross-sectional area at distance r parallel to the axis, and ρ is the density of a binary gas mixture. If the gas is ideal, then one has $\rho = p/RT$. By multiplying $2\pi Z/RT$ in (3.3.6) and integrating the equation, one obtains

$$\gamma\left[r\frac{dx}{dr} + \frac{\Delta M' x}{T}(1-x)r^2\right] = Lx(0) - Lx(r) \tag{3.3.8}$$

where

$$\gamma = \frac{2\pi Z p D_{12}}{RT} \tag{3.3.9}$$

$$\Delta M' = \frac{(M_2 - M_1)\omega^2}{R} \tag{3.3.10}$$

The integrating constant was evaluated by noting that the left side of (3.3.8) vanishes at $r = 0$. Now rewriting the mole fraction x in terms of the abundance ratio ξ given in (3.1.1), x becomes

$$x = \frac{\xi}{1 + \xi} \tag{3.3.11}$$

And if (3.3.8) is written in terms of ξ as a dependent variable, then

$$\gamma \left[r \frac{d\xi}{dr} + \frac{\Delta M'}{T} \xi r^2 \right] = L \left[\frac{\xi(0)}{1 + \xi(0)} - \frac{\xi(r)}{1 + \xi(r)} \right] [1 + \xi(r)]^2$$

$$= L[\xi(0) - \xi] \left[1 + \frac{\xi - \xi(0)}{1 + \xi(0)} \right] \tag{3.3.12}$$

To a first approximation, the term $(\Delta M'/T)\xi$ may be treated as a constant. Such an assumption is comparable to the assumption of $\alpha - 1 \ll 1$ and is valid for the small separations produced by a single centrifugation. Furthermore, $[\xi - \xi(0)]/[1 + \xi(0)]$ may be neglected with respect to unity. Then, after simplification, (3.3.7) becomes

$$\frac{d\xi}{dr} + \frac{L}{\gamma r} \xi = \frac{L\xi(0)}{\gamma r} - \frac{\Delta M r}{T} \tag{3.3.13}$$

where $\Delta M/T = (\Delta M'/T)\xi$. Equation 3.3.13 is a linear first-order ordinary differential equation. Using the boundary condition at $r = 0$, $\xi = \xi(0)$, the solution is readily found to be

$$\xi = \xi(0) - \left(\frac{2\gamma}{L + 2\gamma} \right) \frac{\Delta M' \xi}{2T} r^2 \tag{3.3.14}$$

Letting $\Delta \xi = \xi - \xi(0)$ and rewrite (3.3.14) in terms of the separation factor α, one has

$$\frac{\Delta \xi}{\xi} = \left(\frac{2\gamma}{L + 2\gamma}\right) \ln \alpha \tag{3.3.15}$$

If an equilibrium is established, L vanishes. The difference in the abundance ratio indicates a separation. The quantity ϵ defined as

$$\epsilon = \frac{2\gamma}{L + 2\gamma} \tag{3.3.16}$$

is a measure of the equilibrium, or rather the degree of departure from equilibrium. The effect of disturbing the equilibrium is to introduce a value of ϵ less than 1. As one would expect, removing of the vapor slowly through a shaft along the axis decreases the separation obtained; it has the effect of reducing the separation factor α to a value α', where

$$\ln \alpha' = \ln \alpha^\epsilon \qquad (\epsilon < 1) \tag{3.3.17}$$

In the evaporative centrifuge, if the product is only a small fraction of the charge, or if successive batches of product are kept separate, one may approximate $\xi \cong \xi''$; therefore, the head separating factor β may be replaced by the state separating factor α^ϵ. Hence, the separative power for the evaporative centrifuge may be written from (3.1.64) as

$$\delta U = \frac{L}{2} \frac{\theta}{1 - \theta} [\alpha^{(2\gamma/2\gamma + L)} - 1]^2 \tag{3.3.18}$$

$$\delta U \cong \frac{L}{2} \frac{\theta}{1 - \theta} \left(\frac{2\gamma}{2\gamma + L}\right)^2 (\ln \alpha)^2 \tag{3.3.19}$$

The relationship

$$a^x = 1 + x \ln a + \frac{(x \ln a)^2}{2!} + \frac{(x \ln a)^3}{3!} + \cdots \tag{3.3.20}$$

was used in (3.3.20) to obtain the approximate expression. The value of δU is maximum when $L = 2\gamma$:

$$(\delta U)_{max} = \frac{\gamma}{4} \frac{\theta}{1 - \theta} (\ln \alpha)^2 = \frac{D_{12} P}{RT} (\ln \alpha)^2 \frac{\pi Z}{2} \frac{\theta}{1 - \theta} \tag{3.3.21}$$

In the development, the gas was assumed to be isothermal. This is an oversimplification. When a vapor is withdrawn at the axis, the gas cools by expansion and a temperature gradient is induced. Although the evaporative centrifuge, with $L = 2\gamma$, delivers the maximum separative power, it is not easily adapted to continuous operations.

The Concurrent Centrifuge

In this device, axial flow is used to obtain an axial as well as a radial concentration gradient. As shown in Fig. 3.2, the gas enters in two streams at one end of the rotor and flows axially to the other end, where the streams are removed separately. During the passage through the centrifuge, the streams tend to assume a radial equilibrium distribution. The flow pattern adopted for the concurrent centrifuge consists of two thin cylindrical streams located at radii r_1 and r_2 (periphery) and flowing parallel to the z-axis. There is no radial mass flow.

A rotor with entering streams of two different concentrations has been studied by Cohen [2]. The radial velocity component $\mathbf{v}_r$ is zero, and the flows are generally so large that back diffusion is negligible. The equation describing the concurrent centrifuge for the steady state then reduces to (because no confusion will arise, the subscript 1 has been dropped):

$$\frac{1}{r} \frac{\partial}{\partial r} \left\{ pD_{12} \left[r \frac{\partial x}{\partial r} + \frac{\Delta M'}{T} r^2 x(1 - x) \right] \right\} - \frac{\partial}{\partial z} (px\mathbf{v}_z) = 0 \quad (3.3.22)$$

In the region between $r = r_1$ and $r = r_2$, $\mathbf{v}_z = 0$, so that the net change of desired component due to diffusion in a radial direction is equal to the change in an axial direction. Thus, one writes the material balance of desired component from (3.3.22) giving

$$\frac{-D_{12}\,p}{RT} 2\pi \left[r \frac{dx}{dr} + \frac{\Delta M'}{T} r^2 x(1 - x) \right] = P(z) \quad (3.3.23)$$

where $P(z)$ is the transport of desired component across a cylindrical element of unit between the two streams at $r = r_1$ and $r = r_2$. Also, the two streams have the following property:

$$P(z) = L_2 \frac{d}{dz} x(r_2,z) = -L_1 \frac{d}{dz} x(r_1,z) \quad (3.3.24)$$

and

$$L_2 x(r_2,z) + L_1 x(r_1,z) = L_2 x(r_2,0) + L_1 x(x_1,0) = B = \text{constant} \quad (3.3.25)$$

in which $L_i = L(r_i)$ is a flow rate in an axial direction of the two streams.

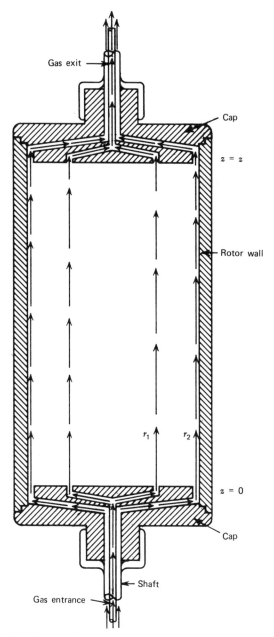

Fig. 3.2 Schematic diagram of simple continuous concurrent flow gas centrifuge. *Source:* Cohen, *The Theory of Isotope Separation,* Ref. 2, by permission of Department of Energy, U.S. Atomic Commission.

Integration of (3.3.23) between r_1 and r_2 assuming the term $(\Delta M'/T)x(1 - x)$ constant yields

$$\frac{D_{12}\,p}{RT}\,2\pi\left[x(r_2,z) - x(r_1,z) + \frac{\Delta M'}{2T}(r_2^2 - r_1^2)x(1 - x)\right] = P(z)\ln\frac{r_2}{r_1}$$

(3.3.26)

Rewriting the above equation using the relationship of (3.3.25) gives

$$\frac{D_{12}\,p}{RT}\,2\pi\left[\frac{B}{L_2} - x\left(1 + \frac{L_1}{L_2}\right) + \frac{\Delta M'}{2T}(r_2^2 - r_1^2)x(1 - x)\right]$$

$$= L_1\frac{d}{dz}x(r_1,z)\ln\frac{r_2}{r_1} \quad (3.3.27)$$

Equation 3.3.27 is a linear differential equation of the first order in $x(r_1,z)$ assuming the term $(\Delta M'/2)(r_2^2 - r_1^2)x(1 - x)$ is constant. Integration with the boundary condition $z \to 0$, $x = x(r_1,0)$ yields

$$\frac{x(r_1,z) - (1 - \theta)\left[\dfrac{B}{L_2} + \dfrac{\Delta M'}{2T}(r_2^2 - r_1^2)x(1 - x)\right]}{x(r_1,0) - (1 - \theta)\left[\dfrac{B}{L_2} + \dfrac{\Delta M'}{2T}(r_2^2 - r_1^2)x(1 - x)\right]} = e^{-bz} \quad (3.3.28)$$

where

$$\theta = \frac{L_1}{L_1 + L_2} \tag{3.3.29}$$

$$b = \frac{2\pi p D_{12}}{RT}\,\frac{1}{\theta(1 - \theta)(\ln r_2/r_1)(L_1 + L_2)} \tag{3.3.30}$$

Using the second expression in (3.3.25) for B, one can rearrange (3.3.28) into

$$x(r_1,z) - x(r_1,0)$$

$$= (1 - \theta)\left[x(r_2,0) - x(r_1,0) + \frac{\Delta M'}{2T}(r_2^2 - r_1^2)x(1 - x)\right](1 - e^{-bz})$$

(3.3.31)

In a similar manner, one also obtains

$$x(r_2,z) - x(r_2,0)$$

$$= -\theta \left[x(r_2,0) - x(r_1,0) + \frac{\Delta M'}{2T}(r_2{}^2 - r_1{}^2)x(1 - x) \right](1 - e^{-bz})$$

$$(3.3.32)$$

With the definition given in (3.1.13), the separation factor α, using the formula

$$\ln x = (x - 1) - \tfrac{1}{2}(x - 1)^2 + \tfrac{1}{3}(x - 1)^3 - \cdots \qquad (2 > x > 0)$$

neglecting the terms higher than the second order, and defining $x'(z) = x(r_1,z)$ and $x'' = x(r_2,z)$, the phases enriching the light and the heavy component, respectively, one can also rewrite (3.3.31) and (3.3.32) as

$$x'(z) - x'(0) = (1 - \theta)[x''(0) - x'(0) + (\alpha - 1)x'(1 - x')](1 - e^{-bz})$$

$$(3.3.33)$$

$$x''(z) - x''(0) = -\theta[x''(0) - x'(0) + (\alpha - 1)x'(1 - x')](1 - e^{-bz})$$

$$(3.3.34)$$

When $z \rightarrow \infty$, the last term in (3.3.33) and (3.3.34) approaches unity, so that

$$\epsilon = 1 - e^{-bz} \qquad (3.3.35)$$

which is a measure of the departure from equilibrium.

As described in (3.1.58), the net variation of the separative duty, called the separative power for the concurrent centrifuge, may be written as

$$\delta U = L_1 \phi[x'(z)] - L_1 \phi[x'(0)] + L_2 \phi[x''(z)] - L_2 \phi[x''(0)] \qquad (3.3.36)$$

Expanding the separating potentials $\phi[x'(z)]$, $\phi[x''(z)]$, and $\phi[x''(0)]$ about $\phi[x'(0)]$, the coefficients of $\phi[x'(0)]$ and $d\phi/dx'(0)$ vanish by the conservation of mass given in (3.3.25). Thus, neglecting orders higher than the second-order terms, one obtains the separative power as

$$\delta U = \frac{d^2\phi}{[dx'(0)]^2} \left\{ \frac{L_1}{2}[x'(z) - x'(0)]^2 + \frac{L_2}{2}[x''(z) - x'(0)]^2 \right.$$

$$\left. - \frac{L_2}{2}[x''(0) - x'(0)]^2 \right\} \qquad (3.3.37)$$

Again using the relation of (3.3.25), the above can be rewritten as

$$\delta U = L_1[x'(z) - x(0)] \frac{d^2\phi}{[dx'(0)]^2} \{\frac{1}{2}[x'(z) - x'(0)]$$

$$+ \frac{1}{2}[x''(0) - x''(z)] + x'(0) - x''(0)\} \quad (3.3.38)$$

Substituting (3.3.33) and (3.1.63) into (3.3.38), the separative power becomes

$$\delta U = \frac{L_1(1 - \theta)\epsilon(2 - \epsilon)}{2[x'(0)]^2[1 - x'(0)]^2} [(\alpha - 1)x'(1 - x') - x'(0) + x''(0)]$$

$$\times \left[\frac{\epsilon}{2 - \epsilon} (\alpha - 1)x'(1 - x') + x'(0) - x''(0) \right] \quad (3.3.39)$$

It is seen that δU is a function of ϵ and $x'(0)$ and $x''(0)$.

If the gas enters in two streams at one end of the rotor, the maximum separative power is obtained by differentiating (3.3.39) with respect to $x'(0) - x''(0)$. It is found to be a maximum when

$$x'(0) - x''(0) = \frac{1 - \epsilon}{2 - \epsilon}(\alpha - 1)x'(0)[1 - x'(0)] \quad (3.3.40)$$

Substituting (3.3.40) into (3.3.39) together with (3.3.35), the maximum separative power per unit length of the device with respect to $[x'(0) - x''(0)]$, the difference in inlet streams concentration is

$$\left(\frac{\delta U}{Z} \right)_{\text{max}}^{\Delta x} = \frac{-\epsilon(\alpha - 1)^2}{(2 - \epsilon) \ln (1 - \epsilon)} \frac{\pi p D_{12}}{RT \ln (r_2/r_1)} \quad (3.3.41)$$

If one takes the limiting value of the separation factor and lets $\Delta M' r_2^2/2 = \ln \alpha_0$ together with (3.3.35), the above may be rewritten as

$$\left(\frac{\delta U}{Z} \right)_{\text{max}}^{\Delta x} = \frac{(\ln \alpha_0)^2(1 - e^{-bz})}{bz(1 + e^{-bz})} \frac{\pi p D_{12}}{RT} \frac{[1 - (r_1/r_2)^2]^2}{\ln (r_2/r_1)} \quad (3.3.42)$$

Differentiating with respect to r_1/r_2 for any value of bz, the maximum separating power per unit length is found when $(r_1/r_2)_{\text{opt}} = 0.534$. The maximum separating power then becomes

$$\left(\frac{\delta U}{Z} \right)_{\text{max}}^{\Delta x, r} = 0.4073 \frac{\pi p D_{12}}{RT} \frac{(\ln \alpha_0)^2}{bz} \left(\frac{1 - e^{-bz}}{1 + e^{-bz}} \right) \quad (3.3.43)$$

If one lets $x'(0) = x''(0)$, which corresponds to no mixing for the single gas stream entering the centrifuge, the separating power may be obtained by replacing L_1 by $(L_1 + L_2)\theta$ from (3.3.29), which gives

$$\delta U = \tfrac{1}{2}(L_1 + L_2)\theta(1 - \theta)(\alpha - 1)^2\epsilon^2 \qquad (3.3.44)$$

The separating power per unit length for single-stream entry at the end of the rotor per unit length may be obtained by using (3.3.35) and (3.3.30) for an expression for a length Z to yield

$$\frac{\delta U}{Z} = -\frac{\pi p D_{12}}{RT} \frac{(\alpha - 1)^2}{\ln (r_2/r_1)} \frac{\epsilon^2}{\ln (1 - \epsilon)} \qquad (3.3.45)$$

which has a maximum at $\epsilon = 1 - e^{-bz} = 0.715$, or $bz = 1.2553$. Thus, one obtains

$$\left(\frac{\delta U}{Z}\right)^\epsilon_{\max} = 0.4073 \frac{\pi p D_{12}}{RT} \frac{(\alpha - 1)^2}{\ln (r_2/r_1)} \qquad (3.3.46)$$

If $(\Delta M'/2)r_2^2 = \ln \alpha_0$ is used, (3.3.45) may be rewritten as

$$\frac{\delta U}{Z} = \frac{(\ln \alpha_0)^2(1 - e^{-bz})^2}{bz} \frac{\pi p D_{12}}{RT} \frac{[1 - (r_1/r_2)^2]^2}{\ln (r_2/r_1)} \qquad (3.3.47)$$

The maximum with respect to r_1/r_2 is the same as $(r_1/r_2)_{\text{opt}} = 0.534$. Substituting $\epsilon = 0.715$ into (3.3.41), one obtains

$$\left(\frac{\delta U}{Z}\right)^\epsilon_{\max} = 0.4433 \frac{\pi p D_{12}}{RT} \frac{(\alpha - 1)^2}{\ln (r_2/r_1)} \qquad (3.3.48)$$

It is seen that the two-stream flow permits an increase in separative power of 8.8% over the single-stream, no-mixing case.

The condition $bz = 1.2553$ gives a definite relation between the length of the rotor and the flow at the condition of the maximum separating power, which is obtained by rearranging (3.3.30),

$$L_1 + L_2 = \frac{2\pi p D_{12}}{1.575RT} \frac{Z}{\theta(1 - \theta)} \qquad (3.3.49)$$

Note that b contains the factor $\theta(1 - \theta)(L_1 + L_2)$. Decreasing $\theta(1 - \theta)$ and increasing $L_1 + L_2$ leaves this factor unchanged; but small total flows are desirable, and $\theta(1 - \theta)$ should obviously be made as large as possible, which requires $\theta = \tfrac{1}{2}$.

The Countercurrent Centrifuge

Countercurrent contacting devices in general have the property of multiplying the separation factor many times in a single unit. Since a large separation can be obtained in one unit, much of the recycling between units is avoided. Furthermore, the problem of cascade operation is considerably simplified since the number of units in series required to effect a given fractionation decreases enormously. Countercurrent flow is also efficient from a process standpoint because it is possible to maintain the maximum separative power through the entire unit. This is contrast to concurrent devices, in which $\partial x/\partial r$ changes with height Z, so that if the separative power is an optimum in one place, it is naturally less in every other height. For all these reasons, countercurrent processes are preferred, and many investigations on improving the gas centrifuge are focused on the countercurrent processes. A simple schematic diagram of a countercurrent centrifuge is presented in Fig. 3.3.

The differential equation describing the countercurrent gas flow centrifuge is

$$p\frac{\partial x}{\partial t} = \frac{1}{r}\frac{\partial}{\partial r}\left\{pD_{12}\left[r\frac{\partial x}{\partial r} + \frac{\Delta M'}{T}x(1-x)r^2\right]\right\}$$

$$- p\mathbf{v}_z(r)\frac{\partial x}{\partial z} + pD_{12}\frac{\partial^2 x}{\partial z^2} = 0 \quad (3.3.50)$$

where $\mathbf{v}_z(r)$ is the axial stream velocity. The solution of (3.3.50) was first obtained by Cohen [2] with a rather drastic simplification of the radial concentration gradient, neglecting the effect of longitudinal back diffusion on the radial concentration gradient. Also, assuming that the centrifuge rotor is so long that end effects in a rotor may be negligible, one may obtain a solution to (3.3.50) as follows:

For a steady-state $\partial x/\partial t = 0$, if a new radial variable defined by

$$\zeta = \frac{r^2 - r_0^2}{r_2^2 - r_0^2} \qquad 0 \le \zeta \le 1 \qquad (3.3.51)$$

is used, (3.3.50) for a steady state may be written as

$$4\frac{\partial}{\partial \zeta}\left\{pD_{12}(\zeta + \eta)\left[\frac{\partial x}{\partial \zeta} + \frac{\Delta M''}{T}x(1-x)\right]\right\}$$

$$- V_z^*(\zeta)\frac{\partial x}{\partial z} + D^*\frac{\partial^2 x}{\partial z^2} = 0 \quad (3.3.52)$$

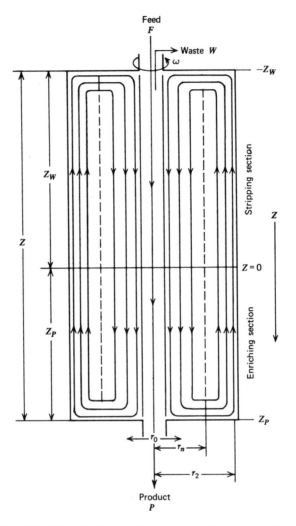

Fig. 3.3   Schematic diagram of countercurrent flow gas centrifuge.

where r_0 is the outside radius of the rotor core and r_2 is the inside radius of the rotor wall. The various quantities are defined as

$$\Delta M'' = \frac{(M_2 - M_1)\omega^2 r_2^2(1 - \sigma^2)}{2R} = \frac{\Delta M'}{2} r_2^2(1 - \sigma^2) \qquad (3.3.53)$$

$$V_z^* = \pi r_2^2(1 - \sigma^2)\rho v_z \qquad \rho = \frac{M}{RT} p \qquad (3.3.54)$$

$$D^* = \pi r_2^2 (1 - \sigma^2) \rho D_{12} \tag{3.3.55}$$

$$\sigma = \frac{r_0}{r_2} \tag{3.3.56}$$

$$\eta = \frac{\sigma^2}{1 - \sigma^2} \tag{3.3.57}$$

The radial boundary conditions are

$$\frac{\partial x}{\partial \zeta} + \frac{\Delta M''}{T} x(1 - x) = 0 \quad \text{at} \quad \zeta = 0 \quad \text{and} \quad \zeta = 1 \tag{3.3.58}$$

which specify that there be no net radial diffusion at solid boundaries. The first integration of (3.3.52) with the boundary condition at $\zeta = 0$ yields

$$\frac{\partial x}{\partial \zeta} = \left[\int_0^\zeta V_z^* \frac{\partial x}{\partial z} \, d\zeta \right] \left[\frac{1}{4\rho D_{12}(\zeta + \eta)} \right] - \frac{\Delta M''}{T} x(1 - x) \tag{3.3.59}$$

The assumption used by Cohen [2] was that the second-order term $D^*(\partial^2 x/\partial z^2)$ in (3.3.52) is so small that it is negligible.

Integration of (3.3.52) across the whole radius, that is, from $\zeta = 0$ to $\zeta = 1$ with the boundary conditions specified in (3.3.58), and integration with respect to the axial coordinate z gives

$$- \int_0^1 V_z^* x \, d\zeta + \frac{d}{dz} \int_0^1 D^* x \, d\zeta = \text{constant} \tag{3.3.60}$$

This expression is the net down flow of the desired component, and the constant can be determined from

$$\text{constant} = \begin{cases} P x_p & \text{for } z > 0 \\ -W x_w & \text{for } z < 0 \end{cases} \tag{3.3.61}$$

The quantity P is the product rate, W is the waste rate, and x_p and x_w are the mole fractions of the desired component in the product and waste streams, respectively. If the first integral in (3.3.60) is expressed by parts, one obtains for the region above the feed point

$$-x(1,z) \int_0^1 V_z^* \, d\zeta + \int_0^1 \frac{\partial x}{\partial \zeta} \int_0^\zeta V_z^* \, d\zeta \, d\zeta + \frac{d}{dz} \int_0^1 D^* x \, d\zeta = P x_p \tag{3.3.62}$$

Substituting (3.3.59) into (3.3.62), one obtains

$$\frac{d}{dz} \int_0^1 D^*x \, d\zeta - \Delta M'' \int_0^1 \frac{x(1-x)I_1}{T} \, d\zeta + \frac{1}{4\pi} \int_0^1 \int_0^\zeta \frac{I_1 \, dt}{\rho D_{12}(t+\eta)}$$

$$\times \left[V_z^* \frac{\partial x}{\partial z} \right] d\zeta - x(1,z) \int_0^1 V_z^* \, d\zeta = Px_p \quad (3.3.63)$$

in which

$$I_1 = \int_0^\zeta V_z^* \, dt \quad (3.3.64)$$

Equation 3.3.63 can be rearranged to give

$$(c_1 + c_2) \frac{dx}{dz} - c_3 x - c_4 x(1-x) = Px_p \quad (3.3.65)$$

where

$$c_1 = \int_0^1 D^* \, d\zeta = D^* \quad (3.3.66)$$

$$c_2 = \frac{1}{4\pi} \int_0^1 V_z^* \int_0^\zeta \frac{I_1 \, dt}{\rho D_{12}(t+\eta)} \quad (3.3.67)$$

$$c_3 = \int_0^1 V_z^* \, d\zeta = P \quad (3.3.68)$$

$$c_4 = \Delta M'' \int_0^1 \frac{I_1}{T} \, d\zeta \quad (3.3.69)$$

In the special case when both $x_0 \ll 1$ and $x_p \ll 1$, $x(1-x) \doteq x$, approximation may be used to integrate (3.3.65) with the boundary condition $x = x_0$ at $z = 0$ to yield

$$\frac{x_p}{x_0} = \frac{(1+\psi) \exp[2\epsilon Z(1+\psi)]}{1 + \psi \exp[2\epsilon Z(1+\psi)]} \quad (3.3.70)$$

where

$$\psi = \frac{c_3}{c_4} = \frac{\displaystyle\int_0^1 V_z^* \, d\zeta}{\Delta M'' \displaystyle\int_0^1 \frac{I_1}{T} \, d\zeta} \quad (3.3.71)$$

$$2\epsilon = \frac{c_4}{c_1 + c_2} = \frac{\Delta M'' \int_0^1 \frac{I_1}{T} d\zeta}{\int_0^1 D^* d\zeta + \frac{1}{4\pi} \int_0^1 \frac{d\zeta}{\rho D_{12}(\zeta + \eta)} I_1^2} \tag{3.3.72}$$

The quantities c_2 and c_4 depend on the function I_1, which measures the countercurrent flow profile upon the separation, and c_1 and c_3 represent the extracting flow rate and the circulation flow rate, respectively. They depend on the flow profile, on the absolute magnitude of the flow, and on the locations and the relative size of the stream. The flow profile is built into a centrifuge rotor by the position of the inlet orifice in the axial core. The magnitude of the flow is an operating variable and can be adjusted within the operation range. The dimensionless quantity ψ is thought of as a normalized rate of production and ϵ as the rate of back diffusion in the radial direction.

The maximum possible value of x_p/x_0 or x'/x for a given length of centrifuge rotor and for a specified type of countercurrent flow occurs when the rate of production is zero, which is analogous to total reflux in a fractional distillation, that is, $\psi = 0$. Then,

$$\left(\frac{x'}{x}\right)_{\max} = \left(\frac{x_p}{x_0}\right)_{\max} = \exp(2\epsilon_0 Z_p) \tag{3.3.73}$$

where $2\epsilon_0$ is the maximum of 2ϵ with respect to I_1:

$$2\epsilon_0 = \frac{c_4}{2(c_1 c_2)^{1/2}} \tag{3.3.74}$$

The corresponding value of I_1 is

$$I_1^0 = \left(\frac{c_1}{c_2}\right)^{1/2} \tag{3.3.75}$$

The existence of a maximum in ϵ is due to the phenomenon of back diffusion. The concentration gradient is proportional to ϵ. When I_1, the total flow in the centrifuge, is large and the concentration gradient is small, back diffusion is negligible ($c_1 \ll c_2$) and ϵ varies inversely with I_1. If I_1 is small and ϵ and the concentration gradient are large, back diffusion plays an important role, finally limiting the attainable fractionation.

It is convenient to characterize all of the stream flow by using the terms of I_1^0 and ψ and by defining

$$m = \frac{I_1}{I_1^0} = \left(\frac{c_2}{c_1}\right)^{1/2} \tag{3.3.76}$$

and

$$m\psi = \frac{c_3}{2c_1 2\epsilon_0} = \frac{L}{L_0} \tag{3.3.77}$$

where

$$L = \int_0^1 V_z^* \, d\zeta \tag{3.3.78}$$

$$L_0 = 4D^* \epsilon_0 \tag{3.3.79}$$

If P is the product flow rate in the enriching section and W is the waste flow rate in the stripping section, L has to be replaced by P and $-W$ in each section, respectively. Also, the signs of c_3 and ψ are positive for the enriching section and negative for the stripping section.

Substituting (3.3.76) into (3.3.71) and (3.3.72) and then into (3.3.70), one obtains

In the enriching section ($Z > 0$):

$$\frac{x'}{x} = \frac{x_p}{x_0} = \frac{(1 + \psi) \exp\left[2\epsilon_0 Z_p \left(\frac{2m}{1 + m^2}\right)(1 + \psi)\right]}{1 + \psi \exp\left[2\epsilon_0 Z_p \left(\frac{2m}{1 + m^2}\right)(1 + \psi)\right]} \tag{3.3.80}$$

In the stripping section ($Z < 0$):

$$\frac{x}{x''} = \frac{x_0}{x_w} = \frac{\psi + \exp\left[2\epsilon_0 Z_w \left(\frac{2m}{1 + m^2}\right)(1 + \psi)\right]}{1 + \psi} \quad \text{for } 1 + \psi \neq 0 \tag{3.3.81}$$

$$\frac{x}{x''} = \frac{x_0}{x_w} = 1 + 2\epsilon_0 Z_w \left(\frac{2m}{1 + m^2}\right) \quad \text{for } 1 + \psi = 0 \tag{3.3.82}$$

If values of x_p, x_0, m, and ψ are assigned, (3.3.80) through (3.3.82) are explicit equations which determine the necessary number of units or length of a countercurrent centrifuge rotor.

Under the condition that a countercurrent gas centrifuge is rotating uniformly at a steady state, the gas is isothermal, and there are no pressure gradients axially and azimuthally, the separative power per unit length of the centrifuge is the product of the net change of mass flux and the separative potential, (3.1.55), in the radial direction. Therefore, one has

$$dU = \left\{ \rho D_{12}(\zeta + \eta) \left[\frac{dx}{d\zeta} + \frac{\Delta M''}{T} x(1 - x)\right] \frac{d^2\phi}{dx^2} \frac{dx}{d\zeta} \right\} \tag{3.3.83}$$

If one takes the concentration gradient as the only variable, the maximum separative power can be obtained by differentiating (3.3.83) with respect to the concentration gradient $dx/d\zeta$ to obtain

$$(dU)_{max} = \rho D_{12}(\zeta + \eta)\left[\frac{\Delta M''}{T}\right]^2 \tag{3.3.84}$$

with

$$\frac{dx}{d\zeta} = -\frac{\Delta M''}{2T}x(1 - x) \tag{3.3.85}$$

The maximum total separative power which can be performed by a centrifuge is obtained by integrating over the entire volume:

$$(\Delta U)_{max} = 2\pi \int_0^z \int_0^1 \rho D_{12}(\zeta + \eta)\left[\frac{\Delta M''}{2T}\right] d\zeta\, dZ$$

$$= \rho D_{12}\left[\frac{\Delta M''}{T}\right]^2 \frac{\pi Z}{4}(1 + 2\eta) \tag{3.3.86}$$

Since $\Delta M'' \propto \omega^2 r_2^2$, and the value of $(\Delta U)_{max}$ depends on the fourth power of the peripheral velocity and is also proportional to the length of the rotor, it is easy to see from (3.3.73) that the maximum separation factor occurs at $\psi = 0$, where the rate of production is zero. Therefore, this is clearly not an ideal situation from the point of view of the separative power. The back diffusion in the radial direction is very important in countercurrent flow, since it is related to the concentration gradient in the radial direction and to the flow pattern in the axial direction.

4 VELOCITY PROFILES OF GAS FLOW IN COUNTERCURRENT CENTRIFUGE ROTORS

For the estimation of the separative power of a centrifuge, it is indispensable to know the profile of gas flow in the rotor and the resultant mole fraction distribution of the constituent components. Various developments of these two areas are discussed below.

In their study of the radial concentration gradient and the countercurrent velocity profile, Los and Kistemaker [24] have suggested that an ideal axial velocity profile must satisfy the relation

$$I_1 \propto \zeta \tag{3.4.1}$$

which implies that all axial mass velocities at any radial location must be the same. To characterize the velocity profile, Groth and Welge [25] have defined the flow pattern coefficient K:

$$K = \frac{\left(\displaystyle\int_0^1 I_1 \, d\zeta\right)^2}{\displaystyle\int_0^1 \frac{I_1^2}{\zeta} \, d\zeta} \qquad (3.4.2)$$

in such a way as to measure the deviation of a flow pattern from an ideal profile as suggested in (3.4.1). The flow pattern coefficient K depends only upon the flow profile I_1 in terms of $V_z^*(\zeta)$ and has a maximum value of unity for the flow profile which satisfies (3.4.1) and values of less than unity for all other profiles. The separative power of the countercurrent centrifuge may be written in terms of the flow pattern coefficient K and the maximum total separative power given in (3.3.86):

$$\delta U = K \frac{m^2}{1 + m^2} (\Delta U)_{max} \qquad (3.4.3)$$

It has been shown that the maximum separation factor x_p/x_0 occurs when the rate of production is zero, that is, $\psi = 0$ for any value of m; however, the separative power for this case is also zero. This, of course, is not a desirable operation. If one replaces m in (3.3.80) or (3.3.81) by $L/(L_0\psi)$ from rearrangement of (3.3.77) and differentiates (3.3.80) or (3.3.81) with respect to L, one obtains

$$m = \frac{L}{L_0\psi} = 1$$

to have the maximum for (3.3.80) or (3.3.81). Therefore, one can conclude that for maximum separation at a given circulation flow rate ψ, m has to be unity.

Several models of the axial velocity profile have been suggested in conjunction with the study of the flow pattern coefficient and the separative power. Kanagawa and Oyama [3] have suggested an axial velocity profile in which $V_z^*(\zeta)$ values are constants and unity at the inner and outer radii of the rotor, respectively, and at the radius of the neutral layer, which is defined as $\zeta_n = r_n/r_2$ where $V_z^*(\zeta_n) = 0$. Then, the velocity profiles can be given respectively as

$$V_z^*(\zeta) = 1 \qquad 0 < \zeta < \zeta_n \qquad (3.4.4)$$

$$V_z^*(\zeta) = -\left(\frac{\zeta_n - I_1}{1 - \zeta_n}\right) \qquad \zeta_n < \zeta < 1 \qquad (3.4.5)$$

By substituting these velocity profiles into (3.4.2), the flow pattern coefficient for the model becomes

$$K = \frac{(1 - \zeta_n)^2[\zeta_n + I_1(1 - \zeta_n)]^2}{[2\zeta_n^2(1 - I_1)^2\ln(1/\zeta_n)] - (1 - \zeta_n)[2\zeta_n^2 - 2I_1\zeta_n(1 + \zeta_n) - I_1^2(1 - 3\zeta_n)]}$$

$$(3.4.6)$$

The quantities I_1 in the enriching and in the stripping sections are given respectively as

$$I_1 = \frac{P}{V_z^*(0)} \quad \text{and} \quad I_1 = \frac{-W}{V_z^*(0)} \qquad (3.4.7)$$

The variations of the flow pattern coefficient K with respect to the radius of the neutral layer ζ_n and the quantity I_1 are presented in graphic form in Fig. 3.4. It is obvious from the figure that the K values approach unity when the neutral layer radius is closer to the outer wall of the rotor. When ζ_n is close to unity, the contribution of the flow pattern in the outer wall region is so small that it may be neglected, and one can use only the inner flow pattern to estimate the K value.

To evaluate the K value, Martin [26] has used the following velocity profile for the inner stream:

$$\int_0^\zeta V_z^* \, d\zeta \propto \zeta^{1/4} \exp\frac{M\omega^2 r_2^2}{4RT}\zeta \qquad (3.4.8)$$

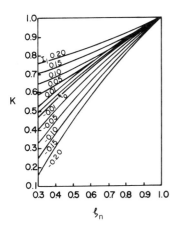

Fig. 3.4 Variation of flow pattern coefficient with respect to the neutral layer radius ζ_n and flow parameter I_1 for Kanagawa and Oyama's flow pattern. *Source:* Reproduced from Ref. 3.

If the neutral layer is very close to the rotor wall, the K value can be evaluated by using the inner stream velocity profile without the profile of the outer stream. Then the expression for K becomes

$$K = \frac{8}{\chi^4} \frac{\left(\int_0^\chi \xi^{3/2} (\exp \xi^2) \, d\xi \right)^2}{\int_0^\chi (\exp 2\xi^2) \, d\xi} \tag{3.4.9}$$

where

$$\xi^2 = \frac{M\omega^2 r_2^2}{4RT} \, \zeta \tag{3.4.10}$$

$$\chi^2 = \left(\frac{M\omega^2 r_2^2}{4RT} \right) \tag{3.4.11}$$

The value of K depends only on χ, and the variation of K values with respect to χ^2 is shown in Fig. 3.5. The K value has a maximum of about 0.96 when χ^2 is 1.85. When the separation is performed for UF_6, argon, and xenon, the periphery velocities of the rotor corresponding to $\chi^2 = 1.85$ that yield the maximum K value are 245, 680, and 375 m/sec at 27.2°C, respectively. For UF_6 gas, the periphery velocities ωr_2 that will yield the values of K for 0.96, 0.90, 0.79, and 0.67 are 245, 300, 350, and 400 m/sec, respectively.

Steenbeck [5] has analyzed the gas flow pattern in the scoop-type rotor by ignoring detailed inlet and outlet end effects of a rotor, linearizing equations of motion and energy, and solving numerically the resulting set of simultaneous, coupled, total differential equations by a perturbation technique. He as-

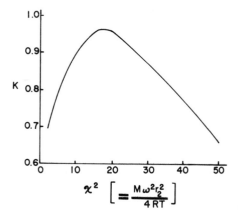

$$\chi^2 \left[= \frac{M\omega^2 r_2^2}{4RT} \right]$$

Fig. 3.5 Flow pattern coefficient K as a function of χ for Martin's flow pattern. Source: Reproduced from Ref. 3.

sumed that at a distance sufficiently far from the driving mechanism, the velocity field is essentially independent of the details of the driving mechanism. He presented an example of $M\omega^2 r_2^2 / 2RT = 2$ in his numerical result, for which $K = 0.60$, also. The reduced radius of the neutral layer ζ_n is 0.48. Comparing this with Kanagawa and Oyama's model, one will note that at $\zeta_n = 0.48$ and $I_1 = 0$, the K value is 0.60 from Fig. 3.4. Therefore, Steenbeck's result by the hydrodynamic approach is very close to the result obtained using the Kanagawa and Oyama model.

Parker and co-workers [27,28] have extended the work of Steenbeck by including terms in the equations of motion and energy which Steenbeck had ignored. The technique for obtaining a solution was similar to that used by Steenbeck, and Parker's numerical solution of the eigenvalue problem agrees with that obtained by Steenbeck at low peripheral velocities. At higher peripheral velocities, Parker's results differ somewhat from those obtained by Steenbeck.

Ging [29] has investigated the solution to the countercurrent flow problem at the limit of high angular velocity using a technique similar to that used by Dirac [30], but including effects due to heat conduction. At sufficiently large peripheral velocities, Ging's results agree with those obtained by Parker et al. [28].

Berman [7] has presented a simple model to describe the internal flow in a long countercurrent centrifuge. In his model, a radial temperature gradient is applied to a long, rapidly rotating cylinder. An external force such as a gravitational field parallel to the axis then produces convective flow. The axial velocity and the temperature are functions of the radial coordinate only. The radial velocity is assumed to be zero everywhere, and the tangential velocity is assumed to be proportional to the radial coordinate. The angular velocity ω is assumed to be a constant. Furthermore, the gas is also assumed to be ideal and to have a viscosity and thermal conductivity independent of pressure and temperature. In the energy balance, the viscous dissipation and the compression–expansion work due to axial flow are negligible. The equations of motion and energy for the model become

$$\mu \frac{1}{r} \frac{\partial}{\partial r} \left(r \frac{\partial v_z}{\partial r} \right) = \frac{\partial p}{\partial z} + \rho g^* \qquad (3.4.12)$$

$$\frac{\partial p}{\partial r} = \rho \omega^2 r \qquad (3.4.13)$$

$$\frac{1}{r} \frac{\partial}{\partial r} \left(r \frac{\partial T}{\partial r} \right) = 0 \qquad (3.4.14)$$

in which μ is the viscosity of the gas and the term ρg^* is not to be taken as the gravitational force, but rather an adjustable driving force for creating countercurrent flow. The magnitude of ρg^* does not affect the shape of the axial velocity profile but does determine the magnitude of the axial velocity. Furthermore, this term is such that the countercurrent flow vanishes when the radial temperature gradient vanishes. Using the ideal gas law, the density ρ in the above equations can be replaced by $\rho = (M/RT)p$. Since T is a function of r only, (3.4.13) can be differentiated with respect to z, after replacing ρ by $(M/RT)p$ and dividing through by p, to give

$$\frac{\partial}{\partial r}\left(\frac{\partial \ln p}{\partial z}\right) \equiv \frac{\partial \Xi}{\partial r} = 0 \qquad (3.4.15)$$

in which $\Xi = (\partial \ln p/\partial z)$ and is independent of r. Equation 3.4.12 may be rewritten as

$$\frac{Mg^*}{RT(r)} + \Xi = \frac{1}{p}\frac{\mu}{r}\frac{\partial}{\partial r}\left(r \frac{\partial v_z}{\partial r}\right) \qquad (3.4.16)$$

It is now assumed that the axial variation of the pressure is very small compared with the radial variation in centrifugation. Thus, one may write

$$p(r,z) \doteq \bar{p}(r) = \frac{1}{Z}\int_0^Z p(r,z)\,dz \qquad (3.4.17)$$

where Z is the height of the centrifuge rotor. Replacing ρ in (3.4.13) by $(M/RT)p$, using reduced quantities defined in (3.3.51), and integrating the rearranged (3.3.13), one obtains the pressure (or density) distribution

$$\frac{p(\zeta)}{p_0} = \exp\left[\frac{M\omega^2 r_2^2}{R}\int_0^\zeta \frac{d\zeta}{T(\zeta)}\right] \qquad (3.4.18)$$

in which $p_0 = p(r_0) = (p)_{\zeta=0}$

The energy equation, (3.4.14), can also be integrated subject to the boundary conditions

$$T(r_0) = (T)_{\zeta=0} = T_0 \quad \text{and} \quad T(r_2) = (T)_{\zeta=1} = T_2 \qquad T_2 \leq T_0$$

$$(3.4.19)$$

The reduced form of the temperature distribution becomes

$$\frac{T_0 - T(\zeta)}{T_0 - T_2} = -\frac{\ln[(\zeta + \eta)/\eta]}{\ln \sigma^2} \qquad (3.4.20)$$

With the pressure and temperature distributions obtained, one can substitute these expressions into (3.4.12) to solve for the axial velocity distribution with the boundary conditions

$$v_z(r_0) = v_z(r_2) = 0 \quad \text{or} \quad V_z^*(0) = V_z^*(1) = 0 \qquad (3.4.21)$$

These boundary conditions imply that there are no slip conditions at the rotor core and the rotor outer wall. Further, one has to satisfy another condition on the material balance, which is

$$2\pi \int_{r_0}^{r_2} \rho v_z r \, dr = P \quad \text{or} \quad \frac{M r_2^2 (1 - \sigma^2)}{R} \int_0^1 \frac{p v_z}{T} \, d\zeta = P \quad (3.4.22)$$

in which P is the product rate in grams per second. If one defines the parameters

$$\phi = \frac{M g^* (1 - \sigma^2) r_2^2}{4 \mu R} \qquad (3.4.23)$$

$$\Psi = \frac{\Xi (1 - \sigma^2) r_2^2}{4 \mu} \qquad (3.4.24)$$

(3.4.12) can be rewritten as follows:

$$\frac{d}{d\zeta} \left[(\zeta + \eta) \frac{d v_z}{d\zeta} \right] = \frac{\phi p}{T} + \Psi p \qquad (3.4.25)$$

Integrating twice and using the boundary condition at the rotor core, one obtains

$$v_z(\zeta) = \phi \int_0^\zeta \frac{dt}{t + \eta} \int_0^t \frac{p}{T} \, ds + \Psi \int_0^\zeta \frac{dt}{t + \eta} \int_0^t p \, ds + C_1 \ln \left(\frac{\zeta + \eta}{\eta} \right)$$

$$(3.4.26)$$

The integration constant C_1 can be evaluated using the boundary condition at the rotor wall. Then one has

$$C_1 = \frac{\phi}{\ln \sigma^2} \int_0^1 \frac{dt}{t + \eta} \int_0^t \frac{p}{T} \, ds + \frac{\Psi}{\ln \sigma^2} \int_0^1 \frac{dt}{t + \eta} \int_0^t p \, ds$$

$$(3.4.27)$$

With several integrations by parts, (3.4.26) and (3.4.27) can be rewritten in the following form to give the axial velocity distribution in a countercurrent gas centrifuge:

$$v_z(\zeta) = \phi[\ln(\zeta + \eta) \int_0^\zeta \frac{p}{T}\, dt - \int_0^\zeta \frac{p}{T} \ln(t + \eta) dt]$$

$$+ \Psi[\ln(\zeta + \eta) \int_0^\zeta p\, dt - \int_0^\zeta p \ln(t + \eta) dt] + C_1 \ln\left(\frac{\zeta + \eta}{\eta}\right) \qquad (3.4.28)$$

$$C_1 = \frac{\phi}{\ln \sigma^2}[\ln(1 + \eta) \int_0^1 \frac{p}{T}\, d\zeta - \int_0^1 \frac{p}{T} \ln(\zeta + \eta) d\zeta]$$

$$+ \frac{\Psi}{\ln \sigma^2}[\ln(1 + \eta) \int_0^1 p\, d\zeta - \int_0^1 p \ln(\zeta + \eta) d\zeta] \qquad (3.4.29)$$

Berman [7] has used a numerical quadrature method to evaluate the integrals in (3.4.28) and (3.4.29) and to obtain the axial velocity profile in the countercurrent gas centrifugation. The values of ϕ and Ψ were provided with the ratio ϕ/Ψ obtained from (3.4.20). He has presented some typical results and compared these with similar solutions which were obtained by a complicated eigenvalue method by Parker et al. [28]. The comparison shows that Berman's results agree very well with Parker's. Their axial velocity profiles in a rotor are reproduced and presented in Fig. 3.6. One notices that the loca-

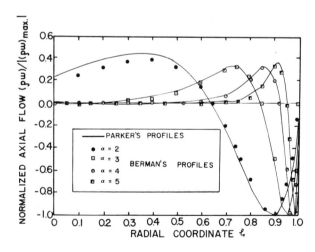

Fig. 3.6   Axial flow profiles in a countercurrent gas centrifuge for $\sigma = 0.01$ and $\Delta T = 1.0$, $\alpha = \omega^2 r_2^2/(2RT/M)$. *Source:* Reproduced from Ref. 8.

tion of the neutral layer, where $v_z(\zeta_n) = 0$, moves closer to the centrifuge wall with increasing angular velocity and the flow in the central region of the rotor rapidly decreases with increasing angular velocity.

5 CONCENTRATION DISTRIBUTIONS IN COUNTERCURRENT CENTRIFUGE ROTORS

The theory of Cohen was extended by Berman [8] to provide a better model for the analysis of centrifugal performance under the usual operating conditions of reasonably high ratio of product to internal flow. As described before, Berman considered the axial back diffusion term on the radial concentration gradient. Thus, the first integration of (3.3.52) with the boundary condition (3.3.58) becomes

$$
\frac{\partial x}{\partial \zeta} = \left[\int_0^{\zeta} V_z^* \frac{\partial x}{\partial z} \, d\zeta - \int_0^{\zeta} D^* \frac{\partial^2 x}{\partial x^2} d\zeta \right] \left[\frac{1}{4\rho D_{12}(\zeta + \eta)} \right] - \frac{\Delta M'' x(1 - x)}{T}
$$

(3.5.1)

Integration of (3.3.52) across the whole radius, that is, from $\zeta = 0$ to $\zeta = 1$, subjecting to the boundary condition (3.3.58), and integration with respect to the axial coordinate z subjecting to the boundary condition (3.3.61), and following the same procedures in Cohen's development, the equivalence of (3.3.63) with the inclusion of the axial back diffusion term becomes

$$
\frac{d}{dz} \int_0^1 D^* x \, d\zeta - \Delta M'' \int_0^1 \frac{x(1 - x)I_1}{T} \cdot d\zeta + \frac{1}{4\pi} \int_0^1 \int_0^{\zeta} \frac{I_1 \, d\zeta}{\rho D_{12}(\zeta + \eta)}
$$

$$
\times \left[V_z^* \frac{\partial x}{\partial z} - D^* \frac{\partial^2 x}{\partial z^2} \right] d\zeta - x(1, z) \int_0^1 V_z^* \, d\zeta = x_p P
$$

(3.5.2)

in which I_1 is defined in (3.3.64). Equation 3.5.2 was solved subject to (3.3.60).

In Cohen's theory, $x(\zeta, z)$ and its derivatives were removed from under the integral signs, the composition $x(1, z)$ at the wall was assigned the same average value, and (3.5.2) became a total differential equation for the axial variation of the average light concentration $x(z)$. Berman used the following simplification to account for a radial dependence on the mole fraction for the desired component:

$$
x(\zeta, z) = \begin{cases} \bar{x}'(z) & 0 \leq \zeta < \zeta_n \\ \bar{x}''(z) & \zeta_n < \zeta \leq 1 \end{cases}
$$

(3.5.3)

where

$$\bar{x}'(z) = \frac{\int_0^{\zeta_n} x \, d\zeta}{\int_0^{\zeta_n} d\zeta} \tag{3.5.4}$$

$$\bar{x}''(z) = \frac{\int_{\zeta_n}^1 x \, d\zeta}{\int_{\zeta_n}^1 d\zeta} \tag{3.5.5}$$

and ζ_n is the value of the radial coordinate at which the axial velocity vanishes, that is, the location of the neutral layer. A more suitable form of $x(\zeta,z)$ for numerical work was chosen as

$$x(\zeta,z) = a_1(z) + a_2(z)\zeta \tag{3.5.6}$$

Then, evaluating a_1 and a_2 in terms of $\bar{x}'(z)$ and $\bar{x}''(z)$ from (3.5.4) and (3.5.5), one obtains

$$x(\zeta,z) = (1 + \zeta_n - 2\zeta)\bar{x}'(z) + (2\zeta - \zeta_n)\bar{x}''(z) \tag{3.5.7}$$

The product composition x_p and the average compositions $\bar{x}'$ and $\bar{x}''$ are related by

$$x_p = \int_0^1 x(\zeta,z)d\zeta = \zeta_n\bar{x}' + (1 - \zeta_n)\bar{x}''; \qquad z = Z \tag{3.5.8}$$

Equations 3.5.7 and 3.5.8 were substituted into (3.5.3) and (3.3.60) together with the definition in (3.5.8); and after algebraic manipulation, one obtains

$$\ddot{\bar{x}}'[(1 + \zeta_n)K_7 - 2K_8] + \ddot{\bar{x}}''[2K_8 - \zeta_n K_7]$$

$$+ \dot{\bar{x}}'[2K_9 - (1 + \zeta_n)P] + \dot{\bar{x}}''[\zeta_n P - 2K_9] = 0 \tag{3.5.9}$$

$$\ddot{\bar{x}}'[(1 + \zeta_n)K_3 - 2K_4] + \ddot{\bar{x}}'[2K_4 - \zeta_n K_3] + \dot{\bar{x}}'[2K_2 - (1 + \zeta_n)K_1]$$

$$+ \dot{\bar{x}}''[\zeta_n K_1 - 2K_2] + \bar{x}'[2K_6 - 2K_9 - (1 + \zeta_n)K_5 + 2P]$$

$$+ \bar{x}''[\zeta_n K_5 - 2K_6 + 2K_9 - 2P] = 0 \tag{3.5.10}$$

In the above equations, the dots represent differentiation with respect to axial coordinate z, and the K_i's are integrals involving the axial flow profile, the diffusion coefficient, and the temperature. The K integrals are defined as follows:

$$K_1 = \frac{1}{4\pi} \int_0^1 V_z^* I_2 \, d\zeta = C_2 \tag{3.5.11}$$

$$K_2 = \frac{1}{4\pi} \int_0^1 V_z^* \zeta I_2 \, d\zeta \tag{3.5.12}$$

$$K_3 = \frac{1}{4\pi} \int_0^1 D^* I_2 \, d\zeta \tag{3.5.13}$$

$$K_4 = \frac{1}{4\pi} \int_0^1 D^* \zeta I_2 \, d\zeta \tag{3.5.14}$$

$$K_5 = \Delta M'' \int_0^1 \frac{I_1}{T} \, d\zeta = C_4 \tag{3.5.15}$$

$$K_6 = \Delta M'' \int_0^1 \frac{\zeta I_1}{T} \, d\zeta \tag{3.5.16}$$

$$K_7 = \int_0^1 D^* \, d\zeta = C_1 \tag{3.5.17}$$

$$K_8 = \int_0^1 D^* \zeta \, d\zeta \tag{3.5.18}$$

$$K_9 = \int_0^1 V_z^* \zeta \, d\zeta \tag{3.5.19}$$

$$I_1(\zeta) = \int_0^\zeta V_z^* \, dt \tag{3.5.20}$$

$$I_2(\zeta) = \int_0^\zeta \frac{I_1 \, dt}{\rho D_{12}(t + \eta)} \tag{3.5.21}$$

In order to evaluate these integrals, the radial variation of axial velocity, temperature, and ρD_{12} have to be specified. Equations 3.5.9 and 3.5.10 are linear, homogenous, total differential equations with constant coefficients, and the characteristic equation for the simultaneously, coupled equations (3.5.9) and (3.5.10) is

$$\alpha^4[K_3 K_2 - K_4 K_7] + \alpha^3[K_2 K_7 + K_4 P - K_1 K_8 - K_3 K_9]$$
$$+ \alpha^2[K_6 K_7 - K_1 K_9 + K_7 K_9 + P K_7 - P K_2 - K_5 K_8]$$
$$+ \alpha[K_9 K_5 + K_9 P - K_6 P - P^2] = 0 \tag{3.5.22}$$

Consistent with his hydrodynamic model of a long, countercurrent gas centrifuge, Berman [8] retained the two smallest eigenvalues for an approximation, since the two larger eigenvalues affect the solution primarily in the neighborhood of the centrifuge ends. One of the characteristic roots is zero, and the other is given approximately by

$$\alpha \doteq \frac{K_5 K_9 - P(P + K_6 - K_9)}{K_5 K_8 + K_1 K_9 - K_7(K_6 - K_9) - P(K_7 - K_2)} \qquad (3.5.23)$$

The approximation given by (3.5.23) can be improved by iteration in (3.5.22). The solution for $\bar{x}'$ and $\bar{x}''$ is

$$\bar{x}'(z) = \gamma_1 + \gamma_2 \exp(\alpha z) \qquad (3.5.24)$$

$$\bar{x}''(z) = \Lambda_1 \gamma_1 + \Lambda_2 \gamma_2 \exp(\alpha z) \qquad (3.5.25)$$

with

$$\Lambda_1 = \frac{l_3 + K_5}{l_3} \qquad (3.5.26a)$$

$$\Lambda_2 = \frac{l_3 + K_5 + (l_2 + K_1)\alpha - (K_3 - K_1)\alpha^2}{l_1 \alpha^2 + l_2 \alpha + l_3} \qquad (3.5.26b)$$

and

$$l_1 = 2K_4 - \zeta_n K_3 \qquad (3.5.27a)$$

$$l_2 = \zeta_n K_1 - 2K_2 \qquad (3.5.27b)$$

$$l_3 = \zeta_n K_5 - 2K_6 + 2K_9 - 2P \qquad (3.5.27c)$$

The integration constants γ_1 and γ_2 have been evaluated by the axial boundary conditions. In practice, there are actually two cases, four equations, to be determined: one pair valid in the region above the feed point (stripping section) and another pair valid below the feed point (enriching section). The axial boundary conditions for the centrifuge were:

1. Mass balance of the desired component at the product end; $z = Z_p$:

$$-\int_0^1 (V_z^* x)_e \, d\zeta + \frac{d}{dz} \int_0^1 D^* x_e \, d\zeta = P x_p \qquad z = Z_p \quad (3.5.28)$$

2. Mass balance of the desired component at the waste end; $z = Z_w$:

$$-\int_0^1 (V_z^* x)_s \, d\zeta + \frac{d}{dz} \int_0^1 D^* x_s \, d\zeta = -W x_w \qquad z = -Z_w$$

$$(3.5.29)$$

3. Desired component mass balance in the downflowing stream at the feed plane; $z = 0$:

$$\int_0^{\zeta_{ns}} (V_z^*x)_s \, d\zeta - \int_0^{\zeta_{ne}} (V_z^*x)_e \, d\zeta + \frac{d}{dz} \int_0^{\zeta_{ne}} D^*x_e \, d\zeta$$

$$- \frac{d}{dz} \int_0^{\zeta_{ns}} D^*x_s \, d\zeta = - F_1 x_f \qquad (3.5.30)$$

4. Desired component mass balance in the upflowing stream at the feed plane; $z = 0$:

$$\int_{\zeta_{ns}}^1 (V_z^*x)_s \, d\zeta - \int_{\zeta_{ne}}^1 (V_z^*x) \, d\zeta + \frac{d}{dz} \int_{\zeta_{ne}}^1 D^*x_e \, d\zeta$$

$$- \frac{d}{dz} \int_{\zeta_{ns}}^1 D^*x_s \, d\zeta = - F_2 x_f \qquad (3.5.31)$$

The subscripts s and e denote quantities evaluated in the region above and below the feed point, P and W are the product and waste rate (in moles/sec), F_1 and F_2 are the feed rates (in moles/sec) into the downflowing and the upflowing stream, and x_f is the composition of feed material.

With the use of (3.5.7), (3.5.8), and the expression corresponding to (3.5.8) for the waste composition, the axial boundary conditions (3.5.28) through (3.5.31) can be transformed into a set of simultaneous total differential equations involving $\bar{x}_e{}'$, $\bar{x}_s{}'$, $\bar{x}_e{}''$, $\bar{x}_s{}''$, and the K integrals defined previously. Subscripts e and s refer to the enricher and stripper portions of the centrifuge, respectively. With (3.5.25) appropriately subscripted for enriching and stripping sections, the differential equations involving the four average compositions $\bar{x}_e{}'$, $\bar{x}_s{}'$, $\bar{x}_e{}''$, and $\bar{x}_s{}''$ are transformed into a set of four simultaneous linear algebraic equations for the four unknowns γ_{1e}, γ_{2e}, γ_{1s}, and γ_{2s}. The solution of these equations provides the axial variation of average desired component mole fractions in the four streams in the centrifuge. They are

$$\bar{x}_e{}'(z) = \gamma_{1e} + \gamma_{2e} \exp(\alpha Z_p) \qquad (3.5.32a)$$

$$\bar{x}_e{}''(z) = \Lambda_{1e}\gamma_{1e} + \Lambda_{2e}\gamma_{2e} \exp(\alpha Z_p) \qquad (3.5.32b)$$

$$\bar{x}_s{}'(z) = \gamma_{1s} + \gamma_{2s} \exp(-\beta Z_w) \qquad (3.5.32c)$$

$$\bar{x}_s{}''(z) = \Lambda_{1s}\gamma_{1s} + \Lambda_{2s}\gamma_{2s} \exp(-\beta Z_w) \qquad (3.5.32d)$$

The eigenvalues for the enriching section α were obtained from (3.5.23) for the two smallest values, and one of the characteristic roots was zero and the other was given in (3.5.24) by adding the subscript e to the K's. The corresponding eigenvalues for the stripping section β can be obtained by replacing P with $-W$ and adding the subscript s to the K's in (3.5.23). The detailed equations and a complete list of equations needed for numerical calculations are given in Berman's report [8].

Once the axial variation of desired component mole fraction is known for each stream, the separation factors for each section or for the whole centrifuge can be calculated easily. These are given by

$$\frac{x_p}{x_f} = [\zeta_{ne} + (1 - \zeta_{ne})K_{1e}]\gamma_{1e} + [\zeta_{ne} + (1 - \zeta_{ne})K_{2e}]\gamma_{2e}\exp(\alpha Z_p)$$

$$(3.5.33a)$$

$$\frac{x_w}{x_f} = [\zeta_{ns} + (1 - \zeta_{ns})K_{1s}]\gamma_{1s} + [\zeta_{ns} + (1 - \zeta_{ns})K_{2s}]\gamma_{2s}\exp(-\beta Z_w)$$

$$(3.5.33b)$$

$$\frac{x_p}{x_w} = \frac{x_p/x_f}{x_w/x_f} \qquad (3.5.33c)$$

In using the boundary conditions, it is not required that the composition be continuous across the feed plane. The mismatch between the compositions at the feed plane and the composition of the feed material can be obtained by replacing x_p and x_w with $\bar{x}_e(z_0)$ and $\bar{x}_s(z_0)$ and the rotor length Z_p and $-Z_w$ with z_0 and $-z_0$ in (3.5.33a) and (3.5.33b), respectively.

If a centrifuge operates at isothermal and total reflux conditions, that is, $P = 0$, the separation factor by the Cohen theory is given in (3.3.73) or

$$\frac{x_p}{x_w} = \exp[2\epsilon(Z_p + Z_w)] \qquad (3.5.34a)$$

and by Berman's theory, that is, for $P = W = 0$, $\alpha = \beta$, as

$$\frac{x_p}{x_w} = \exp[\alpha(Z_p + Z_w)] \qquad (3.5.34b)$$

with

$$2\epsilon = \frac{C_4}{C_1 + C_2} = \frac{K_5}{K_7 + K_1} \qquad (3.5.35)$$

$$\alpha \cong \frac{K_5}{[K_7 + K_1 + (K_5 K_8 - K_7 K_6)/K_9]} \qquad (3.5.36)$$

In evaluating the K integrals, the following ideal flow profile was used:

$$V_z^*(\zeta) = V_{z1}^* + V_{z2}^* \, \delta(\zeta - 1) \tag{3.5.37}$$

in which the function $\delta(\zeta - 1)$ is such that

$$\int_0^{\zeta} \delta(\zeta - 1)d\zeta = H(\zeta - 1) \tag{3.5.38}$$

$$H(t) = 1 \quad \text{for} \quad t = 0 \tag{3.5.39a}$$

$$H(t) = 0 \quad \text{for} \quad t < 0 \tag{3.5.39b}$$

and where $V_{z1}^* + V_{z2}^* = P$, the product flow (in moles/sec). With this profile, (3.3.64) or (3.5.20) becomes

$$I_1(\zeta) = \int_0^{\zeta} V_z^* \, d\zeta = V_{z1}^* \, \zeta + V_{z2}^* H(\zeta - 1) \quad \text{and} \quad I_1(1) = P \tag{3.5.40}$$

Then, (3.5.35) and (3.5.36) become

$$2\epsilon = \frac{\Delta M'' V_{z1}^*}{T\left[2D^* + \dfrac{V_{z1}^*}{(4\rho D)}\right]} \tag{3.5.41}$$

$$\alpha = \frac{\Delta M'' V_{z1}^*}{T\left[2D^* + \dfrac{\Delta M''}{T}\dfrac{D^*}{3} + \dfrac{V_{z1}^{*2}}{4\rho D}\right]} \tag{3.5.42}$$

Hence, at total reflux, Berman's theory predicts somewhat lower enrichment than the Cohen theory, and the difference becomes greater at high peripheral velocities, that is, higher $\Delta M''$.

6 COUPLING OF THERMAL AND MASS FLOWS IN GAS CENTRIFUGES

Recently, Nakayama and Usui [31] have theoretically studied thermal convection and weak forced axial flows in a gas centrifuge and obtained the mass velocity distributions in a rotor. Compressibility of the gas in the form of a density stratification in the radial direction and the effect of the Coriolis force in the flow field were taken into account in their analysis.

A steady motion of a compressible gas in a cylinder of radius R and height $2H$, rotation about its own axis with a constant angular velocity ω, was considered in their analysis with the assumptions that (1) the rotating speed of the rotor is so high that the gravitational acceleration is negligible compared to the centrifugal acceleration $\omega^2 R$, (2) the walls of the rotor are thermally conductive, (3) the gas is a perfect gas, and (4) the viscosity, the thermal conductivity, and the specific heats are all constants.

Assuming the gas is rotating as a rigid body under an isothermal condition subject to no perturbation from the outside, the distributions of dimensionless density and pressure can be obtained by modifying (3.1.10) with the definitions of the speed of sound, $c = \sqrt{\gamma R T}$, and the Mach number, $M = v/c$, which gives

$$\rho_o = p_o/RT_o = \exp\left\{-\frac{1}{2}\gamma M^2(1 - r^2)\right\} \tag{3.6.1}$$

where γ is the ratio of the heat capacities and M is the rotational Mach number:

$$M = \frac{v}{c} = \frac{\omega^2 R}{\sqrt{\gamma R T_0}} \tag{3.6.2}$$

The dimensionless density and pressure are defined as

$$\rho^* = \frac{\rho - \rho_0}{\rho_0} \tag{3.6.3}$$

$$p^* = \frac{p - p_0}{p_0} \tag{3.6.4}$$

The subscript 0 denotes the mean quantity in the rotor.

The reduced conservation equations for a compressible gas for the model becomes

$$\nabla \cdot \{\rho_0(1 + \rho^*)\mathbf{q}\} = 0 \tag{3.6.5}$$

$$\mathbf{Ro}\left\{\frac{1}{2}\nabla(\mathbf{q}\cdot\mathbf{q}) + (\nabla \times \mathbf{q}) \times \mathbf{q}\right\} + 2\mathbf{k} \times \mathbf{q} - \frac{(\Xi)}{\mathbf{Ro}}\theta\mathbf{k} \times (\mathbf{k} \times \mathbf{r})$$

$$= -\frac{1}{\gamma M^2 \mathbf{Ro}(1 + \rho^*)}\nabla p^* + \frac{2\mathbf{E}}{\rho_0(1 + \rho^*)}\nabla \cdot \overline{\overline{\tau}} \tag{3.6.6}$$

$$- \mathbf{Ro}^2\left\{\frac{1}{2}\mathbf{q}\cdot\nabla(\mathbf{q}\cdot\mathbf{q})\right\} - \mathbf{q}\cdot\{\mathbf{k} \times (\mathbf{k} \times \mathbf{r})\}$$

$$+ \frac{2\mathbf{Ro}\cdot\mathbf{E}}{\rho_0(1 + \rho^*)}\{\mathbf{q}\cdot(\nabla \cdot \overline{\overline{\tau}}) + \overline{\overline{\tau}} : \nabla\mathbf{q}\}$$

$$= \frac{(\Xi)}{(\gamma - 1)M^2}\mathbf{q}\cdot\nabla\theta - \frac{E(\Xi)}{(\gamma - 1)M^2\mathbf{Pr}\,\mathbf{Ro}}\frac{1}{\rho_0(1 + \rho^*)}\nabla^2\theta \tag{3.6.7}$$

where $\mathbf{q}$ is the dimensionless velocity vector; θ is the reduced temperature $[\theta = (T - T_0)/\Delta T$, where ΔT, is the temperature variation scale]; $\overline{\overline{\tau}}$ is the shear rate tensor; $\mathbf{k}$ is a unit vector in the axial direction of rotation; $\mathbf{Pr}$ is the Prandtle number ($\mathbf{Pr} = c_p\mu/k$); $\mathbf{Ro}$ is the Rossby number ($\mathbf{Ro} = v_0/\omega R$); $\mathbf{E}$ is the Ekman number ($\mathbf{E} = \mu/\rho_z\omega R^2$); and ($\Xi$) is the temperature fraction $[(\Xi) = \Delta T/T_0]$.

The above conservation equations together with the equation of state for the gas completely specify the system. The equation of state for an ideal gas written in dimensionless variables p^*, ρ^*, θ, and (Ξ) yields

$$p^* = \rho^* + (\Xi)(1 + \rho^*)\theta \qquad (3.6.8)$$

In what follows, the existence of a nonuniform temperature distribution induces a corresponding nonuniform density distribution in a cylindrical rotor. The nonuniformity in density in turn causes a body force which induces the convection. The driving force for this phenomenon is represented by the last term in the left-hand side of (3.6.5), and the ratio (Ξ)/$\mathbf{Ro}$ associated with the term determines whether the Coriolis force in the same equation, $2\mathbf{k} \times \mathbf{q}$, is significant or not in the gas flow profile in a countercurrent centrifuge. Nakayama and Usui [31] have performed comparisons of the order of magnitude of each term in those conservation equations, the so-called scaling analysis, to examine which term may be negligible or insignificant in obtaining the flow profile.

Actually, in the rotor, the fraction of temperature gradient defined as (Ξ) is a very small value: (Ξ) $\ll$ 1. The Rossby number is also a very small number in a centrifuge. However, there are two possibilities existing between the Rossby number $\mathbf{Ro}$ and the fraction of temperature gradient (Ξ):

$$\mathbf{Ro} \ll (\Xi) \ll 1 \qquad (3.6.9a)$$

and

$$\mathbf{Ro} \sim (\Xi) \ll 1 \qquad (3.6.9b)$$

The order of magnitude for (Ξ) $\ll$ 1 in (3.6.9a) indicates that the initial term, the first term in (3.6.6), is negligible. In the case of (3.6.9a), the Coriolis force term may be negligible compared to the convective driving force, the pressure gradient, and the viscous shear stress terms. Moreover, if one takes the dot product of $\mathbf{k}$ and (3.6.6), an axial direction momentum balance equation is obtained:

$$-\frac{1}{\gamma M^2 \mathbf{Ro}}\frac{\partial p^*}{\partial z} + \frac{\mathbf{E}}{\rho_0}\frac{1}{r}\frac{\partial}{\partial r}\left(r\frac{\partial v_z}{\partial r}\right) = 0 \qquad (3.6.10)$$

If (3.6.9b) is to be applied in a gas centrifuge, the coefficient in each term must be in the same order of magnitude:

$$\frac{E}{\rho_0} \sim \frac{1}{\gamma M^2 Ro} \qquad (3.6.11)$$

Under all practical applications of centrifugation, the Mach number M is in the order of magnitude $M \sim 0(1)$, so that $\gamma M^2 Ro$ rarely exceeds $0(1)$. As to the order of magnitude of the Ekman number, it is usually $E \ll 1$, and hence, the condition (3.6.11) is realized only when $\rho_0 \ll 1$. From (3.6.1), one sees that ρ_0 decreases as $r \rightarrow 0$ and $M > 1$. The characteristics of a very low-density region can be best demonstrated by rewriting (3.6.11) in terms of the flow Mach number $M_f = v_0/\sqrt{\gamma R T_0}$ and the local Reynolds number based on the local value of ρ_0, $Re_o = \rho_0 v_0 R/\mu$:

$$\frac{\gamma M_f^2}{Re_o} \sim 0(1) \qquad (3.6.12)$$

This reveals that the local Knudsen number (ratio of length of mean free path to characteristic dimension of the system) should be of order of unity under the condition of (3.6.11). Obviously, the flow cannot be regarded as a continuous flow or as a slip flow. The mass velocity in such a region is very small, so that it contributes very little to the separating process when the condition of (3.6.11) is possible in a centrifuge with such a high peripheral velocity. For this reason, the case of (3.6.9a) can be discarded.

In case of (3.6.9b), one lets $Ro = (\Xi)$, whence one can use

$$v_0 = \frac{\omega R \Delta T}{T_0} \qquad (3.6.13)$$

as the velocity scale. The small value of E/ρ_0 suggests that the viscous stress term is insignificant in determining the field of flow in most regions of the rotor. Thus, from (3.6.11) one obtains that the axial pressure gradient is also insignificant in that the pressure p^* is a function in the radial direction only. The region where the effect of viscosity is negligible is called an inviscid core in the rotor.

In accordance with the elimination of the viscous effect terms in (3.6.5) through (3.6.8), further simplification can be made in the equations describing the model. Since p^* depends on r, it can be interpreted as a variation of rigid rotation pressure caused by a radial temperature variation; thus, it may be referred to as a nonisothermal rigid rotation pressure. As long as $(\Xi) \ll 1$, the difference between the isothermal and nonisothermal rigid rotation pressures can be uncoupled from the equations for thermal convection. Hence, one can let

$$p^* = 0 \qquad (3.6.14)$$

On eliminating the term of order less than $0[(\Xi)]$ from the reduced equation of state, (3.6.8), one has

$$\rho^* = -(\Xi)\theta \qquad (3.6.15)$$

Thus, one may approximate the density distribution by ρ_0 since the departure from the rigid rotation distribution is so small. Eliminating all the negligible terms in (3.6.6), one obtains

$$2v_\theta = r\theta \qquad (3.6.16)$$

Equation 3.6.16 signifies that the Coriolis force dominates the flow field and that the azimuthal component v_θ is of the order of unity. Nakayama and Usui called this azimuthal flow a "thermal wind."

To find the order of magnitude of the radial velocity component v_r, if one takes the dot product of the unit vector in the azimuthal direction ϕ and (3.6.6), then one obtains

$$2v_r = \frac{E}{\rho_0}\left[\frac{\partial}{\partial r}\left(\frac{1}{r}\frac{\partial}{\partial r}\right)rv_\theta + \frac{\partial^2 v_\theta}{\partial z^2}\right] \qquad (3.6.17)$$

Hence, one finds that $v_r \sim 0(E/\rho_0)$.

The order of magnitude of the various terms in (3.6.7) can be estimated by using the results of scaling analysis on the momentum equation (3.6.6). Results of the scaling analysis of (3.6.6) have been summarized in terms of the radio of the respective terms to the conduction term:

Kinetic energy transport	$\mathbf{M}^{*2}(\Xi)\lambda^*$
Centrifugal pressure work	$\mathbf{M}^{*2}$
Dissipation	$\mathbf{M}^{*2}\lambda^*E/\rho_0$
Convection	λ^*
Conduction	1

where

$$\lambda^* \equiv \frac{(\Xi)}{(E/\rho_0)^{1/2}} \quad \text{and} \quad \mathbf{M}^{*2} = (\gamma - 1)\mathbf{M}^2\mathbf{Pr}$$

In isotope centrifuge operations, $\mathbf{M}^{*2}$ is usually smaller than unity. This is due to the fact that, while the gases of heavy molecules such as UF_6 yield large $\mathbf{M}$, the values of the specific heat ratios γ for those high-molecular-weight substances are very close to unity. Even with those gases having a large γ value, the $\mathbf{M}$ value could rarely exceed unity in a centrifugal operation at the present time.

Since (Ξ) and E/ρ_0 are very small, the kinetic energy transport and dissipation terms may be eliminated from the energy equation. The term for centrifugal work is obtained by combining (3.6.16) and (3.6.17) expressed in terms of

$$\mathbf{q} \cdot \{\mathbf{k} \times (\mathbf{k} \times \mathbf{r})\} = \frac{-1}{4} \frac{E}{\rho_0} \left(r \, \nabla^2 \theta + 2 \frac{\partial \theta}{\partial r} \right) \qquad (3.6.18)$$

Then, (3.6.7), the energy equation, can be simplified to

$$\left(1 + \frac{1}{4} M^{*2} r \right) \nabla^2 \theta + \frac{1}{2} M^{*2} \frac{\partial \theta}{\partial r} - \lambda^* \frac{\mathbf{q}}{(E/\rho_0)^{1/2}} \cdot \nabla \theta = 0 \quad (3.6.19)$$

In the case of $\lambda^* \ll 1$, (3.6.19) can be further reduced to

$$\nabla^2 \theta + \frac{M^{*2}}{2[1 + (M^{*2} r/4)]} \frac{\partial \theta}{\partial r} = 0 \qquad (3.6.20)$$

The state of rigid rotation is also perturbed by the feed and extraction of the gas into and out of the rotor. The inlet and outlet are usually provided on the end plates in the form of a narrow circular slit or several small holes arranged in a circle. Near a narrow inlet, the gas velocity is relatively high, so that the inertia effect may become important. However, Nakayama and Usui [31] have reviewed various operating conditions governing the running of actual centrifuges; they concluded that in almost all cases the mass flow rate is so small that the inertia effect may be neglected without causing errors in the analysis. For this reason, the characteristic velocity can be given by the mass flow rate divided by the density at the inlet and the cross-sectional area of the rotor. The characteristic velocity $\mathbf{v}_0$ in the Rossby number is obtained by this method, so that in the actual centrifuge $\mathbf{Ro} \ll 1$.

The temperature scale associated with the forced flow is that of an adiabatic temperature change caused by the gas motion. Hence, an appropriate definition of (Ξ) may be found from (3.6.6):

$$(\Xi) = \mathbf{Ro} M^{*2} \qquad (3.6.21)$$

Since $M^{*2} < 1$ and $\mathbf{Ro} \ll 1$, (Ξ) is usually very small. The momentum equation, (3.6.6), for the case may be reduced to

$$2\mathbf{k} \times \mathbf{q} - M^{*2} \theta \mathbf{k} \times (\mathbf{k} \times \mathbf{r}) = -\frac{1}{\gamma M^2 \mathbf{Ro}} \nabla p^* \qquad (3.6.22)$$

If one takes the dot product of $\mathbf{k}$ and (3.6.22), one finds

$$\mathbf{k} \cdot \nabla p^* = 0 \qquad (3.6.23)$$

Therefore, $\nabla p*$ has only a radial component in the inviscid core, and (3.6.23) implies that flow and pressure fields vary radially and are uniform in the axial direction. This is precisely what is stipulated by the Proundam–Taylor theorem [32,33].

To transfer a certain amount of gas from an inlet in the central core to the discharge location close to the rotor wall, a pressure difference must exist to overcome the viscous resistance of the boundary layer close to the rotor wall. The pressure difference would be higher than that produced by the rigid rotation at the corresponding radial locations. From (3.6.23), one sees that this externally applied pressure difference is maintained uniformly in the axial direction. The radial pressure difference, on the other hand, induces an azimuthal flow which is balanced by the Coriolis force, as can be seen from (3.6.22). In the viscinity of the end plate, the azimuthal velocity is decelerated, until it is reduced to zero on the end plate surface due to the viscous stress. This deceleration of the azimuthal velocity is limited to a thin viscous layer close to the rotor wall, and this is where the radial transport of mass takes place. Therefore, the pressure difference between the inlet and outlet is determined by viscous resistance against the radial flow occurring within the viscous layer.

When thermal convection coexists with the forced flow, two types of temperature parameter have to be considered: (1) (Ξ), the temperature fraction for the thermal convection, and (2) $\mathbf{Ro}M*^2$, the physical characteristics for the forced flow. These two parameters are related to each other by a factor $\mathbf{M}*^2$ in magnitude, since $\mathbf{Ro}$ and (Ξ) were found to be in the same order of magnitude in most gas centrifugations.

The Ekman number $\mathbf{E}$ is a gross measure of how the typical viscous force compares to the Coriolis force and is, in essence, the inverse Reynolds number for the flow. Likewise, the Rossby number $\mathbf{Ro}$, a ratio of the convective acceleration to the Coriolis force, provides an overall estimate of the relative importance of nonlinear terms. The Rossby number $\mathbf{Ro}$ is of unit magnitude or less; linear theories presume an infinitesimal value. The gas flow in a centrifuge is of low Rossby number, and, consequently, the fundamental equations are linear. Thus, the complete flow pattern can be obtained by superposing the solutions obtained separately for thermal convection and forced flow, provided they satisfy all the boundary conditions. From the discussion of scaling analysis, Nakayama and Usui's model is that the fluid in a rotor is separated into two regions which are composed of an inviscid, rigid core separated from a viscous layer concentric to the axis of rotation and an enveloped viscous layer, which is a very narrow region, lining the side and end faces of the rotor. A typical axial velocity distribution in a countercurrent centrifuge rotor is reproduced in Fig. 3.7. The analytical procedure for evaluating the whole flow pattern is given elsewhere [30].

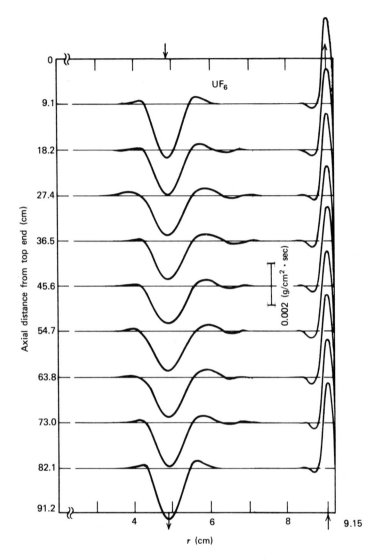

Fig. 3.7 Axial mass velocity distribution in countercurrent flow gas centrifuge. *Source:* Reproduced from Ref. 31. (*a*) UF_6—R = 9.15 cm; $2l''$ = 91.2 cm; inner inlet and outlet radius = 4.88 cm; outer inlet and outlet radius = 9.0 cm; feed rate = 2.97 g/min on both inner and outer inlets; P_w = 100 torr; T_0 = 348°K; WR = 270 m/sec. Physical properties of UF_6 used here are μ = 1.90 × 10^{-4} poise, γ = 1.067 (*above*). (*b*) SF_6—R = 10 cm; $2l''$ = 80 cm; inlet radius = 5.0 cm; outlet radii = 5.0, 9.8 cm; feed rate = 6.0 g/min; cut θ = 0.5; P_w = 150 torr; T = 320°K; ΔT = 5°C; WR = 260 m/sec (*right*).

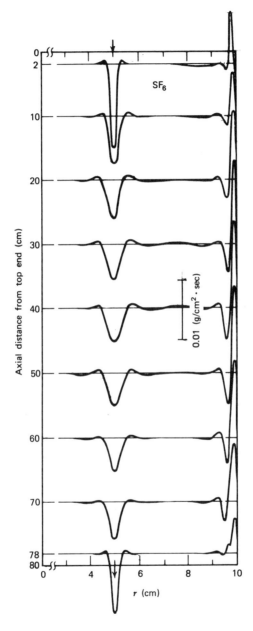

Fig. 3.7 (b).

SYMBOLS

A	Cross-sectional area
b	Quantity as defined in (3.3.30)
c	Velocity of sound $(=\sqrt{\gamma R T})$
c_1, c_2, c_3, c_4	Constants as defined in (3.3.66) through (3.3.69)
c_p	Heat capacity measured at constant pressure
c_v	Heat capacity measured at constant volume
d_p	Quantity as defined in (3.2.4)
D_{ij}	Multicomponent diffusivity of i to j
D^T	Thermal diffusion coefficient
D^*	Reduced diffusivity as defined in (3.3.55)
E	Ekman number $(=\mu/\rho\omega^2 R)$
f	Feed plate location
f_D	Correction factor of ordinary diffusion
F	Molar flow rate of feed
H	Height of centrifugal cylinder
I_1	Reduced molar flow rate as defined in (3.3.64) and (3.4.7)
I_2	Reduced quantity as defined in (3.5.21)
I_1^0	Ratio of reduced quantity as defined in (3.3.75)
J	Total molar flow rate in entire cascade
$J_p{}'$	Total molar heads flow rate in enriching section
$J_p{}''$	Total molar tails flow rate in enriching section
$J_w{}'$	Total molar heads flow rate in stripping section
$J_w{}''$	Total molar tails flow rate in stripping section
K	Flow pattern coefficient defined in (3.4.2), (3.4.6), and (3.4.9)
K_1, K_2, K_3, K_4	Constants as defined in (3.5.11) ~ (3.5.14)
k	Thermal conductivity
l_1, l_2, l_3	Constants as defined in (3.5.27a), (3.5.27b), and (5.5.27c)
L_i	Stream molar flow rate in the ith stage
L	Molar flow rate of stage feed
L'	Molar flow rate of stage heads
L''	Molar flow rate of stage tails
M	Number of moles
$\langle M \rangle$	Average molecular weight of mixture

$\Delta M'$	Quantity as defined in (3.3.10)
$\Delta M''$	Quantity as defined in (3.3.53)
m	Coefficient as defined in (3.3.76)
$\mathbf{M}$	Mach number ($= v/c = \omega^2 R/\sqrt{\gamma R T_0}$)
$\mathbf{M}^*$	Dimensionless parameter defined as $= \sqrt{(\gamma - 1)\mathbf{Pr}\cdot\mathbf{M}}$
n	Number of stages in cascade
n_w	Number of stages in stripping section
N_i	Total number of moles of component i
N	Molar flow rate
P	Cascade molar product rate
p^*	Reduced quantity as defined in (3.6.4)
$\mathbf{Pr}$	Prandtle number ($= C_p \mu/k$)
$\mathbf{q}$	Dimensionless velocity vector
R	Gas constant
R	Radius of centrifuge cylinder
$\mathbf{Ro}$	Rossby number ($= v_0/\omega R$)
t	Time
T	Temperature
T_{ij}^*	Reduced temperature ($= kT/\epsilon_{ij}$)
u_i	Velocity of the ith component relative to the mass-average velocity
U	Separative duty of cascade as defined in (3.1.54)
$v_x, v_y, v_z, v_r, v_\theta$	Velocity components in x, y, z, r, and θ directions
V_z^*	Reduced axial velocity component as defined in (3.3.54)
w	Weight fraction of desired product
W	Cascade molar waste rate
x	Atom fraction of desired product
x_F	Atom fraction in cascade feed
x_0	Atom fraction in waste of optimum composition
x_p	Atom fraction in cascade product
x_w	Atom fraction in cascade waste
Z	Length of centrifuge rotor

Greek Symbols

α	Stage separation factor as defined in (3.1.8)
α	Parameter as defined in (3.5.36)
β	Heads separation factor as defined in (3.1.14)

γ	Quantity as defined in (3.3.9)
γ	Heat capacity ratio ($= c_p/c_v$)
ϵ	Quantity as defined in (3.3.16)
ϵ, ϵ_0	Constant ratio as defined in (3.3.72) and (3.3.74)
ϵ_{ij}	Intermolecular potential parameter of energy
ζ	Reduced radial distance as defined in (3.3.51)
η	Reduced distance as defined in (3.3.57)
θ	Cut, head to feed ratio as defined in (3.1.5)
θ	Quantity as defined in (3.3.29)
θ	Reduced temperature $[= (T - T_0)/\Delta T]$
(Ξ)	Temperature fraction $[= (T - T_0)/T_0]$
κ	Boltzmann constant
λ^*	Dimensionless quantity $[= (\Xi)/(\mathbf{E}/\rho_0)]$
Λ_1, Λ_2	Quantity as defined in (3.4.25a) and (3.4.25b)
μ	Dynamic viscosity
ξ	Abundance ratio as defined in (3.1.1)
ξ	Reduced quantity as defined in (3.4.10)
Ξ	Quantity defined as $\partial \ln p / \partial z$
ρ	Density
ρ^*	Reduced density as defined in (3.6.3)
σ	Reduced quantity as defined in (3.3.56)
σ_{ij}	Intermolecular potential parameter of diameter
$\overline{\overline{\tau}}$	Shear stress tensor
ϕ	Parameter as defined in (3.4.23)
ϕ_i	Separative potential as defined in (3.1.55)
χ	Reduced parameter as defined in (3.4.11)
ψ	Reduced parameter as defined in (3.4.24)
ω, Ω	Angular velocity

Superscripts

$'$	Stage heads stream
$''$	Stage tails stream
$-$	Average Quantity

Subscripts

f	Feed location of cascade
i	Stage number in enriching section counted from waste end of cascade

j	Stage number in stripping section counted from waste end of cascade
p	Cascade product, or enriching section
w	Cascade waste, or stripping section; values evaluated at wall
n	Highest stage of enriching section
n_w	Highest stage of stripping section
min	Minimum
opt	Optimum, resulting in minimum total flow
0	Evaluated at zero position or an arbitrary reference position
e	Enriching section
s	Stripping section

References

1. M. Benedict and T. H. Pigford, *Nuclear Chemical Engineering*, McGraw-Hill, New York, 1957, Chap. 10.
2. K. Cohen, *The Theory of Isotope Separation*, McGraw-Hill, New York, 1951, Chap. 6.
3. A. Kanagawa and Y. Oyama, *J. Soc. Atomic Energy (Japan)*, **3**, 868, 918 (1961).
4. A. Kanagawa and Y. Takashima, *J. Soc. Atomic Energy (Japan)*, **6**, 20, 511 (1964).
5. M. Steenbeck, *Kernenergie*, **1**, 921 (1958).
6. A. Kanagawa, Y. Takashima, and Y. Oyama, *J. Soc. Atomic Energy (Japan)*, **10**, 118 (1968).
7. A. S. Berman, A Simplified Model for the Axial Flow in a Long Countercurrent Gas Centrifuge, USAEC Report K-1535, Oak Ridge Gaseous Diffusion Plant, 1963.
8. A. S. Berman, A Theory of Isotope Separation in a Long Countercurrent Centrifuge, USAEC Report K-1536, Oak Ridge Gaseous Diffusion Plant, 1962.
9. A. Kanagawa, *Atomic Energy Ind. (Japan)*, **14** (8), 4 (1968).
10. H. Mikami, *J. Nucl. Sci. Tech.*, **10** (7), 396 (1973).
11. H. Mikami, *J. Nucl. Sci. Tech.*, **10** (9), 580 (1973).
12. T. Matsuda, *J. Nucl. Sci. Tech.*, **12** (8), 512 (1975).
13. T. Matsuda, *J. Nucl. Sci. Tech.*, **13** (2), 92 (1976).
14. O. Okada and T. Dodo, *J. Nucl. Sci. Tech.*, **10** (10), 626 (1973).
15. K. Cohen, *The Theory of Isotope Separation*, McGraw-Hill, New York, 1951, Chap. 1.
16. M. R. Fenske, *Ind. Eng. Chem.*, **24**, 482 (1931).
17. J. O. Hirschfelder, C. F. Curtiss, and R. B. Bird, *Molecular Theory of Gases and Liquids*, Wiley, New York, 1954, p. 518.

18. R. S. Mulliken, *J. Am. Chem. Soc.*, **44**, 1033 (1922).
19. J. W. Beams, *Rev. Mod. Phys.*, **10**, 245 (1938); *Rev. Sci. Instrum.*, **9**, 413 (1938).
20. C. Skarstorm, H. E. Carr, and J. W. Beam, *Phys. Rev.*, **55**, 591 (1939).
21. J. W. Beam and C. Skarstorm, *Phys. Rev.*, **56**, 266 (1939).
22. R. F. Humphreys, *Phys. Rev.*, **56**, 684 (1939).
23. H. C. Urey, *Rep. Progr. Phys.*, **VI**, 72 (1939).
24. J. Los and J. Kistermaker, *Proceedings of the Symposium on Isotope Separation*, Amsterdam Interscience, New York, 1958, p. 696.
25. W. Groth and K. H. Welge, *Z. Phys. Chem. (Neue Folge)*, **19**, 1 (1959).
26. H. Martin, *Z. Elektrochem.*, **54**, 120 (1950).
27. H. M. Parker and T. T. Mayo, *University of Virginia, Charottsville Report No. EP-42-174-61S*, August 1961.
28. H. M. Parker, *University of Virginia, Charottsville, Report No. EP-4422-188-61S*, November 1961.
29. J. L. Ging, *University of Virginia, Charottsville, Report No. EP-442-198-62S*, January 1962.
30. P. A. M. Dirac, General Electric, General Engineering Laboratories, *Report No. Br. 42*, Schenectady, N.Y., 1940.
31. W. Nakayama and S. Usui, *J. Nucl. Sci. Tech.*, **11** (6), 242 (1974).
32. J. Proudman, *Proc. Roy. Soc.* (London), **A92**, 408 (1916).
33. G. J. Taylor, *Proc. Roy. Soc.* (London), **A100**, 114 (1921).

Chapter IV
PREPARATIVE LIQUID CENTRIFUGES

A liquid centrifuge is a device for separating particles from a solution. The term "particles" is understood to cover dissolved substances and particles of both microscopic and macroscopic dimensions, that is, everything except the suspending fluid, which is usually water. In biology, the particles are usually cells, subcellular organelles, or large molecules. In chemistry, they are usually dissolved macromolecular solutes. The objectives of preparative centrifuges are to separate specific particles and to concentrate separated particles from a dilute solution. These two objectives are related. An optimum separation and purification is achieved by having a high resolution in each separated zone or band, with the concentration increasing as the density of the separated particles in a band increases, further contributing to achievement of high resolution in a separated zone.

The most common and crude separation method is that of differential centrifugation, or pelleting. In this method, the centrifuge tube or the rotor is filled initially with a uniform mixture of sample solution. Through centrifugation, one obtains a separation of two fractions: a pellet containing the sedimented particle and a supernatant solution of the unsedimented fraction of the solution. Any particular component in the mixture may end up in the supernatant or in the pellet, or it may be distributed in both fractions, depending upon its size and/or the conditions of centrifugation. The pellet is a mixture of all of the sedimented components, and it is contaminated with whatever unsedimented particles were in the bottom of the tube or on the rotor wall initially. The only component that is in purified form is the most slowly sedimenting one, but its yield is often very low. The two fractions are recovered by decanting the supernatant solution from the pellet. The supernatant can be recentrifuged at higher speeds to obtain further purification, with formation of a new pellet and supernatant. The pellet can be recentrifuged after resuspension in a small volume of a suitable solvent.

Another method of separation is by density-gradient centrifugation, a method somewhat more complicated than differential centrifugation, but one which has compensating advantages. Not only does the density-gradient method permit the complete separation of several or all of the components in a

111

mixture, it also permits analytical measurements to be made. The density-gradient method involves a supporting column of fluid whose density increases toward the bottom of the tube or whose density increases with distance r from the axis of rotation in a centrifuge rotor. The density-gradient fluid consists of a suitable low-molecular-weight solute in a solvent in which the sample particles can be suspended.

The mass of particulate material that may be fractionated using gradient techniques is generally one or more orders of magnitude lower than that which can be handled in the same centrifuge using differential centrifugation. It is therefore often necessary to prefractionate samples before gradient techniques are used. This requirement has led to the development of precision differential centrifugation [1], which may be used for preliminary characterization of particle sedimentation-rate distributions and for partial resolution of a mixture. Most particulate suspensions encountered in biological studies contain a spectrum of particle sizes too broad to be resolved in a single density-gradient run.

It should be noted that the distinction between analytical and preparative centrifugation is made largely on the basis of intent. In analytical studies, regardless of the quantity of sample used, the objective is to make accurate measurements of sedimentation rate, banding density, molecular weight, and diffusivity, or of anomalies associated with attempts to measure these. In preparative work, the objective is purification, and each aspect of the separation is examined only from the viewpoint of how it affects resolution and separative capacity.

There are two basic methods of density-gradient centrifugation: the rate zonal (sedimentation velocity) method and the isopycnic (sedimentation equilibrium) method.

Rate Zonal Method

In the rate zonal method, a sample solution containing particles to be separated is layered on the top of the density-gradient column. Under centrifugal force, the particles will begin sedimenting through the gradient in separate zones, each zone consisting of particles characterized by their sedimentation rate. To achieve a rate zonal separation, the density of the sample particles must be greater than the density of the liquid at any specific point along the gradient column. The run must be terminated before any of the separated zones reaches the position in the centrifuge tube at which the gradient density is equal to its own density.

Isopycnic Method

In the isopycnic method, the density-gradient column encompassses the whole range of densities of the sample particles. Each particle will sediment only to the position in the centrifuge tube at which the gradient density is equal to its own density, and there it will remain. The isopycnic method, therefore, separates particles into zones solely on the basis of differences of their densities, independent of time.

In the isopycnic procedure with salt solutions, it is not always convient to perform a gradient and to layer the sample solution on top. It is sometimes easier to start with a uniform solution of the sample and the gradient material. Under the influence of centrifugal force, the gradient material redistributes in the tube so as to form the required concentration (or density) gradient. Meanwhile, sample particles, which are initially distributed throughout the tube, sediment or float to their isopycnic positions. This self-generating gradient technique often requires long hours of centrifugation. Isopycnic banding of DNA, for example, takes 36 to 48 hr in a self-generating cesium chloride gradient. It is important to note that the run time cannot be shortened by increasing the rotor speed; this only results in changing the positions of the zones in the tube since the gradient material will redistribute farther down the tube (or rotor) under greater centrifugal force. Sedimenting of the gradient-forming solute beyond its saturation concentration at the bottom of the tube will cause crystal precipitation. Overstressing of the rotor may result, and rotor failure may occur.

As one has seen, the density-gradient method permits separations of sample particles based either on differences in sedimentation rate or differences in density. However, sample size is limited by the tube diameter and hence the area over which the sample spreads. The only practical way to significantly increase the quantity of sample handled in a density-gradient separation is to increase the area of the sample cavity. The vessel which would allow large volumes to be centrifuged at high speed is a hollow cylindrical pressure vessel spinning on its own axis. Cylindrical or bowl-shaped zonal rotors now commonly used in research work have capacities of 50 to 100 times that of the typical swinging bucket rotor and are highly efficient tools for isolating and purifying a variety of particles [2]. The procedures used and the equipment required for large-scale centrifugation are quite different from those used for similar experiments in swinging bucket rotors.

Zonal rotors are large cylindrical containers or bowls in which the rotor volume varies with the square of the radial distance from the center of the rotation. The cylindrical cavity of the bowl is divided into sector-shaped compartments by vanes attached to the core, similar to those used in the analytical ultracentrifuge [2, 3]. The rotor is enclosed by a threaded lid. A rotating seal assembly, which is fixed or removable, allows fluid to be pumped in and out of the cavity while the rotor is spinning.

The design of zonal centrifuges requires the simultaneous solution of a number of problems. The diameter at a given speed is a function of the strength of the materials of construction, and detailed information on the optimum wall thickness and internal diameter as a function of speed is available for aluminum, titanium, and steel rotors [3]. Rotor length is limited by the first critical frequency of the rotor on the one hand and considerations of rotor stability on the other. Rotors are not operated above or close to their first critical frequency. Rotor stability is markedly affected by the ratio of its length to its diameter. If the L/D ratio approaches unity ($L/D = 1$), the rotor tends to be instable. As the ratio is increased or decreased, increasing stability is observed. Rotors are constructed as either cylinders ($L/D > 1$) or as disks ($L/D < 1$). Low-speed rotors have generally been of the disk type, while high-speed rotors have generally been made in both configurations. Disk rotors can be supported from above or from below on a single shaft. Cylindrical rotors operated at high speed require shafts and damper bearings at both ends. Regardless of speed, the cylindrical configuration is chosen when maximum volume is required.

1   SEPARATIONS BY ZONAL CENTRIFUGES

As shown in Table 1.3, many types of zonal rotors have been developed and manufactured commercially. For small particles with S values below 20, such as most proteins, hormones, and enzymes, a high centrifugal force is desirable. The J-1 rotor has a volume of 825 ml and generates up to 994,000 g at 100,000 rpm. Particles with S values from 20 to 100, such as macroglobulins or ribosomal subunits, are most effectively separated with the B-XIV titanium rotor, which generates up to 102,000 g at 37,000 rpm. It has a total volume of 630 ml. For large particles, such as mammalian viruses and subcellular organelles from plant or animal tissue homogenates, the B-XV titanium ultracentrifuge zonal rotor is preferable, since it develops 77,000 g at 28,000 rpm with a volume of 1665 ml.

The three rotors mentioned are made of titanium to withstand the highest force fields and for good corrosion resistance to salt and most other solutions within the range of pH 4 to 10. Where lower forces suffice, and where only noncorrosive solutions will be used, an aluminum rotor is more economical.

Two types of cores are available for zonal rotors. The choice of cores depends mainly on the method used to load and unload the rotor. The standard core permits loading and unloading the rotor while it is spinning (dynamic loading). The other core is designed to load and unload the rotor while it is at rest (static loading) using the reorienting-gradient technique. In the standard core, there are modifications to add versatility to the operation of zonal rotors by allowing fraction recovery at the edge of the rotor as well as at its center, as for B-XXIII- and B-XXIX-type rotors.

Density-Gradient Solutions

The shape of the gradient is important for achieving the desired separation and for determining properties such as buoyant density and sedimentation coefficient. Gradients linear in density and volume are the most commonly used and offer the best resolution of components such as proteins, enzymes, hormones, ribosomal subunits, and some plant viruses.

Concave gradients are best used for lipoproteins and other samples that require separation by flotation.

Discontinuous or step gradients are best suited for separating whole cells or subcellular organelles from plant or animal tissue homogenates, and for purifying some mammalian and insect viruses. Multiple steps are sometimes used with high-viscosity gradients made with Ficoll* solutions.

The "cushion," a small volume of high-density solution often placed at the bottom of the rotor, also acts as a step in the gradient and may be useful in effecting part of the separation. A cushion makes it easier to resuspend any sedimented material at the end of the run and prevents damage to particles that may not withstand pelleting. In particular, some viruses lose viability when pelleted.

There is no ideal all-purpose gradient material. Sucrose is used for most rate separations, and cesium chloride is often used for isopycnic separations. The basic requirement is that the gradient permit the desired type of separation. Additional considerations in selecting a gradient material include the following [4]:

1. Its density range should be sufficient to permit separation of the particles of interest by the chosen density-gradient technique, without overstressing the rotor.

2. It should not affect the biological activity of the sample.

3. It should be neither hyperosmotic nor hypoosmotic when the sample is composed of sensitive organelles.

4. It should not interfere with the assay technique.

5. It should be removable from the purified product.

6. It should not absorb in the ultraviolet or visible range.

7. It should be inexpensive and readily available; more expensive materials should be recoverable for reuse.

8. It should be sterilizable.

9. It should not be corrosive to the rotor, particularly for zonal or continuous flow operation.

10. It should not be flammable or toxic to the extent that its aerosols could be hazardous.

Table 4.1 lists some commonly used gradient materials with their solvents and densities at 20°C [4].

Gradient solutions are formed before centrifugation either by hand-layering the different concentrations into the rotor or by the use of a gradient former. A

---

*Polysucrose, Pharmacia Fine Chemicals, Sweden.

Table 4.1 Commonly Used Gradient Materials with Their Solvents [4]

Materials	Solvent	Maximum Density at 20°C
Sucrose (66%)	H_2O	1.32
Sucrose (65%)	D_2O	1.37
Silica sols	H_2O	1.30
Diodon	H_2O	1.37
Glycerol	H_2O	1.26
Cesium chloride	H_2O	1.91
	D_2O	1.98
Cesium formate	H_2O	2.10
Cesium acetate	H_2O	2.00
Rubidium chloride	H_2O	1.49
Rubidium formate	H_2O	1.85
Rubidium bromide	H_2O	1.63
Potassium acetate	H_2O	1.41
Potassium formate	H_2O	1.57
	D_2O	1.63
Sodium formate	H_2O	1.32
	D_2O	1.40
Lithium bromide	H_2O	1.83
Lithium chloride	D_2O	1.33
Albumin	H_2O	1.35
Sorbitol	H_2O	1.39
Ficoll	H_2O	1.17
Metrizamide	H_2O	1.46

Source: Reprinted from Ref. 4 by permission of Beckman Instruments, Spinco Division.

sucrose or viscous gradient formed by the layering process can either be run as a step gradient or it can be made into a linear gradient by allowing it to diffuse. The higher the viscosity, the longer the time for each layer to diffuse into a smooth gradient solution.

Many gradient-forming devices are available, commercially and custom-made. To produce a gradient of a specific shape, a gradient pump which can be programmed to deliver linear and nonlinear gradients is highly useful. Commercially available pumps have this capacity. A gradient pump's versatility and the ease with which it can be made to produce the necessary gradient shapes are important in zonal separations.

For a discontinuous gradient, an ordinary peristaltic pump can be used, or the gradient shape can be programmed if a programmable pump is used.

Although most gradient shapes used in zonal rotors are the same as those used in swinging bucket rotors, allowances must be made for the fact that the shape of a gradient changes when it is loaded into a cylindrical or bowl-shaped rotor.

To duplicate experiments performed with linear gradients in swinging bucket rotors, it is necessary to introduce into the zonal rotor a gradient that is linear in volume and concentration with respect to rotor radius. Such a gradient offers the best resolution of components from samples of proteins, enzymes, hormones, ribosomal subunits, and most plant viruses.

A gradient prepared so that it is linear in concentration with respect to rotor volume becomes a concave in the zonal rotor. This gradient is found to be useful for lipoprotein isolation and for other samples requiring separation by flotation.

The discontinuous or step gradients used in zonal rotors are also similar to those used in swinging bucket rotors. The concentration at each step is determined from the separation observed in previous experiments using linear gradients. This type of gradient is best suited for separations of subcellular and cytoplasmic organelles and purification of some mammalian insect viruses. Many components from tissue homogenates band isopycnically in sucrose gradients; hence, the discontinuity in the gradient facilitates better purification of the individual component.

Laboratory experience has shown that preformed continuous gradients with starting and ending concentrations differing by a factor of 3, especially when the starting concentration is less than 10% w/w, produce better resolution of components. These concentration limits are especially important when shallow or low concentration gradients are used.

To duplicate experiments reported in the literature, one must know whether the gradient solutions were mixed according to weight by weight or weight by volume. A 20% w/w sucrose solution, for example, has the same density (1.081 g/cm^3) as a 21.5% w/v sucrose solution, while a 20% w/v sucrose solution has the same density (1.0748 g/cm^3) as an 18.6% w/w solution. An error of this magnitude could produce undesirable results when separations are made on a density basis. The difference is even greater with CsCl, since a 20% w/w CsCl solution has the same density as a 23.8% w/v CsCl solution, namely, 1.758 g/cm^3 [4].

Loading of Zonal Rotors

Zonal experiments are usually carried out in the cold, at least below 15°C. Thus, the rotor is normally chilled before the filling process is started, and the gradient solutions as well are usually cold. Some cooling of the rotor and its contents can be accomplished in the centrifuge itself during the loading process.

Dynamic Loading

With a standard core, the rotor is loaded and unloaded dynamically at a speed of 2000 to 3000 rpm and requires fluid lines connected to both the center and the edge of the rotor (see Fig. 4.4). Hydrostatic pressure is equalized by leading both lines back as close as possible to the axis of rotation. When a single seal is used, the two lines are coaxial, with the line connected to the rotor center in the center of the seal. The edge line therefore must come into the seal a few millimeters from the axis, thereby creating a small difference in hydrostatic pressure between the two lines during rotation. This difference is largely equalized by the presence of fluid in the edge lines, which is denser than that found in the center line, the density difference being the difference between the density of the underlay, or cushion, under the gradient and the density of the overlay solution on the sample.

When loading starts, the light end of the gradient is introduced to the edge of the rotor during rotation. Centrifugal force at this stage in the loading cycle makes the rotor act as a pump which draws the gradient in. When the lightest part of the gradient enters the rotor cavity, it forms a thin cylindrical layer on the wall, due to centrifugal force. As each succeeding element of the gradient solution enters the cavity, a new layer is formed which displaces the former layer, so that the first element introduced is gradually pushed to the center of the rotor cavity. The gradient filling process is continued by introducing gradient solution of increasing density and concentration until the entire gradient volume is pumped into the cavity. The volume of the gradient is usually 60 to 70% of the rotor cavity volume. Since most gradient engines pump at a constant rate, dissolved air in the gradient tends to form air bubbles in the edge line due to the negative pressure in this line. The air bubbles cause little difficulty until the rotor is nearly full. At that time, the hydrostatic pressure due to the fluid in the main rotor chamber is only partially balanced by fluid in the edge line, resulting in a high back pressure. This problem is resolved by use of a gradient-producing device which forms the gradient as fast as it is drawn into the rotor. The remainder of the rotor volume is then filled with a cushion, a solution identical to the heavy end of the gradient or, in some cases, considerably denser.

In all dynamically loading rotors, tapered surfaces exist which serve to funnel particle zones into the center exit line during the unloading and also to minimize mixing of the sample with the gradient and the overlay during loading. To leave the sample in contact with these slanted surfaces would result in a starting sample of uneven thickness which is too close to the axis of rotation, that is, in a low centrifugal field. Hence, a solution having a density slightly less than that of the sample layer, termed the "overlay," is added to move the sample layer out into the rotor and free of the core surface. At this juncture, flow is reversed several times, and small volumes of fluid are run alternately to the center and the edge to ensure that all entrapped air has been expelled. This illustrates an additional function of the overlay and the underlay, which is to provide the fluid volume necessary to allow the sample and gradient to be moved back and forth radially in the spinning rotor. The completion of the gradient filling process is signaled by the appearance of fluid at the centerline connection to the rotor.

Static Loading

Static loading is generally done through a core connecting a tube to the bottom of the centrifuge rotor (see Fig. 4.4). The overlay, sample, and gradient may be introduced in that order, or the sample and overlay may be carefully added from the top after the gradient is in position. An alternate method is to fill the rotor with dense underlay and then, by withdrawing part of the underlay from the bottom, draw in the gradient, sample layer, and overlay from the top.

To date, the practical use of gradient reorientation has been chiefly continuous-sample-flow-with-banding centrifugation. However, it is a potentially useful method for large-scale, low-speed gradient centrifugation and also for centrifugation at very high speeds where continuously attached seals cannot be used. Use of detachable seals for high-speed rotors also presents problems because they are light in weight and have insufficient rotational momentum to allow seals to be attached without creating turbulence. As discussed later, the turnover effect can markedly decrease resolution in rotors loaded at rest. However, it does not affect gradients after a separation has been made.

Sample Concentration and Loading Capacity

Sample concentration in zonal rotors is just as important as in swinging bucket rotors. Svenson et al. [5] pointed out that the theoretical capacity of the gradient is a function of its density slope. Later, Brakke [6] discovered experimentally that only a few percent of the theoretical sample load could be supported in swinging bucket rotor tubes. Based on Brakke's work, it has been demonstrated [4] that gradients in swinging bucket rotors can support most samples if the ratio between sample concentration (% w/w) and starting gradient concentration (% w/w) is 1:10. If the sample concentration on the gradient is too high, streaming (turnover) will result. Even if streaming is not evident, too high a sample concentration may overload the gradient causing broadening of the separated zones and loss of resolution. If the sample concentration is too low, it may be difficult to identify the separated zones.

For zonal rotors, the total sample concentration (excluding the additional sucrose), when observed in the Bausch & Lomb hand refractometer, should be approximately 40% less than the starting gradient concentration to prevent density inversion or turnover effect. Berman [7] and Spragg and Rankine [8] discuss the maximum sample loading capacity for gradients. Laboratory investigations confirmed that the measurements observed in the Bausch & Lomb refractometer are in reasonable agreement with Berman's formula for gradient capacity.

Prior to applying the sample to the gradient, the sample concentration and volume should be determined. Typical loading and unloading volumes and flow rates for zonal experiments with Beckman's four zonal rotors are shown in Table 4.2. Sample volumes smaller than those shown are not recommended because there will be a resultant loss of resolution due to diffusion. Larger sample volumes than those shown may be used, but there will be a reduction in the number of components that can be effectively separated in the remaining radial distance.

Table 4.2 Suggested Typical Loading Volumes and Flow Rates for Four Beckman Zonal Rotors [4]

Z-60 *Standard Core*	*Al/Ti-14*	
	Standard Core	*B-29 Core*
20 ml overlay	30 ml overlay	30 ml overlay
20 ml sample	20 ml sample	20 ml sample
250 ml gradient	450 ml gradient	450 ml gradient
40 ml cushion	150 ml cushion	50 ml cushion
330 ml total volume	650 ml total volume	550 ml total volume

Flow rates
Loading: sample and overlay 10–12 ml/min; gradient and cushion 20–25 ml/min
Unloading: same as loading gradient

JCF-Z *Standard or* *Reograd Core*	*Al/Ti-15*	
	Standard Core	*B-29 Core*
100 ml overlay	50 ml overlay	50 ml overlay
50 ml sample	50 ml sample	50 ml sample
1500 ml gradient	1200 ml gradient	1200 ml gradient
250 ml cushion	350 ml cushion	50 ml cushion
1900 ml total volume	1650 ml total volume	1350 ml total volume

Flow Rates
Loading: sample and overlay 15–25 ml/min; gradient and cushion 30–50 ml/min
Unloading: same as loading gradient

Source: Spinco DS-468, by O. M. Griffith (by permission of the Beckman Spinco Division).

Experimental Runs

After the gradient and sample are in the rotor, the seal assembly is removed and the rotor accelerated to 5000 rpm before attaching the vacuum cap. This increase in speed causes the rotor to expand slightly, allowing any air remaining in the rotor to be displaced to the center. The trapped air in the rotor will aid in removing the vacuum cap at 2000 rpm at the end of the run. After capping temperature settings and proper chamber vacuum are achieved, the rotor is accelerated to the operating speed. It should be noted that the centrifugal force has a very considerable stabilizing effect on a liquid density gradient.

Selection of rotor speed and run time depend upon the kind of experiment to be conducted, the physical properties of the sample components such as size and density, and upon the density and viscosity of the gradient solution. Some typical experimental conditions suggested by Spinco Division of Beckman Instruments Co., using their rotors, are presented in Table 4.3. If the properties of the sample components are unknown, or if they include a wide range of sizes and densities, then the most appropriate rotor speed and time of operation may be best determined experimentally in a series of runs. In addition, if some information is available about the properties of the sample, suitable sample concentrations or gradient limits may be calculated or some experimental designs based on mathematical theories to be discussed later are recommended.

Unloading and Sample Recovery

In all zonal experiments, the rotor can be decelerated with the brake to 2000 rpm. At this speed, the chamber is opened and the vacuum cap removed. The seal assembly, which should be immersed in water during the run, is now primed with the displacement solution up to the edge-port fitting. After attaching the seal assembly to the rotor, the pump is actuated and the rotor contents are displaced by introducing heavy displacement fluid, typically 60% w/w sucrose, at the periphery and collecting fractions through the center-port fitting of the seal assembly. If a modified core is used for B-XXIII- or B-XXIX-type rotors, the rotor loading is in the same manner as the regular standard core, but unloading can be accomplished by introducing buffer or distilled water at the center and collecting fractions exiting through the edge port of the core. For both types of rotors, the rotor cavities are made with a tapered wall, so that the area perpendicular to the radius occupied by a separated particle zone decreases with the radius at the tapered wall zone, which gives a funnel effect analogous to the effective collection of fractions at the center.

With edge unloading, it is not necessary to use concentrated gradient material to replace fractions, thus realizing a considerable savings in preparation time and material expenses. Besides its economy, a tapered wall zonal

Table 4.3 Typical Parameters for Experiments and Runs Using Beckman Zonal Rotors [4]

Sample	S Value	Rotor	Gradient (% w/w)	Rotor Speed (rpm)	Time (hr)	Temperature (°C)
Serum proteins	4 S, 7 S, 19 S	Z-60	5-40	60,000	16	5
Ribosomal subunits	30 S and 50 S	Ti-14	10-40	48,000	6	5
Tobacco mosaic	180 S	Ti-14	10-40	48,000	1.5	5
Mammalian virus	700 S	Ti-15	10-50[a]	32,000	1	5
Mitochondria	—	Ti-15	10-50[a]	32,000	0.75	5
Subcellular components (rat liver)	—	JCF-Z	10-50	20,000	1.5	5
Subcellular components (brain)	—	Ti-15	10-55	30,000	0.5	5

Source: Sprinco DS-468, by O. M. Griffith (by permission of the Beckman Spinco Division).
[a]Discontinuous gradients are suggested for these experiments.

rotor is especially useful in applications such as flotation separations (sample introduced at the edge, buoyant particles recovered at the center) and flota-tion–sedimentation separations (sample introduced between the center and the edge to let buoyant particles be recovered at the center and heavy particles sedimented to the edge) to increase the dimension of separations.

The special design of the rotor core and fluid passages allows the relative order and distribution of material in the rotor at the conclusion of the high-speed run to be retained in the effluent stream of the rotor. Usually the ef-fluent is continuously monitored for ultraviolet absorption at a specific wavelength while the separation profile is automatically recorded. About 20 to 40 equal fractions of the rotor volume are collected during the unloading. The collected fractions may be analyzed for concentration of gradient material (e.g., using a refractometer reading directly in weight percent sucrose), for specific biological activity, for radioactivity, or for some other property re-quired for the experiment.

Commercial systems are available which have a flow-cell attachment for on-line monitoring of the gradients from zonal rotors. There are, however, two im-portant factors to be observed in choosing a monitoring system: (1) the tubing from the rotor seal assembly and (2) the cuvette (flow cell) must clear air bub-bles readily.

An Oak Ridge B-XV rotor, a partially assembled and the completely assembled rotor with the removable upper seal in place, are shown in Figs. 4.2a and 4.2b. In Fig. 4.3, the rotor and seal in place in a Spinco Model L preparative ultracentrifuge are shown. A side view of the assembled rotor in position in the centrifuge is presented in Fig. 4.4. Several schematic diagrams of typical zonal operations that apply to B-XIV and B-XV zonal centrifuge rotors are shown in Fig. 4.1.

Continuous-Flow Zonal Centrifuges

In order to avoid pelleting of the separated particles, continuous-flow cen-trifuges are often used. In this type of centrifuge the rotor is filled with a heavy density-gradient solution, and a light sample stream containing the particles to be separated (axially along the length of the rotor) is made to flow continuously through the centrifuge [9]. The heavy density-gradient solution confines the feed stream to a thin layer near the core, and the particles sediment from the feed layer to the outlying gradient as the sample feed stream moves along the core and can eventually be banded in the gradient.

The rotors are similar to zonal rotors, but the core is different due to the fluid flow patterns in the rotor. In operation, the sample is pumped in con-tinuously through the centerline of the seal assembly while the rotor is spinning at operating speed. The sample flows along the bottom of the core and moves over the centripetal surface of a density gradient or other solution. The cen-

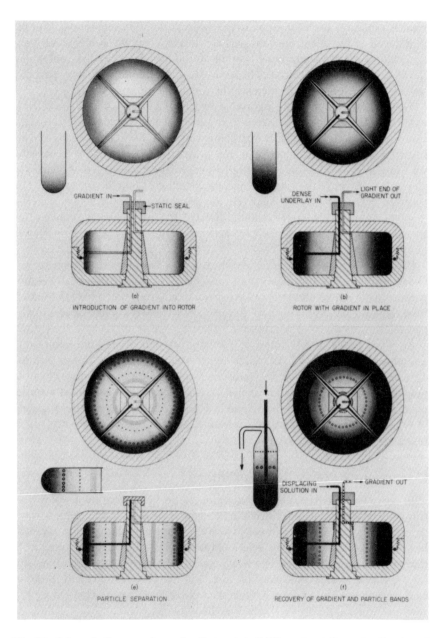

Fig. 4.1 Schematic diagrams of operation that apply to B-XV zonal centrifuge rotor. Courtesy of Oak Ridge National Laboratory.

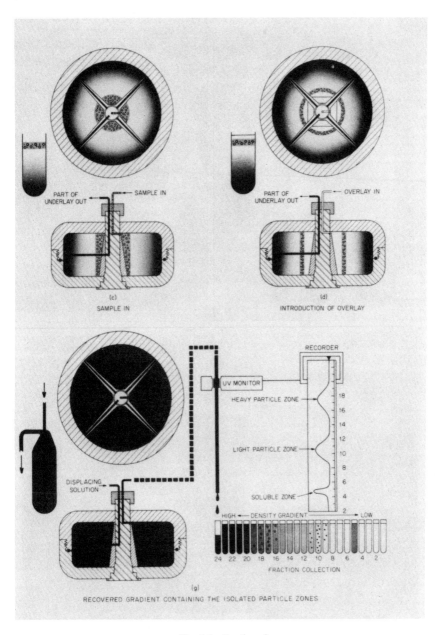

PART OF UNDERLAY OUT — SAMPLE IN

(c)

SAMPLE IN

PART OF UNDERLAY OUT — OVERLAY IN

(d)

INTRODUCTION OF OVERLAY

RECORDER

UV MONITOR

HEAVY PARTICLE ZONE

18

16

14

12

LIGHT PARTICLE ZONE

10

8

6

SOLUBLE ZONE

4

DISPLACING SOLUTION

2

HIGH ← DENSITY GRADIENT → LOW

24 22 20 18 16 14 12 10 8 6 4 2

FRACTION COLLECTION

(g)

RECOVERED GRADIENT CONTAINING THE ISOLATED PARTICLE ZONES

Fig. 4.1 Continued

Fig. 4.2 B-XV Rotor. (*a*) Partially assembled. (*b*) Completely assembled with removable upper seal in place. Courtesy of Oak Ridge National Laboratory.

Fig. 4.3   Rotor and seal in place in Spinco Model L preparative ultracentrifuge. Courtesy of Oak Ridge National Laboratory.

trifugal separation therefore accounts for two fractions: a sedimenting particle that moves out into the rotor cavity and a supernatant fraction that continues to flow along the core and over the centripetal surface of the gradient and then out of the rotor via the outlet lines. Various flow rotors are currently available from various manufacturers. The K- and J-series rotors are effectively in use for these purposes.

Continuous-flow processing finds particular use in large-scale purification of viruses (plant, mammalian, or insect) from tissue culture media or other solutions and of some subcellular organelles from large volumes of tissue homogenates. The technique is also useful for harvesting bacteria and for separating clay particles in water pollution studies.

Many viruses, especially the myxoviruses, lose biological activity when pelleted. In addition, sedimented particles often do not resuspend properly, and, in subsequent rate and banding purifications, infectivity may be observed in a spectrum of particles, discrete virus particles, and clumped or aggregated particles of many different sizes. A major contribution in the development of these continuous-flow methods is in the application of producing effective vaccines which contain only those antigenic proteins necessary to produce immunity [9].

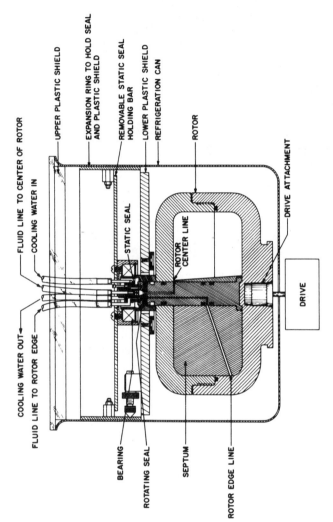

Fig. 4.4 Side view of assembled rotor in position in centrifuge during loading. Courtesy of Oak Ridge National Laboratory.

Continuous-flow operations are normally carried out in the cold, and the rotor is chilled before placing it in the centrifuge. In addition, since the run may continue for several hours, some provision must be made for keeping the sample supply cold. A simple peristaltic pump may be used which does not heat up the solution it pumps. The pump should be capable of flow rates ranging from 5 to approximately 200 ml/min and not have to be stopped to change flow rates. Alternately, the cold sample solution can be forced from its container (for example, a pressure vessel placed in an ice or refrigerated bath) into the rotor by air or nitrogen gas pressure.

In a continuous-flow procedure, a gradient of sufficiently high density is needed to band the particles isopycnically. Because the radial distance from the core to the rotor wall is so short (approximately 1 cm), it is not necessary to build a programmed linear gradient into continuous-flow rotors. Diffusion of the solutions at each step in the gradient is enough to produce a continuous gradient during the loading of the rotor.

After the rotor is filled, the rotor is accelerated to operating speed while the buffer or distilled water is kept flowing into the center of the rotor at a constant rate of approximately 10% higher than the flow rate for the run. When operating speed is achieved, the pump is stopped and the sample medium is pumped into the center of the rotor replacing the buffer or the distilled water.

Before a given gradient reaches its maximum capacity in an isopycnic banding run, the continuous-flow zonal rotor is unloaded by decelerating it to 2000 rpm and displacing the rotor contents with a heavier (unloading) solution, as in batch zonal rotor operation with the standard core. There is, however, one special consideration that makes continuous-flow operation different from batch zonal rotor operation: the continuous-flow zonal rotor cores have an additional passage to allow the particle-free effluent to flow out of the rotor. This passage has a common connection with the edge-fluid line of the rotor. Hence, prior to injecting the unloading solution into the edge-fluid line, a small volume of air is pumped into the rotor ahead of the unloading solution. This air blocks off the additional passage in the core so that the unloading solution can flow only to the rotor edge for normal unloading.

Static unloading is accomplished by bringing the rotor to complete rest and withdrawing the rotor contents, namely, the banded particle zone and particle-free effluent. This is achieved by gravitational flow from the rotor bottom for K-type rotors or by pumping high-density fluid through the edge-fluid line to replace the rotor contents.

A typical disassembled K-series centrifuge rotor for continuous-flow isopycnic banding zonal runs is presented in Fig. 4.5a. A safe rotor chamber, armor shielding, and the K-series centrifuge enclosed in an armor shielding ready for operation are shown in Figs. 4.5b and 4.5c, respectively.

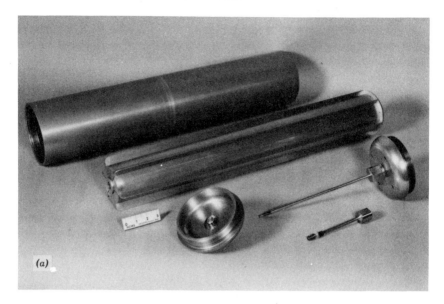

Fig. 4.5   Typical K-series rotor and its operation system. (*a*) Disassembled K-type centrifuge rotor. (*b*) Armor shielding for K-rotor. (*c*) K-Series rotor closed in armor shielding ready for operation. Courtesy of Oak Ridge National Laboratory.

Problems in Zonal Runs

There are some operating problems that are frequently encountered in zonal separations. Precautions must be taken to minimize errors due to these. They are discussed below.

Seal Assembly Leaks

Gradient or sample may be lost from the system during passage through the rotating seal assembly due to leaks because of the following reasons: (1) failure to lubricate O-rings on the rotor core and on the seal housing; (2) failure to tighten the threaded center fitting; (3) failure to maintain the seal assembly after using; (dried gradient, in particular sucrose, causes the bearings to seize, resulting in overheating of the seal assembly or scratching the surfaces of the rotating seal; after each use, the seal assembly must be washed in warm water and the bearings must be lubricated with silicone grease and the seal assembly must be cleaned); (4) failure to select a proper tubing system, so that the seal assembly is subjected to excess pressure.

Fig. 4.5 Continued

Air Bubbles

Occasional air bubbles introduced into the rotor during loading will not cause problems, but large blocks of air indicate a leak in the tubing or seal assembly. These air bubbles cause mixing of the gradient. In continuous-flow operations, air bubbles introduced into the rotor during sample flow will cause the back pressure to rise. If the pressure is above 25 psi, the following steps are

Fig. 4.5 Continued

suggested [4]: (1) turn off the pump; (2) decelerate the rotor to 2000 rpm; (3) replace sample flow with buffer; (4) turn on pump to a flow rate 10% higher than previous sample flow (this flow rate will dislodge the air block); (5) accelerate to operating speed; (6) at speed, turn off pump and replace buffer flow with sample flow; (7) reset pump flow rate for correct sample flow; and (8) turn on pump. If air bubbles are observed at the exit line during sample flow, these bubbles are caused by sample and/or gradient outgassing. If the pressure does not exceed 20 psi, the experiment may be continued without additional problems.

Gradient Loading

Fast gradient loading can cause mixing of the gradient and subsequent loss of resolution. Usually, flow rates for loading the gradient solution into the rotor should not exceed 50 ml/min. Charging gradient in the wrong order or charging it into the center of the rotor instead of the periphery are examples of improper loading. Loading or unloading of rotors with a solution that is not at the same temperature as the rotor contents can produce thermal gradients or outgassing. Therefore, the gradient solution and the sample solution should be kept at the same temperature as the rotor before or during the loading.

Gradient Shape

All centrifugation techniques depend on the shape of the gradient and on the gradient properties which are variable with respect to the space coordinate in terms of concentration. The shape of the gradient is important in achieving the desired separation. Thus, one has to know the exact shape of the density-gradient solution at a given rotor location and at a specified time for either sedimentation–velocity (rate zonal) technique or sedimentation–equilibrium (isopycnic) technique. A wrong density for the ending gradient solution can cause pelleting of the sample against the rotor wall. Likewise, a wrong density profile can overload the sample in the gradient solution causing a turnover effect which decreases the resolution. In continuous-flow runs, a wrong density profile and an improper sample flow rate may cause the loss of the sample attributing to the sample elution from the rotor.

2 ANALYTICAL DEVELOPMENTS IN ZONAL CENTRIFUGATION

In batch zonal centrifugation, with the simultaneous occurrence of sedimentation and diffusion in any isothermal–nonreacting fluid mixtures, the continuity equation can be written from (2.1.18) and (2.8.3) to give

$$\frac{\partial \rho_i}{\partial t} = \frac{1}{r}\frac{\partial}{\partial r}\left[r\sum_{j=1}^{\nu}(D_{ij})_v\frac{\partial \rho_i}{\partial r} - (S_i)_v\omega^2 r\rho_i\right] \qquad (i = 1, 2, \ldots, \nu) \qquad (4.2.1)$$

In obtaining (4.2.1), constant partial specific volumes of all components are assumed, and only the centrifugal force field exists as an external force field. The quantities S_i and D_{ij} are the sedimentation coefficients of component i and the diffusivity of i into j in a multicomponent mixture ($\nu > 2$), where the subscript v denotes that both S_i and D_{ij} are to be referred to the volume-fixed frame. Neither S_i nor D_{ij} depend on the concentration gradients, but they can be functions of the local state variables such as temperature, pressure, and composition. The sedimentation coefficient of component i, S_i, is simply a measure of sedimentation velocity of component i per unit centrifugal force field; it is defined by Svedberg [10, 11] as

$$S_i = \frac{\mathbf{v}_{ri}}{\omega^2 r} = \frac{d\mathbf{r}_i/dt}{\omega^2 r} \quad \text{(in Svedberg units} = \sec \times 10^{-13}\text{)} \quad (4.2.2)$$

Equation 4.2.1 is generally known as the generalized Lamm equation. The time in (4.2.1) or (4.2.2) is measured from the instant of band formation, which is assumed to occur after the rotor has attained a certain velocity, that is, ω reaches a constant specified value. Neglecting the Coriolis effect, the equation of motion can be adequately expressed by a radial component only. Thus, for an incompressible liquid density-gradient system, one has equations of motion from (2.9.9):

$$\frac{d\mathbf{v}_{ri}}{dt} + \frac{18\mu(r)}{\rho_i D_{pi}^2}\mathbf{v}_{ri} = \frac{\omega^2 r}{\rho_i}[\rho_i - \rho(r)] \quad (i = 1, 2, \ldots, \nu) \quad (4.2.3)$$

The quantities $\mu(r)$ and $\rho(r)$ are respectively the viscosity and density of the gradient solute as a function of radial distance from the rotating axis, which are the characteristics of a gradient solution. A mathematical technique to predict $\mu(r)$ and $\rho(r)$ for the diffusion and sedimentation of gradient-forming solutes has been developed by Breillatt et al. [12]. The quantities ρ_i and D_{pi} are the ith particle's density and size, respectively, and $\mathbf{v}_{ri}$ is the sedimentation velocity of the ith particle. In the derivation of (4.2.3), the particles are all assumed to be spherical, and the drag force on the particles during the sedimentation can be approximated by Stokes' resistance law. For the nonspherical particles, the equation can be modified by the introduction of appropriate shape correction factors in the usual fashion.

In theory, (4.2.1) and (4.2.3), together with appropriate initial and boundary conditions, completely specify the zonal centrifugal systems. Solutions to these equations describe the transport phenomena of zonal centrifugations.

Sedimentation–Velocity Method

For this method, a centrifugal field of specified force is chosen for the system under investigation. Because the rotor is operated at high speed, the solute particles are forced to settle at appreciable rates toward the wall of the rotor. Thus, the sedimentation–velocity method is also known as rate centrifugation and is basically a transport method; the parameter used to describe the transport of particles is the "sedimentation coefficient" given in (4.2.2).

The rate of particle movement is a function of molecular weight as well as particle size and shape, the difference in density between the particles and the local sustaining liquid density, the local sustaining liquid viscosity, and the centrifugal force field strength. In order to make the characteristics unique, the observed sedimentation coefficient is customarily converted to the sedimentation medium of water at 20°C for the particle. Thus, the conversion is

$$S_{H_2O, 20°C} = S_{obs} \left(\frac{\mu_{GT}}{\mu_{H_2O, 20°C}} \right) \cdot \left(\frac{\rho_p - \rho_{H_2O, 20°C}}{\rho_p - \rho_{G,T}} \right) \qquad (4.2.4)$$

where $\mu_{H_2O, 20°C}$ is the viscosity of water at 20°C; $\mu_{G,T}$ is the local viscosity of a sustaining gradient solution at temperature T; and ρ_p, $\rho_{G,T}$, and $\rho_{H_2O, 20°C}$ are the densities of a solute particle, the local gradient solution, and water at 20°C, respectively.

The separation using the sedimentation-velocity method requires building an optimum density-gradient in a rotor and estimating an optimum centrifugation time, so that the resolution of separation (the difference between the positions of the two different particles) is the maximum, and also to evaluate each particle's position in a rotor when centrifugation is stopped, so that the separated particles can be collected adequately.

Sedimentation–Equilibrium Method

Using the sedimentation–equilibrium method (isopycnic banding), the density distribution of molecules at sedimentation–equilibrium positions is used in the fractionation. The banding solute i moves outward from the center of the rotor with velocity $\mathbf{v}_{r_i}$ until the solute reaches its respective location of isodensity in the gradient; then $\mathbf{v}_{r_i}$ vanishes and an equilibrium state is established. There are many observable properties that are characteristic of systems in a state of internal equilibrium.

The mathematical relationship describing sedimentation equilibrium can be obtained from equations which express the criteria of equilibrium in the system. Thus, the general conditions for sedimentation equilibrium in a continuous system are:

1. Thermal equilibrium:

$$\text{grad } T = 0 \qquad (4.2.5)$$

2. Mechanical equilibrium:

$$\text{grad } P = \rho_i \omega^2 r \qquad (4.2.6)$$

3. Equilibrium distribution of concentration of various components:

$$M_i[1 - \bar{v}_i(r) \, \rho(r)] \omega^2 r = \sum_{k=2}^{\nu} \mu_{ik}(r) \, \text{grad } c_k(r)|_{T,P} \qquad (i = 2, 3, \ldots, \nu) \quad (4.2.7)$$

where M_i and $\overline{v}_i$ are respectively the molecular weight and partial specific volume of component i; ρ is the density of the system; ω is the angular velocity; r is the distance from the center of rotation; and (r) represents the values evaluated at this distance; subscript T and P are constant temperature and pressure; c_k is the molar concentration of component k (component 1 is the principal solvent); $\mu_{ik} = (\partial\mu_i/\partial c_k)_{p,T,c_l}$, where μ_i is the chemical potential of component i and the subscript c_l signifies constancy of all molarities except that indicated in the differentiation. Furthermore, $\mu_{ik} = \mu_{ki}$ due to the Maxwell relation. Equation 4.2.7 results from (2.8.3) for equilibrium, with vanishing of fluxes.

Using the sedimentation–equilibrium method for centrifugal separation, one would like to know how the liquid density gradient changes with time and what the maximum amount of particles is that can be loaded into a given density-gradient without loss of resolution during the settling of the particles in an isodensity zone.

Approach-to-Equilibrium Method

The third centrifugal method, approach-to-equilibrium (pseudosedimentation–equilibrium), preserves several advantages of the equilibrium method while shortening the time. The approach-to-equilibrium method has proved to be of great value in the separation of many enzymes and proteins.

When the banding solutes move outward from the center to the wall in a radial direction, they reach a pseudoequilibrium state before each solute reaches its isopycnic position in the gradient solution. The state is established by mutual compensation of sedimentation and back diffusion, and the total outward fluxes vanish.

The simultaneous occurrence of sedimentation and diffusion in any isothermal fluid mixture, the flux of component i, also from (2.8.3), may be written as

$$\mathbf{J}_i = \rho_i\mathbf{v}_{ri} - \sum_{j=1}^{\nu} D_{ij} \, \text{grad} \, \rho_j \qquad (j = 1, 2, \ldots, \nu) \qquad (4.2.8)$$

In the sedimentation–equilibrium method, $\mathbf{v}_{ri}$ vanishes; while in the approach-to-equilibrium method, $\mathbf{v}_{ri} = S_i\omega^2 r$ is almost vanishing but is still finite, and the total flux vanishes due to the counterbalances of the two types of transport phenomena, and the flux of solute particles vanishes everywhere. Thus, one obtains for the pseudoequilibrium condition with the relation $\rho_i = c_i(1 - \overline{v}_i/\overline{v}_j)$ and assuming all $\overline{v}_i$ are constants, as from (4.2.8),

$$c_i S_i \omega^2 r = \sum_{j=1}^{\nu-1} D_{ij} \, \text{grad} \, c_j \qquad (i = 1, 2, \ldots, \nu) \qquad (4.2.9)$$

On the other hand, the thermodynamics for sedimentation equilibrium given in (4.2.7) must also hold. Therefore, the general relationship between sedimentation coefficients S_i and multicomponent diffusion coefficients D_{ij} in a density-gradient solution can be obtained by solving (4.2.7) for grad c_j and substituting the expressions for grad c_j into (4.2.9). The solution can be used in estimating the location of the pseudoequilibrium for the ith particle in a rotor for a specified gradient solution. The approach-to-equilibrium method is, in many respects, directly related to the sedimentation–equilibrium method. Therefore, similar procedures are generally used in the design of separational runs.

Particle Sedimentation in Density-Gradient Solutions

The mechanical structure of a zonal centrifuge is basically a cylindrical pressure vessel. Therefore, measurements of quantities such as concentration and sedimentation velocity, which are usually measured by an optical method in an analytical ultracentrifuge, cannot be made during the centrifugation runs. Thus, to predict such quantities, it becomes necessary to use a refined theory of mathematical modeling.

Quantitative estimations of the location of a particle, of its instantaneous sedimentation velocity, and of its instantaneous sedimentation coefficient during a zonal centrifugation run in an arbitrary gradient solution, characterized by the density and viscosity profiles given as functions of radial distance as polynomials, have been made [22]. Thus, the analysis will permit the rational selection of centrifuge conditions for mass separation of new particles in a specified density-gradient solution.

A cylindrical rotor system of radius R filled with a gradient solution whose viscosity and density increase with distance r from the axis is expressed by the polynomials

$$\mu = \mu_0(1 + \lambda_1'r + \lambda_2'r^2 + \lambda_3'r^3 \ldots) \qquad (4.2.10a)$$

$$\rho = \rho_0(1 + \epsilon_1'r + \epsilon_2'r^2 + \epsilon_3'r^3 \ldots) \qquad (4.2.10b)$$

where μ_0 and ρ_0 are the light end of viscosity and density, respectively. The coefficients λ_i' and ϵ_i' are characteristic constants for the viscosity and density profiles in a rotor as a gradient solution.

The cylindrical rotor is to rotate about its own axis at an angular velocity ω. The biomaterials or particles are suspended in the light end of a gradient solution. Before the rotor reaches a given constant angular velocity ω, we assume that the particles ride in the gradient solution and do not exercise their sedimenting action under the centrifugal force field. The sedimenting motion will begin after a pseudosteady state, the state in which the isodense paraboloid interfacial area does not change with the speed [13], has been reached. The particles are all assumed to be spherical. Then, the sedimentation equation of the particle in an incompressible density-gradient solution relative to the rotating coordinate system can be written from (2.9.9) as

$$\frac{d\mathbf{v}_r}{dt} + \frac{18\mu_0(1 + \lambda_1'r + \lambda_2'r^2 + \lambda_3'r^3 + \ldots)}{\rho_p D_p^2}\mathbf{v}_r - \frac{r\omega^2}{\rho_p}$$

$$[\rho_p - \rho_0(1 + \epsilon_1'r + \epsilon_2'r^2 + \epsilon_3'r^3 + \ldots)] = 0 \qquad (4.2.11)$$

$$\frac{d\mathbf{v}_\theta}{dt} + \frac{18\mu_0(1 + \lambda_1'r + \lambda_2'r^2 + \lambda_3'r^3 + \ldots)}{\rho_p D_p^2}\mathbf{v}_\theta + 2\omega\mathbf{v}_r = 0 \qquad (4.2.12)$$

Since the radial velocity component $\mathbf{v}_r$, can be written as $\mathbf{v}_r = dr/dt$, (4.2.11) is a second-order differential equation for the radial coordinate r. It is assumed that particle sedimentation takes place after the rotor reaches a pseudosteady state; hence, (4.2.12) does not directly relate to the present investigation. Equation 4.2.11 can now be written with the reduced variables for the radial coordinates as

$$P\frac{d^2\zeta}{d\tau^2} + 18Q(1 + \lambda_1\zeta + \lambda_2\zeta^2 + \lambda_3\zeta^3 + \ldots)\frac{d\zeta}{d\tau}$$

$$+ [Q(1 + \epsilon_1\zeta + \epsilon_2\zeta^2 + \epsilon_3\zeta^3 + \ldots) - 1]N\zeta = 0 \qquad (4.2.13)$$

in which

$$P = \left(\frac{D_p}{R}\right)^2 \qquad (4.2.14a)$$

$$Q = \frac{\rho_0}{\rho_p} \qquad (4.2.14b)$$

$$N = \left(\frac{\rho_0\omega R D_p}{\mu_0}\right)^2 \qquad (4.2.14c)$$

$$\zeta = \frac{r}{R} \qquad (4.2.14d)$$

$$\tau = \frac{\mu_0 t}{\rho_0 R^2} \qquad (4.2.14e)$$

$$\lambda_i = \lambda_i' R^i \qquad (4.2.14f)$$

$$\epsilon_i = \epsilon_i' R^i \qquad (4.2.14g)$$

For systems of interest in the present study, the largest particle diameter D_p, will be in the order of 1 to 3 μ; the radius of the rotor, R, will be approximately 6 to 8 cm (for B- and K-series rotors); the rotor speed will be approximately 30,000 to 36,000 rpm; and the density and viscosity of the light end of the gradient solution will be approximately 1.1 to 1.3 g/ml and 1.2 to 20 centipoises, respectively. Therefore, the orders of magnitude for the dimensionless coefficients P, Q, N in (4.2.13) are roughly in the order of 10^{-10}, 10^{-1}, and 10^{-2}, respectively. Hence, (4.2.13) can be approximated to the first-order equation by neglecting the second-order term with P as a coefficient without error. Thus, (4.2.13) can be approximated to

$$(1 + \lambda_1 \zeta + \lambda_2 \zeta^2 + \lambda_3 \zeta^3 + \ldots)\frac{d\zeta}{d\tau} - A[1 - \frac{Q}{1 - Q}$$

$$(\epsilon_1 \zeta + \epsilon_2 \zeta^2 + \epsilon_3 \zeta^3 + \ldots)]\zeta = 0 \qquad (4.2.15)$$

where

$$A = \left(\frac{1}{Q} - 1\right)\frac{N}{18} \qquad (4.2.16)$$

The boundary condition for the case is

$$\tau = 0, \qquad \zeta = \xi_i \qquad (4.2.17)$$

where $\xi_i = r_i/R$ is the reduced initial position of the particle, which can be obtained from the loading condition of the sample into the rotor [13]. Usually, the length of the outer edge of the rotor core can be used as the initial position of the particle, r_i.

Equation 4.2.15 was solved by a perturbation method with the boundary condition given in (4.2.17). One assumes that the solution is to be of the following form:

$$\zeta = \zeta_0 + A\zeta_1 + A^2\zeta_2 + A^3\zeta_3 + \ldots \qquad (4.2.18)$$

By substituting (4.2.18) into (4.2.15) and setting the coefficients of each power of A equal to zero, one obtains a set of differential equations which are to be solved for $\zeta_n(n = 0, 1, 2, 3, \ldots)$. The quantity A, (4.2.16), is used as a perturbation parameter. With the boundary condition, (4.2.17), the solution for ζ is as follows:

$$\zeta = \xi_i\left(1 + C_1 A\tau + \frac{C_2}{2!}A^2\tau^2 + \frac{C_3}{3!}A^3\tau^3 + \frac{C_4}{4!}A^4\tau^4 + \ldots\right) \quad (4.2.19)$$

where the C_i's are constants given in terms of the density gradient characteristics λ_i and ϵ_i and the boundary value ξ_i. A detailed expression of these constants is presented in Ref. 14.

It is interesting to note that, if the viscosity and density of the sedimenting medium are constants, that is, λ_i and ϵ_i are all zero, and all C_i's reduce to unity, then (4.2.19) becomes

$$\zeta = \xi_i\left(1 + A\tau + \frac{A^2\tau^2}{2!} + \frac{A^3\tau^3}{3!} + \frac{A^4\tau^4}{4!} + \ldots\right) \qquad (4.2.20a)$$

$$= \xi_i e^{A\tau} \qquad (4.2.20b)$$

For the case of vanishing λ_i and ϵ_i, the solution of (4.2.15) can be easily integrated with the boundary condition (4.2.17) to give (4.2.20b).

In order to evaluate the effects of a density gradient in a centrifugal sedimentation process, the following constants for a hypothetical gradient solution were used:

$$\lambda_1 = a \qquad \lambda_2 = \frac{\lambda_1^2}{2} \qquad \lambda_3 = \frac{\lambda_1^3}{6} \qquad a = -0.5,\ 1.0(0.5),^*\ 3.0$$

$$\epsilon_1 = b \qquad \epsilon_2 = \frac{\epsilon_1^2}{2} \qquad \epsilon_3 = \frac{\epsilon_1^3}{6} \qquad b = 0.1(0.1),^*\ 1.0$$

The profiles of the gradient solution were parts of the natural exponential curves, $e^{a\zeta}$ or $e^{b\zeta}$. Figure 4.6 shows the viscosity and density profiles of a gradient solution in rotors for $\lambda_1 = 1.5,\ 2.0,\ 2.5,$ and 3.0 and $\epsilon_1 = 0.10,\ 0.15,\ 0.20,$ and 0.25.

Figure 4.7 shows the reduced particle position in a rotor as a function of reduced rotation time $(A\tau)$. If the viscosity and density of a sedimenting medium are constants, a particle's position and the rotation time have a commonly known linear relationship, as can be seen from (4.2.20b):

$$\ln \zeta = \ln \xi_i + A\tau \qquad (4.2.20c)$$

In a semilog plot, the above equation gives a straight line, which is shown as a broken line in Fig. 4.7. The curves in the figures show the correction for the gradient solution. The variations of λ_1 from -0.5 to 3.0 for $\epsilon_1 = 0.1$ are shown in Fig. 4.7a, and the variations for ϵ_1 from 0.1 to 0.9 for $\lambda_1 = 1.5$ are shown in Fig. 4.7b.

It is interesting to note that, for a gradient solution with $\lambda_1 = -0.5$ and $\epsilon_1 = 0.1$, in which the viscosity of a gradient solution decreases while the density of a gradient solution increases, a particle sediments much faster

*The numbers in parentheses indicate the interval for the increment.

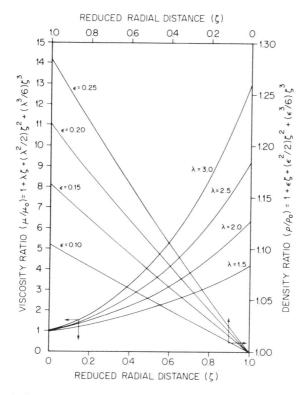

Fig. 4.6 Hypothetical density-gradient profile used in zonal calculations. *Source:* Reproduced from Ref. 14 by permission of Marcel Dekker, Inc.

than in the medium of constant density and viscosity. Therefore, we may conclude from the quantitative studies that:

1. The viscosity of a gradient solution is the controlling factor in the decision of length of time required for a zonal centrifugation run for isopycnic banding and also for velocity-sedimentation separation.

2. The variation of the viscosity profile indicates that the steeper the profile, the longer is the time required for a particle to reach a given position.

3. The centrifugation time for a particle to reach a certain position in a rotor is inversely proportional to the square of the particle diameter and to the square of the rotor speed.

4. The influence of particle diameter and that of the rotor speed are also inversely proportional to each other. If the diameter of a particle is $1/10$ that of a reference particle, an increase in rotor speed to 10 times the original speed will result in the particle reaching the same position at the same time as the reference particle traveling at the original speed.

5. The variation of the density profile also indicates that the steeper the profile, the longer is the time required for a particle to reach a given position.

6. The variation in the density ratio between a particle and the light end of a gradient solution shows that the smaller the ratio, the shorter is the time for a particle to reach its isopycnic position.

The sedimentation velocity of particles in a gradient solution now can be easily obtained from the expression of the position of particles as a function of time. The sedimentation velocity of particles in a reduced form is obtained by differentiating (4.2.20) with respect to the reduced rotation time τ, which gives

$$\mathbf{V}_r^* = \frac{d\zeta}{d\tau} = A(C_1 + AC_2\tau + A^2C_3\tau^2 + A^3C_4\tau^3 + \ldots) \qquad (4.2.21)$$

where the quantity A was defined in (4.2.16) and all the C_i's were constants.

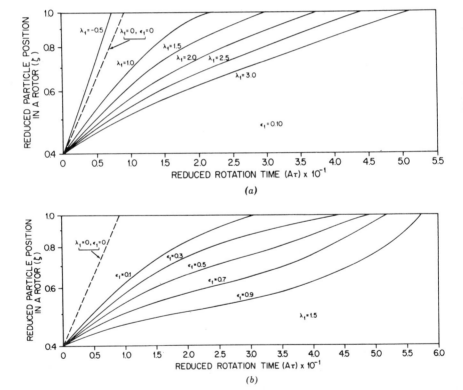

Fig. 4.7 Particle's position in a rotor as a function of rotation time. (a) With $\epsilon_1 = 0.10$ and various values of λ_1. (b) With $\lambda_1 = 1.5$ and various values of ϵ_1. *Source:* Reproduced from Ref. 14 by permission of Marcel Dekker, Inc.

The sedimentation coefficient of a particle, S_i, was defined by Svedberg [10, 11] as in (4.2.2). If the sedimenting medium is a density-gradient solution characterized by (4.2.10a) and (4.2.10b) the observed sedimentation coefficient of a particle, or a macromolecule, can be expressed in a reduced form by

$$Se = \frac{S\omega^2 R^2 \rho_0}{\mu_0} = \frac{V_r^*}{\zeta} \tag{4.2.22}$$

It is emphasized here that (4.2.2) or (4.2.22) gives an expression for an instantaneous observed particle sedimentation coefficient which is not, in a general sense, the sedimentation coefficient customarily given.

The observed sedimentation coefficients as a function of reduced rotation time are shown in Figs. 4.8a and 4.8b. The effects of variations of λ_1 from -0.5 to 3.0 for $\epsilon_1 = 0.1$ on the observed sedimentation coefficients are shown in Fig. 4.8a, and the variations of ϵ_1 from 0.1 to 0.9 for $\lambda_1 = 1.5$ are shown in Fig. 4.8b. It is interesting to note that there is a minimum point in Figs. 4.8a and 4.8b, which suggests the importance of the profiles of viscosity and density in a gradient solution for a zonal run. With a properly designed gradient profile, one can conduct zonal velocity separation according to the specifications. This suggests that investigations on mixed gradient medium solutions are warranted.

Knowledge of shear stress exerted by the particles during the sedimentation which can be obtained from the radial velocity component [13], may be informative to prevent damage to biomaterials in the separation. The shear stress exerted by a particle is the same as that of the sustaining liquid, but opposite in sign:

$$\tau_{rr} = \mu \left\{ 2 \frac{\partial v_r}{\partial r} - \frac{2}{3} \left[\frac{1}{r} \frac{\partial}{\partial r} (r v_r) \right] \right\} \tag{4.2.23a}$$

In a reduced form (4.2.23a) can be rewritten as

$$T_{rr} = \frac{\tau_{rr}}{2S\omega^2\mu_0} = \frac{1}{3}(1 + \lambda_1\zeta + \lambda_2\zeta^2 + \lambda_3\zeta^3 + \ldots)\left[\frac{2}{S} \frac{d \ln V_r^*}{d\tau} - 1\right] \tag{4.2.23b}$$

An example of shear stress distribution of particles is presented in Fig. 4.9. The shear stress reaches a maximum and disappears immediately before the particle reaches the wall with a viscous medium.

The results presented here are based on the assumptions that the particles ride in the gradient solution and do not sediment before the rotor reaches its

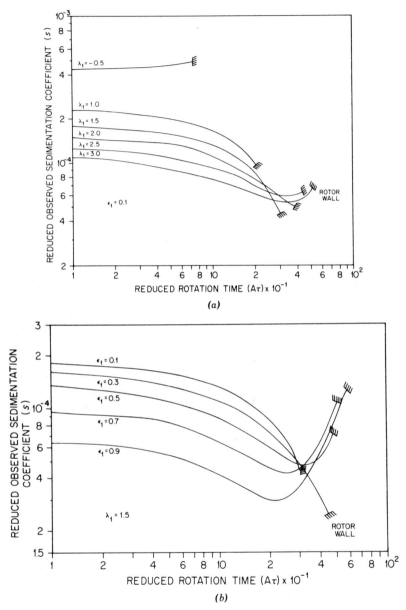

Fig. 4.8 Observed sedimentation coefficient as a function of rotation time. (*a*) With $\epsilon_1 = 0.10$ and various values of λ_1. (*b*) With $\lambda_1 = 1.5$ and various values of ϵ_1. *Source:* Reproduced from Ref. 14 by permission of Marcel Dekker, Inc.

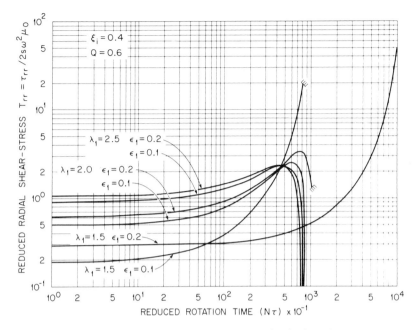

Fig. 4.9 Shear-stress distribution of a particle during sedimentation. *Source:* Reproduced from Ref. 14 by permission of Marcel Dekker, Inc.

given full angular velocity ω. Usually, the dynamic loading for B- and K-rotors is performed at rotor speeds of 2000 to 3000 rpm. For zonal centrifuge separations of the particle sizes in consideration, the sedimentation velocity is negligible around the dynamic loading rotation speed. Therefore, this assumption seems to be reasonable and without error. It is also assumed that all the particles are spherical, with D_p as their particle diameters. The method of handling the nonspherical particles can be modified in the usual fashion by the introduction of appropriate shape factors. During the sedimentation period, the particles will generally orient themselves to the minimum drag position. Therefore, the shortest characteristic length of the particle was used as the particle diameter D_p. For osmometric particles, the equivalent particle diameter can be obtained from the information on the partial specific volume of particles and the partial specific quantity of a gradient solution. The diffusion effect in a gradient solution during the sedimentation period and the Brownian motion of ultrasmall particles in a gradient solution are also neglected. Experimental investigations on these effects are warranted.

For the application of these results, one must first find the viscosity and density profiles of a gradient solution in a rotor by curve fitting to determine

λ_1' and ϵ_1'. The diameter and density of the particles have to be known to use these mathematical results. By using (4.2.19), one can determine where and when the particles will be in a given gradient solution in a rotor. If one is interested in a velocity sedimentation, (4.2.22) together with information on the gradient solution can be used to determine when and where the centrifugal separation will be optimum. With a gradient solution where $\lambda_1 = 1.5$ and $\epsilon_1 = 0.1$, one will find that the maximum separation between $N = 0.1$ and 0.01, when $Q = 0.6$ and $\xi_i = 0.4$ (approximately for the B-XV rotor), will take place at $\tau = 2.5 \times 10^2$. Then, using the definition of τ, the best velocity sedimentation centrifugation time can be found. Furthermore, the location for optimum velocity sedimentation can also be found.

The problem has also been investigated by Thomson and Mikuta [15], Rosenblum and Cox [16], and Duve et al. [17] by modifying the density and viscosity in a sustaining gradient solution. Recently, Pretlow et al. [18] have used linear density and exponential viscosity for the Ficoll gradient to analyze their particle position data.

Stability of Isopycnic Banding and Loading Capacity

The separation of material in zonal centrifugation frequently uses the equilibrium method. When the banding particles move outward from the axis with velocity v_r until particles reach their respective zone of isodensity in the sustaining gradient solution, then v_r vanishes and the equilibrium state, an isopycnic banding, is established. To maintain stable zones of settling particles is of importance since an understanding of them may lead to an estimate of the maximum amount of particles that can be loaded into a given density-gradient without the loss of resolution during the settling of the particle zones. A complete understanding of factors which affect a stable isodensity banding would also lead to a specification of the best set of operating conditions and design of a gradient solution for any separation task.

Previous work in this area is sparse and limited in scope. Unstable phenomena during a banding centrifugation have been observed by Anderson [19] and Brakke [20]. A quantitative experimental study of this instability has been done by Nason et al. [21]. In order to explain this instability Svensson et al. [5] and Schumaker [22] have assumed that instability is due to occurring of the formation of a density inversion within the fluid by diffusion. Sartory [23] has used a small perturbation analysis to determine whether the disturbance grows or decays in time for two layers of stationary diffusion solute in a common solvent in a uniform gravitational field (static loading). In testing Svensson's theory, Brakke [24] concluded that only a fraction of the theoretical zone capacity (i.e., of the maximum particle loading in a zone) could be attained in practice. Meuwissen and Heirwegh [25] have shown experimentally that the stability of zones in liquid density gradients under 1 g depends on the strength of the supporting gradient. Hsu [26] has used a small perturbation analysis to determine the stability criteria of an equilibrium method in a zonal centrifugation run. An infinitesimal disturbance is introduced in the gradient solution and the settling particle, and an analysis is made to determine the necessary condition in which the disturbance decays in time. Such a disturbance might be initiated in practice during the fluctuation in rotor speed or might arise from vibration or from many other sources. He found that the stability is related to the width of the isopycnic band, which can be expressed in terms of various diffusivities between the particles and a gradient solution, a profile of gradient solution, particle size, and rotational speed of a rotor. The results unify the previous observations in sedimentation and theoretical analysis for diffusion phenomena under one gravitational field.

A centrifuge rotor filled with a gradient solution whose viscosity and density increase with radial distance from the axis, which is expressed by $\mu(r)$ and $\rho(r)$, was considered. The cylindrical rotor is to rotate about its own axis at an angular velocity ω. The materials or particles are suspended in the light end of a gradient solution either by dynamic loading or by static loading and reoriented to a steady state. The two solutes, a gradient solute (1) and the macromolecule (2) involved, are assumed to diffuse in a solvent (0). The diffusivities are assumed constant. Variations may not be negligible but are believed to be not essential to the fluctuation phenomena. Thus, the equations governing the sedimentation and diffusion of two solutes are obtained from (4.2.1):

$$\frac{\partial \rho_1}{\partial t} = \frac{1}{r}\frac{\partial}{\partial r}\left[r\left(D_{11}\frac{\partial \rho_1}{\partial r} + D_{12}\frac{\partial \rho_2}{\partial r}\right) - S_1\omega^2 r^2\rho_1 \right] \qquad (4.2.24a)$$

and

$$\frac{\partial \rho_2}{\partial t} = \frac{1}{r}\frac{\partial}{\partial r}\left[r\left(D_{21}\frac{\partial \rho_1}{\partial r} + D_{22}\frac{\partial \rho_2}{\partial r}\right) - S_2\omega^2 r^2\rho_2 \right] \qquad (4.2.24b)$$

where ρ_i are the concentrations of the solute, D_{ij} are the diffusivities, and S_i are the sedimentation coefficient of the ith species and defined by Svedberg [4] as

$$S_i = \frac{dr_i/dt}{\omega^2 r} = \frac{\mathbf{v}_{ri}}{\omega^2 r} \qquad (4.2.25)$$

The sedimentation of a spherical particle in an incompressible gradient solution is given in (4.2.3) as

$$\frac{d\mathbf{v}_{ri}}{dt} + \frac{18\mu(r)}{\rho_i D_{pi}^2}\mathbf{v}_{ri} = \frac{\omega^2 r}{\rho_i}[\rho_i - \rho(r)] \tag{4.2.3}$$

in which D_{pi} is the diameter of the particle i.

We now split the concentration and sedimentation coefficients into unperturbed parts (barred quantity) and a perturbation of infinitesimal amplitude (primed quantity). Let

$$\rho_1 = \bar{\rho}_1 + \rho_1'$$
$$\rho_2 = \bar{\rho}_2 + \rho_2'$$
$$S_1 = \bar{S}_1 + S_1' \tag{4.2.26}$$
$$S_2 = \bar{S}_2 + S_2'$$

The attainment of sedimentation equilibrium, or the isopycnic banding conditions, may be stated by

$$\frac{\partial \bar{\rho}_i}{\partial t} = 0 \quad \text{and} \quad \bar{S}_i = 0 \tag{4.2.27}$$

Thus, the unperturbed concentration at an isopycnic point is

$$\bar{\rho}_1 = \bar{\rho}_2 = \rho(r)|_{\text{iso}} = \text{const.} \tag{4.2.28}$$

where $\rho(r)|_{\text{iso}}$ denotes the density of the gradient solution at the isopycnic point.

The criterion for stability is that these perturbed quantities ρ_1', ρ_2', s_1', and S_2' decay with time. Collecting terms of the first order in the primed quantities for (4.2.24a) and (4.2.24b), we have

$$\frac{\partial \rho_1'}{\partial t} = \frac{1}{r}\frac{\partial}{\partial r}\left[r\left(D_{11}\frac{\partial \rho_1'}{\partial r} + D_{12}\frac{\partial \rho_2'}{\partial r}\right)\right] \tag{4.2.29a}$$

$$\frac{\partial \rho_2'}{\partial t} = \frac{1}{r}\frac{\partial}{\partial r}\left[r\left(D_{21}\frac{\partial \rho_1'}{\partial r} + D_{22}\frac{\partial \rho_2'}{\partial r}\right) - \omega^2 r^2 \bar{\rho}^2 S_2'\right] \tag{4.2.29b}$$

$$\frac{dS_2'}{dt} + \frac{18\mu(r)}{D_{p2}^2 \bar{\rho}_2}S_2' = \frac{\rho_2' - \rho_1'}{\bar{\rho}_2} \tag{4.2.29c}$$

In obtaining (4.2.29a), we have assumed that $S_1' = 0$, since $\bar{S}_1 = 0.024$ (at 20°C, H$_2$O) for a sucrose solution. The assumptions are believed to be adequate, since the unperturbed term $\bar{S}_1$ is so small compared to the other terms.

We now seek a solution to (4.2.29) in terms of normal modes, usually used in a linear stability analysis:

$$\rho_1{}' = \hat{\rho}_1 \exp i(\alpha r + \gamma t) \tag{4.2.30a}$$

$$\rho_2{}' = \hat{\rho}_2 \exp i(\alpha r + \gamma t) \tag{4.2.30b}$$

where $\hat{\rho}_1$ are small (complex) constants; α is the wave number, which is real; and γ is the growth rate constant, which may be complex. Thus, the stability criterion in terms of normal modes given above is that the imaginary part of γ must be positive semidefinite; that is,

$$\text{Im}(\gamma) \geq 0 \quad \text{for stability} \tag{4.2.31}$$

Now, (4.2.29c) is a linear first-order ordinary differential equation which can be easily integrated with the boundary condition that when $t = 0, S_2' = 0$ to give a solution. Substituting the solution of $S_2{}'$ from (4.2.29c) together with (4.2.30a) and (4.2.30b) into (4.2.29a) and (4.2.29b), one finds that the stability criterion is

$$\alpha r \leq \left\{ 1 - \frac{(D_{11} + D_{22})^2}{D_{11}D_{22} - D_{12}D_{21}} \left[\frac{D_{11} + D_{12}}{(D_{11} - D_{22})^2} \frac{\omega^2 D_{p_2}{}^2 r^2 \bar{\rho}_2}{18\mu} \left(3 - \frac{d \ln \mu}{d \ln r} \right) \right. \right.$$

$$\left. \left. + \frac{1}{4} \left(1 - \frac{1}{D_{11} + D_{22}} \frac{\omega^2 D_{p_2}{}^2 r^2 \bar{\rho}_2}{18\mu} \right)^2 \right] \right\}^{1/2} \tag{4.2.32}$$

Equation 4.2.32 implies that the stability of an isopycnic banding is determined by the bandwidth, the maximum band capacity, as obtained above. In obtaining (4.2.32), the solution of $S_2{}'$ was linearized. Since the largest macromolecule in a zonal centrifugation is in the order of 10^{-4} cm, the viscosity of a gradient solution $\mu(r)$ is greater than 1.2 centipoise, and the density of a particle $\bar{\rho}_2$ is 1.2 to 1.5 g/cm^3, one has $[D_{p_2}{}^2\bar{\rho}_2/18\mu] \ll 1$.

The stability criterion given in (4.2.32) presents the unifying theory. The theory obtained from the previous observations during centrifugation, such as density inversion theory, and the theory from the infinitesimal perturbation analysis of diffusional phenomena under one gravitational field are inclusively represented by (4.2.32).

It is interesting to note that Meuwissen and Heirwegh's conclusion [25] that the stability depends on the shape or strength of the supporting gradient is also represented by the terms of $(3 - d \ln \mu/d \ln r) \cdot (\omega^2 D_{p_2}{}^2 r^2 \bar{\rho}_2/18\mu)$ in

(4.2.32). If the gradient increases with the cube of the radius, the term $(3 - d \ln \mu/d \ln r)$ drops out. If the slope of the gradient viscosity is $(d \ln \mu/d \ln r) < 3$, the contribution of that term to stability is negative, and the maximum load capacity is reduced. If $(d \ln \mu/d \ln r) > 3$, the steeper the gradient solution, the more the stability of the system. In this case, the bandwidth increases with the slope of the gradient solution, and the range of the stable region increases. The same conclusion may also be drawn for the density inversion theory. If the shape of the gradient is $(d \ln \mu/d \ln r) < 3$, the second term at the right-hand side of (4.2.32) is positive. Therefore, the following situation has occurred in the gradient: $\bar{\rho}_2 < \rho_{grad} (3 - d \ln \mu/d \ln r)$, since at an isopycnic point, $\bar{\rho}_2 = \bar{\rho}_{grad}$. Thus, density inversion does take place in order to return to the physically stable situation. If $(d \ln \mu/d \ln r) > 3$, that is, $\bar{\rho}_2 > \rho_{grad} (3 - d \ln \mu/d \ln r)$, the system remains unchanged within the region. If a system is under one gravitational field, that is, the terms containing $\omega^2 D_{p_2}{}^2 r^2 \bar{\rho}_2 / 18\mu$ drop out, the stability criterion becomes

$$\alpha r \le \left(1 - \frac{(D_{11} + D_{22})^2}{4(D_{11}D_{22} - D_{12}D_{21})} \right)^{1/2} \tag{4.2.33}$$

The criterion is similar to that obtained by Sartory [23], except for the powers on the diffusivities. The four diffusivities for three-component systems can be obtained from Fujita and Gosting's procedure [27].

An additional characteristic besides previous theories, which appears in (4.2.32), is that one would like to increase the right-hand side of (4.2.32) so that the stability range can be increased. In order to increase the right-hand side of (4.2.32) for a given system, one would like to set the term

$$\frac{\omega^2 D_{p_2}{}^2 r^2 \bar{\rho}_2}{(D_{11} + D_{22})\mu} < 1$$

After rearranging, it becomes

$$\omega < \frac{1}{D_{p_2}{}^2 r} \left(\frac{(D_{11} + D_{22})\mu}{\bar{\rho}_2} \right)^{1/2} \tag{4.2.34}$$

where r is the position where ideal isopycnic banding takes place. Equation 4.2.34 shows that increasing angular velocity in a zonal run does not improve the resolution. The angular velocity has to be constrained by (4.2.34). The physical interpretation of this phenomenon is that with too high an angular velocity, the banding solute penetrates too deeply into the gradient solution; therefore, an instability due to a density inversion or lack of strength in the supporting gradient will take place. Thus, the angular velocity also has to be constrained.

The stability criterion is such a complicated phenomenon that one should consider all the means to increase the right-hand side of (4.2.32). Thus, (4.2.32) shows the unification of all previous theories.

Optimum Density-Gradient Profile for the Sedimentation–Velocity Method

In using the sedimentation–velocity method for particle separation in a zonal centrifuge, a centrifugal force field is chosen for the system. Because the rotor is operated at high speeds, the solute particles are forced to settle at appreciable rates toward the wall of the rotor. The rate of particle movement is a function of the molecular weight as well as the particle size and shape, the difference between the particles and the local sustaining liquid, the local sustaining viscosity, and the centrifugal force field strength. It has been shown that the viscisoty of a density-gradient solution is the major factor in controlling the rate of particle movement [14]. Thus, designing an optimal density-gradient solution is of importance in the sedimentation–velocity separation method.

An analysis to determine the optimal profile of a density-gradient solution characterized by viscosity and density profiles, which are expressed as functions of radial distance in the zonal rotor, has been made [28]. The analysis also provides a means to estimate an optimal centrifugation time, so that the resolution of separation by the sedimentation–velocity method will be maximal and to evaluate each particle's position in a rotor when centrifugation is stopped, so that the separated particles can be collected adequately. Thus, the analysis will permit the rational selection of optimal operating conditions for mass separation by the sedimentation–velocity method in zonal centrifugation.

The two macromolecules, particles 1 and 2, are in the light end of a density-gradient solution at a rotor core position R_c for $t \geq 0$. Then, the equations governing the sedimentation of two macromolecules are exactly the same as those given in (4.2.24). Using (4.2.24), the conventional formula converting the observed sedimentation coefficient to the customarily expressed sedimentation coefficient in the sedimenting medium of water at $20°C$, the equations of (4.2.24) can be rewritten in the following forms:

$$\frac{\partial \rho_1}{\partial t} = \frac{1}{r} \frac{\partial}{\partial r} \left[r \left(D_{11} \frac{\partial \rho_1}{\partial r} + D_{12} \frac{\partial \rho_2}{\partial r} \right) - k_1 \frac{\omega^2 r^2}{\mu(r)} \frac{\rho_1 - \rho(r)}{\rho_2 - \rho_{H_2O,20°C}} \rho_1 \right]$$

$$(4.2.35a)$$

$$\frac{\partial \rho_2}{\partial t} = \frac{1}{r} \frac{\partial}{\partial r} \left[r \left(D_{21} \frac{\partial \rho_1}{\partial r} + D_{22} \frac{\partial \rho_2}{\partial r} \right) - k_2 \frac{\omega^2 r^2}{\mu(r)} \frac{\rho_2 - \rho(r)}{\rho_2 - \rho_{H_2O,20°C}} \rho_2 \right]$$

$$(4.2.35b)$$

where $k_i = S_i^0(H_2O,20°C)\mu(H_2O,20°C) = $ constant.

The optimum separation for the sedimentation velocity method in a zonal centrifugation for particles 1 and 2 may be defined as the maximum distance of separation between the two particles. The maximum distance of separation between particles 1 and 2 in terms of a performance index can be written as

$$\max\left[\Delta R = \int_0^{t_f} \left(\frac{dr_1}{dt} - \frac{dr_2}{dt} \right) dt \right] \tag{4.2.36}$$

The problem is to find a density-gradient solution characterized by viscosity $\mu(r)$ and density $\rho(r)$ which maximize the difference in instantaneous position of two particles in a rotor, so that the resolution of the separation will be maximal. The conditions associated with the performance index (4.2.36) and (4.2.35) are

$$\frac{dr_1}{dt} = \frac{k_1 \omega^2 r}{\mu(r)} \frac{\rho_1 - \rho(r)}{\rho_1 - \rho_{H_2O,20°C}} \tag{4.2.37a}$$

$$\frac{dr_2}{dt} = \frac{k_2 \omega^2 r}{\mu(r)} \frac{\rho_2 - \rho(r)}{\rho_2 - \rho_{H_2O,20°C}} \tag{4.2.37b}$$

Equations 4.2.37 are obtained by combining (4.2.2) and (4.2.4) with $k_i = S_i^0(H_2O, 20°C)\mu(H_2O, 20°C)$.

Solution to (4.2.36) subject to the constraint conditions given in (4.2.35) and (4.2.37) was obtained by Hsu [28] using Pontryagin's maximum principle [29].

The optimum viscosity and density profiles for a sedimentation–velocity method are

$$\mu(r) = \mu_0 \left(\frac{r}{R_c} \right)^{4/3} \tag{4.2.38a}$$

$$\rho(r) = (\rho_0 - \alpha)\left(\frac{r}{R_c} \right)^{2/3} + \alpha \tag{4.2.38b}$$

where $\mu_0 = \mu(R_c)$, $\rho_0 = \rho(R_c)$, and

$$\alpha = \frac{1 - \left(\dfrac{\rho_2}{\rho_1} \right)\left(\dfrac{S_2^0}{S_1^0} \right)\left(\dfrac{\rho_1 - \rho_H}{\rho_2 - \rho_H} \right)}{1 - \left(\dfrac{S_2^0}{S_1^0} \right)\left(\dfrac{\rho_1 - \rho_H}{\rho_2 - \rho_H} \right)} \qquad \rho_H = \rho(H_2O, 20°C)$$

$$\tag{4.2.38c}$$

In obtaining (4.2.38a) and (4.2.38b), $d \ln \rho_i / d \ln r = 0$ has been used. The particles are moving in the rotor as bands; the relation holds true inside and outside the particle bands. Furthermore, the dispersion due to the interaction of diffusion and sedimentation in a high centrifugal force field is almost negligible [30] before an isopycnic point.

Besides obtaining an optimal density-gradient solution in a rotor for a velocity sedimentation, an important problem is when and where the maximum separation will take place, so that one can stop the centrifugation run and collect the separated fractions from a rotor either by draining the gradient out of the bottom of the rotor or by displacing it out the top using a heavier gradient solution. The final positions of particles 1 and 2 are obtained as in Ref. 28:

$$
r_1(t_f) = R_c \left\{ \frac{\frac{4}{6}\left[\frac{S_1^0(\rho_1 - \alpha)}{\rho_1 - \rho_H} - \frac{S_2^0(\rho_2 - \alpha)}{\rho_2 - \rho_H}\right] + \frac{S_1^0(\rho_2 - \rho_1)}{\rho_1 - \rho_H}}{\left[\frac{k_1}{\rho_1 - \rho_H} - \frac{k_2}{\rho_2 - \rho_H}\right](\rho_0 - \alpha)} \right\}^{3/2}
$$

$$(4.2.39a)$$

$$
r_2(t_f) = R_c \left\{ \frac{-\frac{4}{6}\left[\frac{S_1^0(\rho_1 - \alpha)}{\rho_1 - \rho_H} - \frac{S_2^0(\rho_2 - \alpha)}{\rho_2 - \rho_H}\right] + \frac{S_2^0(\rho_2 - \rho_1)}{\rho_2 - \rho_H}}{\left[\frac{S_1^0}{\rho_1 - \rho_H} - \frac{S_2^0}{\rho_2 - \rho_H}\right](\rho_0 - \alpha)} \right\}^{3/2}
$$

$$(4.2.39b)$$

The maximum separation ΔR thus obtained is

$$
(\Delta R)_{max} = R_c \left\{ \frac{\frac{4}{3}\left[\frac{S_1^0(\rho_1 - \alpha)}{\rho_1 - \rho_H} - \frac{S_2^0(\rho_2 - \alpha)}{\rho_2 - \rho_H}\right]}{\left(\frac{S_1^0}{\rho_1 - \rho_H} - \frac{S_2^0}{\rho_2 - \rho_H}\right)(\rho_0 - \alpha)} + \frac{\rho_2 - \rho_1}{\rho_0 - \alpha} \right\}^{3/2}
$$

$$(4.2.40)$$

The optimum centrifugation time in dimensionless variables is obtained as

$$
\tau_f = \frac{1}{AB} \sum_{n=0}^{\infty} \frac{3M^n}{2(n + 2)} \zeta_c^{2(n+2)/3}
$$

$$
\cdot \left[\left\{ \frac{\frac{4}{6}\left[\frac{k_1(\rho_1 - \alpha)}{\rho_1 - \rho_H} - \frac{k_2(\rho_2 - \alpha)}{\rho_2 - \rho_H}\right] + \frac{k_1(\rho_2 - \rho_1)}{\rho_1 - \rho_H}}{\left(\frac{k_1}{\rho_1 - \rho_H} - \frac{k_2}{\rho_2 - \rho_H}\right)(\rho_0 - \alpha)} \right\}^{n+2} - 1 \right]
$$

$$(4.2.41)$$

where

$$\tau = \frac{\mu_0 t}{\rho_0 R^2} \qquad \zeta = \frac{r}{R} \qquad Se_i = \frac{S_i^0 \omega^2 R^2 \rho_0}{\mu_0} \qquad \text{(4.2.41a, b, c)}$$

$$\beta_i = \frac{\rho_i}{\rho_0} \qquad \delta = \frac{\mu_{H_2O,20°C}}{\mu_0} \qquad M = C/B \qquad \text{(4.2.41d, e, f)}$$

$$A = \frac{Se_i}{\beta_i - \beta_H} \qquad B = \beta_i - \alpha \qquad C = \frac{1 - \alpha}{\zeta_c^{2/3}} \qquad \text{(4.2.41g, h, i)}$$

These results are based on the assumptions that particles are still far away from their respective isopycnic points, so that sedimentation is dominant in the mutual interference of diffusion and sedimentation. Hence, changes of the density-gradient solution due to diffusion are negligible. In velocity sedimentation, the centrifugation time is generally very short. Therefore, concentrations of marcromolecules in the density-gradient solution do not disperse appreciably by molecular diffusion. Also, the partial specific volumes of each component are assumed constant. This implies that the solution is regular. The solution may not be regular, but a correction can be made by use of an activity coefficient. These assumptions are considered to be reasonable and permit the development of an analytical solution with relative ease.

The estimates of an optimal centrifugation time and positions of particles serve as a guide for designing velocity–sedimentation runs. It is perhaps very difficult to build an optimal density-gradient solution as specified in (4.2.38a), (4.2.38b), and (4.2.38c) with a single gradient solute. A mixed-solute gradient solution probably can provide the viscosity and density profiles required.

An interesting result is that all the diffusion coefficients in Lamm's equation cancel out and do not show up in the optimum profiles. This may be due to the fact that in the sedimentation–velocity method, the centrifugation time is generally very short. Therefore, dispersion of solute particles is offset by sedimentation velocity in a density-gradient solution.

Fractional Elution in a Continuous-Flow Centrifugation

As mentioned previously, improper operating conditions and a wrong density-gradient profile may cause the loss of sample attributing to the sample elution from the rotor in a continuous-flow zonal centrifugation. Sartory [31] has derived a formula to predict the fraction of particles removed from the sample feed stream as a function of the feed rate and other parameters describing the centrifuge and the particles.

If one assumes that a solution in the rotor is incompressible and the rotor is rotating with a constant angular velocity ω, the overall conservation of mass may be written from (2.1.27) as

$$\frac{1}{r}\frac{\partial(r\mathbf{v}_r)}{\partial r} + \frac{\partial \mathbf{v}_z}{\partial z} = 0 \qquad (4.2.42)$$

in which $\mathbf{v}_r(r, z)$ and $\mathbf{v}_z(r, z)$ are the radial and axial components of the volume-mean velocity in the centrifuge rotor. Since the flow of the sample feed stream is confined to a thin layer near the core, one may assume that $r \simeq r_1$, the core radius, and (4.2.42) can be approximated to

$$\frac{\partial \mathbf{v}_r}{\partial r} + \frac{\partial \mathbf{v}_z}{\partial z} = 0 \qquad (4.2.43)$$

If a stream function $\Psi(r, z)$ is introduced, one has the velocity components such that

$$\mathbf{v}_z = \frac{\partial \Psi}{\partial r} \qquad (4.2.44a)$$

$$\mathbf{v}_r = \frac{-\partial \Psi}{\partial z} \qquad (4.2.44b)$$

Equations 4.2.44a and 4.2.44b define Ψ only up to an additive constant. Since $\mathbf{v}_r$ vanishes at the core wall, Ψ is a constant along the rotor core so $\Psi(r_1, z) = 0$ is chosen as a constant for convenience. Then, $\Psi(r, z)$ is the volumetric flow rate of fluid, per unit of perimeter, which passes between the core wall and the point (r, z).

The macromolecules or particulate materials are carried along with the fluid, a solvent, but also sediment radially. The radial and axial fluxes of particulate materials are then given by

$$\mathbf{J}_r = (\mathbf{v}_r + S\omega^2 r)c = \left(\frac{-\partial \Psi}{\partial z} + S\omega^2 r\right)c \qquad (4.2.45a)$$

$$\mathbf{J}_z = \mathbf{v}_z c = \left(\frac{\partial \Psi}{\partial r}\right)c \qquad (4.2.45b)$$

where c is the concentration (in g/ml solution), S is the sedimentation coefficient of particulate materials, and ω is the angular velocity of the rotor. In a steady-state operation, conservation of particulate material means that the divergence of the mass fluxes vanishes. With the thin layer sample feed stream, $r \doteq r_1$, and the constant sedimentation assumptions, one has the conservation equation

$$\left(-\frac{\partial \Psi}{\partial z} + S\omega^2 r_1\right) \frac{\partial c}{\partial r} + \left(\frac{\partial \Psi}{\partial r}\right) \frac{\partial c}{\partial z} = 0 \qquad (4.2.46)$$

If one defines a modified stream function as

$$\Psi_1 = \psi - S\omega^2 r_1 z \qquad (4.2.47)$$

then (4.2.46) becomes

$$\left(\frac{-\partial \Psi_1}{\partial z}\right)\left(\frac{\partial c}{\partial r}\right) + \left(\frac{\partial \Psi_1}{\partial r}\right)\left(\frac{\partial c}{\partial r}\right) = 0 \qquad (4.2.48)$$

Equation 4.2.48 represents the vanishing of the Jacobin $\partial(\Psi_1, c)/\partial(r, z) = 0$. It follows that there is a functional dependence of c on Ψ_1:

$$c(r, z) = F[\Psi_1(r, z)] \qquad (4.2.49)$$

which, for arbitrary F, is the general solution of (4.2.48).

As a boundary condition, we require that the radial flux of particulate materials should vanish at the impermeable core wall:

$$J_r(r_1, z) = 0 \qquad z \geq 0 \qquad (4.2.50)$$

which, in view of (4.2.45a), gives

$$c(r_1, z) = 0 \qquad z \geq 0 \qquad (4.2.51)$$

It is also required that, at the inlet axial position, $z = 0$, the concentration has its initial value c_0:

$$c(r, 0) = c_0 \qquad r > r_1 \qquad (4.2.52)$$

The solution satisfying (4.2.51) and (4.2.52) is

$$c = c_0 \, S(\Psi_1) \qquad (4.2.53)$$

where S is the step function:

$$S(x) = \begin{cases} 0 & x \leq 0 \\ 1.0 & x > 0 \end{cases} \qquad (4.2.54)$$

Equation 4.2.53 can be written

$$c = c_0 \, S(\Psi - S\omega^2 r_1 z) \qquad (4.2.55)$$

The fractional cleanout of particulate materials is defined by Sartory [31] as

$$f = \frac{\text{influx} - \text{outflux}}{\text{influx}}$$

The feed layer is assumed to be confined to the interval from r_1 to $r_1 + \delta$, where $\delta \ll r_1$. The mass influx of particulates into the feed layer is given by

$$\text{influx} = 2\pi r_1 \int_{r_1}^{r_1+\delta} \mathbf{J}_z |_{z=0}\, dr = 2\pi r_1 \int_{r_1}^{r_1+\delta} [c(r, z)\mathbf{v}_z(r, z)_{z=0}\, dr$$

$$= 2\pi r_1 \int_{r_1}^{r_1+\delta} [c(\partial\Psi/\partial r)]_{z=0}\, dr = 2\pi r_1 \int_0^{\Psi_\delta} c\,|_{z=0}\, d\Psi$$

$$= 2\pi r_1 c_0 \int_0^{\Psi_\delta} S(\Psi - S\omega^2 r_1 z)|_{z=0}\, d\Psi = 2\pi r_1 c_0 \Psi_\delta \qquad (4.2.56)$$

where Ψ_δ is the value of Ψ at the outer edge of the feed layer. Similarly, the outflux of particulates from the feed layer to the exit channels is given by

$$\text{outflux} = 2\pi r_1 \int_{r_1}^{r_1+\delta} J_z|_{z=L}\, dr = 2\pi r_1 c_0 \int_0^{\Psi_\delta} S(\Psi - S\omega^2 r_1 z)|_{z=L}\, d\Psi$$

$$= 2\pi r_1 c_0 \int_{s\omega^2 r_1 L}^{\Psi_\delta} d\Psi = 2\pi r_1 c_0 (\Psi_\delta - S\omega^2 r_1 L) \qquad (4.2.57)$$

where L is the length of the rotor.

The fractional cleanout is then given by

$$f = \frac{S\omega^2 r_1 L}{\Psi_\delta} \qquad (4.2.58)$$

Since the entire feed stream passes between r_1 and $r_1 + \delta_1$,

$$\psi_\delta = \int_{r_1}^{r_1+\delta} \mathbf{v}_z(r, z)\, dr \qquad (4.2.59)$$

is the total volumetric throughput of fluid per unit of perimeter. Using the inner perimeter for a thin layer,

$$\psi_\delta = \frac{Q}{2\pi r_1} \qquad (4.2.60)$$

where Q is the volumetric throughput in cm^2/sec. The fractional clean-out is then given by

$$f = \frac{S\omega^2 2\pi r_1^2 L}{Q} \qquad (4.2.61)$$

where all quantities are expressed in cgs units.

If the flow rate is expressed as Q_1 in liters/hr, the rotor speed $(60\omega/2\pi)$ is given in rpm, and the sedimentation coefficient is expressed as S_1 in Svedberg units, (4.2.61) becomes

$$f = \frac{8\pi^3 r_1^2 L S_1}{10^{16}} \frac{(\text{rpm})^2}{Q_1} \qquad (4.2.62)$$

Then, the fractional elution is

$$f_e = 1 - f \qquad (4.2.62a)$$

In order to reduce the sample loss from the rotor, one has to choose the volumetric flow rate of feed Q and the rotor speed in rpm or the angular velocity ω in such a way as to make f as close to unity as possible. The steady-state assumption used in the derivation is applicable only to the sample feed layer. Macromolecular particulates continue to sediment and to accumulate in the heavy density gradient. The value of the sedimentation coefficient S_1 is the observed sedimentation coefficient in the density gradient at the feed stream r_1 and at that temperature, not the standard value in water at 20°C.

3   PROPERTIES OF DENSITY-GRADIENT SOLUTIONS

The application of density-gradient centrifugation requires information on the shape of the gradient and on the gradient properties which are variable with respect to the space coordinates, in terms of concentration. The shape of the density gradient is important in achieving the desired separation and in determining properties such as buoyant density, sedimentation coefficient, and molecular weight in ultracentrifugal analysis. In order to prescribe an exact profile of a density-gradient solution at a given rotor location and at a specified time, one has to know the diffusivity of a gradient solute in a solvent as a function of its concentration which is a function of space coordinates. In a density-gradient centrifugation, the solvent is usually water.

Sucrose or cesium chloride solutions are often used as density-gradient solutions in zonal centrifugation for the separation of biomaterials. However, for virus separation or isolation, the use of potassium citrate or potassium tartrate as gradient materials has some advantages over sucrose or cesium chloride solutions. A density higher than that of sucrose solutions is obtainable by the use of cesium chloride solutions, but the use of cesium chloride solutions at higher density produces deleterious effects. The structure of a virus protein may be destroyed by the high ionic strength of a cesium chloride solution.

Potassium citrate solutions are chemically more complex than sucrose solutions because potassium citrate is a uni-trivalent electrolyte of a strong base and a weak tribasic acid that has relatively small differences in the ionization constants for the three hydrogen ions. Consequently, the hydrolysis of the pure salt in water produces solutions that are too basic for virus stability, the pH approaching 9.4 at saturation [32]. The addition of citric acid will lower the pH value into the range considered satisfactory for virus isolations, but the amount of citric acid required to obtain a given pH change varies with the potassium citrate concentration. The use of potassium tartrate as gradient solutions reduces those deficiencies to an acceptable range for virus stability.

For bioparticles behaving as osmometers, there have been reports using Ficoll* and methylcellulose as gradient solutions in zonal centrifugation to isolate red blood cells [33–35], mitochondria [36, 37], peroxisomes [38], and lysosomes [39] in a more "native" state. These gradient materials were used because of their low osmotic characteristics and because of the high molecular weights and low contents of dialyzable material. In density-gradient centrifugations, cells and similar particles can therefore be collected at lower densities in those gradient materials than in sucrose gradient solutions. Sorbitol is generally used as a sweetening agent for diabetics and has properties to increase absorption of vitamins and other nutrients in pharmaceutical preparations. It may be used in the zonal centrifugation for purification of cellular materials, from in vivo or in vitro, without altering their metabolic state.

Recently, Metrizamide† has been successfully used in the separation of nuclei [40], mitochondria and lysosomes [41], and various subcellular particles and macromolecules [42]. Metrizamide gives dense solutions of low viscosity and osmolality and would appear particularly well suited for the separation of intact cells.

Concentration-Dependent Diffusivities of Gradient Materials

Densities and viscosities of gradient materials are usually available from the manufacturers. In order to prescribe more accurately the shape of density-gradient solutions, concentration-dependent binary diffusivities of seven gradient materials—potassium citrate, potassium tartrate, sucrose, methylcellulose (M-278), Ficoll, Sorbitol, and Metrizamide—were measured by a microinterferometric method [43–46].

---

*Polysucrose, Pharmacia Fine Chemicals, Sweden.
†A triiodinated benzamido derivative of glucose manufactured by Nyegaart & Co., Norway.

The information is important not only in prescribing the performed density-gradient profile in a zonal rotor but also in the prediction of self-forming gradient solutes in a zonal rotor by diffusion and sedimentation. In improving the separation resolution, diffusivity at various concentrations is the prime information required for evaluating the maximum load capacity [6] or the band-broadening effect [47].

From irreversible thermodynamics considerations, Gosting [48] has introduced a term called the thermodynamic term for nonideal mixtures. Then, an expression for an isothermal, isotropic diffusion coefficient in nonreacting mixtures in the absence of external forces can be written as

$$D = D_0 \left[1 + \frac{d \ln \gamma^{(c)}}{d \ln c} \right] \tag{4.3.1}$$

where D_0 is the diffusivity at infinite dilution by extrapolating experimental data points, in which the activity coefficient $\gamma^{(c)}$ becomes unity. The expression in brackets is the thermodynamic term.

The microinterferometric method used in the measurement of diffusivity is essentially a transient technique. In order to compensate for the frictional coefficient with concentration, following Gosting's approach [48], (4.3.1) was modified to

$$D = D_0 \left[\frac{\mu_0 \bar{v}_0 \rho}{\mu} \right] \left[1 + \frac{d \ln \gamma^{(c)}}{d \ln c} \right] \tag{4.3.2}$$

where μ, $\bar{v}_0$, and ρ are viscosity, specific volume of solvent, and density at a given concentration, respectively, and μ_0 is the viscosity of pure solvent.

The data points obtained from measurements are presented in Tables 4.4 and 4.5. The partial specific volumes $\bar{v}_0$ were determined graphically by the method of intercepts from the densities. The quantity D_0 in Table 4.5 was obtained by extrapolation to an infinite dilution from a diffusivity-versus-concentration plot. The activity coefficient was obtained from the equation given below [50]:

$$\ln \gamma^{(c)} = \int_0^c \left[\frac{D\mu}{D_0 \mu_0 \bar{v}_0 \rho} - 1 \right] d \ln c \tag{4.3.3}$$

The quantities D, ρ, $\bar{v}_0$, and μ are all functions of concentration. Hence, substituting experimentally determined values of each quantity at various concentrations into the integrand and performing a numerical integration to that concentration, the activity coefficient at that concentration was obtained.

Table 4.4 Density, Viscosity, and Partial Specific Volume at 25°C at Various Concentrations

c (g solute/ml)	μ (centipoises)	ρ (g solution/ml)	$\bar{v}_0$ (ml/g solution)
Potassium Citrate			
0.1031	1.065	1.031	1.000
0.2207	1.383	1.104	1.000
0.3530	1.970	1.176	1.000
0.5045	3.496	1.260	0.955
0.6740	5.350	1.345	0.955
Potassium Tartrate			
0.1038	1.022	1.039	1.000
0.2199	1.258	1.100	1.000
0.3546	1.695	1.182	0.969
0.5021	2.500	1.256	0.959
0.6701	4.280	1.340	0.959
Ficoll			
0.050	11.8	1.013	1.010
0.075	12.6	1.021	1.010
0.100	13.2	1.028	1.010
0.220	18.0	1.059	1.010
0.310	23.0	1.077	1.010
0.400	28.0	1.096	1.010
0.550	41.0	1.119	1.010
0.700	60.0	1.139	1.010
Methyl cellulose			
0.050	1.78	1.012	1.001
0.075	6.15	1.017	1.001
0.100	17.89	1.022	1.001
0.125	56.38	1.028	1.001
0.150	158.45	1.034	1.001
0.175	500.00	—	—
0.200	1,690.00	2.136	1.001
Sucrose			
0.3375	2.7384	1.1250	1.000
0.3944	3.5283	1.1464	0.999
0.4726	5.2199	1.1755	0.997
0.5473	7.9508	1.2055	0.994
0.6198	12.7500	1.2322	0.992
Sorbitol			
0.1025	1.2290	1.0250	1.003
0.2120	1.8363	1.0600	1.003

Table 4.4 *(continued)*

c (g *solute*/ml)	μ (centipoises)	ρ (g *solution*/ml)	$\bar{v}_0$ (ml/g *solution*)
0.3305	2.9397	1.1015	1.003
0.4544	4.6915	1.1360	1.003
0.5853	9.4778	1.1705	1.003
Metrizamide			
0.1054	1.2512	1.0570	1.000
0.2213	1.6762	1.1200	1.000
0.3486	2.4001	1.1868	1.000
0.4862	3.4800	1.2600	1.000

Source: Reprinted from Refs. 43–46, by the permission of Marcel Dekker, Inc.

Table 4.5 Diffusion Coefficient and Calculated Physical Parameters at 25°C at Various Concentrations

C (g *solute*/ml)	$D \times 10^6$ (cm^2/sec)	$\mu/\mu_0 \bar{v}_0 \rho$	$\ln \gamma^{(c)}$
Potassium Citrate ($D_0 \times 10^6 = 2.305$)			
0.1031	3.149	1.156	0.4774
0.2207	3.539	1.402	1.0536
0.3530	3.712	1.874	1.8022
0.5045	3.620	3.251	2.8449
0.6740	4.420	4.660	4.4540
Potassium Tartrate ($D_0 \times 10^6 = 1.000$)			
0.1038	2.191	1.100	1.4029
0.2199	3.387	1.280	3.0175
0.3546	4.207	1.656	5.1823
0.5021	5.116	2.322	8.2074
0.6701	6.629	3.724	15.2857
Ficoll ($D_0 \times 10^6 = 1.161$)			
0.100	1.586	0.0702	14.44
0.220	2.415	0.0524	36.59
0.310	2.660	0.0416	51.56
0.400	2.750	0.0349	66.13
0.550	4.390	0.0242	98.54
0.700	5.620	0.0169	150.93
Methyl cellulose ($D_0 \times 10^6 = 1.161$)			
0.075	0.1659	0.147	0.25×10^3
0.100	0.1769	0.115	1.00×10^3

Table 4.5 *(continued)*

C (g *solute*/ml)	$D \times 10^6 (\text{cm}^2/\text{sec})$	$\mu/\mu_0 \bar{\nu}_0 \rho$	$\ln \gamma^{(c)}$
0.125	0.1858	0.162×10^{-1}	2.50×10^3
0.150	0.2140	0.058×10^{-1}	6.00×10^3
0.175	0.2886	—	—
0.200	0.3630	0.112×10^{-2}	43.74×10^3

Sucrose-0.025 M *Sodium Phosphate Buffer* $(D_0 \times 10^6 = 5.266)$

0.3375	8.451	2.724	0.1309
0.3944	—	3.499	0.7946
0.4725	8.790	4.986	2.1232
0.5473	9.300	7.422	3.8938
0.6198	10.900	11.808	5.1377

Sorbital-Distilled Water $(D_0 \times 10^6 = 5.2659)$

0.1025	5.524	1.338	1.3421
0.2120	7.625	1.933	3.8064
0.3305	13.476	2.978	8.2943
0.4544	16.527	4.609	16.1080
0.5853	18.846	9.036	28.9975

Sorbital-0.010 M Sodium Phosphate Buffer

0.1025	2.9935	—	0.7160
0.2120	4.7849	—	1.6115
0.3305	6.1026	—	2.9006
0.4544	4.8847	—	4.7551
0.5853	4.2006	—	7.3174

Sorbital-0.025 M Sodium Phosphate Buffer

0.1025	3.4937	—	0.6490
0.2120	2.0776	—	1.4906
0.3305	5.3158	—	2.7624
0.4540	5.4413	—	4.7225
0.5853	4.9659	—	7.6456

Sorbital-0.05 M Sodium Phosphate Buffer

0.1025	3.2034	—	0.4609
0.2120	4.1099	—	1.0635
0.3305	3.8559	—	1.9849
0.4544	4.7031	—	3.3993
0.5853	3.6613	—	5.4718

Sorbital-0.10 M Sodium Phosphate Buffer

0.1025	3.7102	—	0.4902
0.2120	4.1315	—	1.1333
0.3305	4.5025	—	2.1004

Table 4.5 (continued)

C (g $solute$/ml)	$D \times 10^6$(cm^2/sec)	$\mu/\mu_0 \bar{\nu}_0 \rho$	ln $\gamma^{(c)}$
0.4544	3.1499	—	3.5674
0.5853	4.0863	—	3.6793
Metrizamide–Distilled Water			
0.1054	0.4180	1.1837	1.8000
0.2213	0.6416	1.4966	4.0000
0.3486	0.8512	2.0223	6.9000
0.4862	0.8821	2.7619	10.6100
Metrizamide–0.005 M *Phosphate Buffer at pH 6.2*			
0.1054	0.6337	—	3.1120
0.2213	1.1427	—	7.3000
0.3486	1.7552	—	13.3260
0.4862	2.3069	—	22.2000
Metrizamide–0.5 M *Acetate Buffer at pH 4.6*			
0.1054	0.5004	—	2.1000
0.2213	0.7799	—	5.0230
0.3486	1.2007	—	9.1480
0.4862	1.3658	—	14.6700
Metrizamide–1.0 M *Borate Buffer at pH 8.2*			
0.1054	0.3558	—	1.6990
0.2213	0.6128	—	3.8720
0.3486	1.0810	—	7.0490
0.4862	1.0146	—	11.4860

Source: Reprinted from Refs. 43–46 by permission of Marcel Dekker, Inc.

Empirical formulae for the diffusion coefficient and the activity coefficient as a function of the solute concentration were also obtained. They are listed below:

POTASSIUM CITRATE

$$D \times 10^6 = 2.305 + 10.986c - 30.080c^2 + 27.289c^3 \text{ (cm}^2/\text{sec)} \qquad (0.0556)$$

$$\qquad (4.3.4)$$

$$\ln \gamma^{(c)} = 4.727c - 2.435c^2 + 16.942c^3 - 26.687c^4 + 19.326c^5$$

$$(4.3.5)$$

or

$$\ln \gamma^{(c)} = -0.31 + 1.91\left(0.37 + \frac{2.90c}{1.75 - 0179c}\right)^{1.35} \quad (0.0370)$$

$$(4.3.5a)$$

in wt % of solute (w):

$$\ln \gamma^{(w)} = -0.31 + 1.91\left(0.31 + \frac{1.62w}{1 - w}\right)^{1.35} \quad (4.3.5b)$$

POTASSIUM TARTRATE

$$D \times 10^6 = 1.000 + 13.903c - 19.484c^2 + 16.785c^3 \ (\text{cm}^2/\text{sec}) \quad (0.0588)$$

$$(4.3.6)$$

$$\ln \gamma^{(c)} = 13.903c - 6.827c^2 + 32.613c^3 - 28.398c^4 + 19.571c^5$$

$$(4.3.7)$$

$$\ln \gamma^{(c)} = 1.58\left(0.91 + \frac{4.82c}{1.75 - 0.79c}\right)^{1.64} \quad (0.1077) \quad (4.3.7a)$$

In wt % of solute (w):

$$\ln \gamma^{(w)} = 1.58\left(0.91 + \frac{2.69w}{1 - w}\right)^{1.64} \quad (0.108) \quad (4.3.7b)$$

FICOLL

$$D \times 10^6 = 1.161 + 5.488c - 5.594c^2 + 9.949c^3 \quad (0.195)$$

$$(4.3.8)$$

$$\ln \gamma^{(c)} = 81.25c + 938.00c^2 - 3581.20c^3 + 5485.94c^4 - 2703.55c^5 \quad (0.680)$$

$$(4.3.9)$$

In wt % of solute (w):

$$\ln \gamma^{(w)} = 17.00\left(1.00 + \frac{0.50w}{1.00 - w}\right)^{7.36} \quad (1.67) \quad (4.3.9a)$$

METHYLCELLULOSE

$$D \times 10^6 = 0.110 + 1.715c - 19.527c^2 + 86.851c^3 \quad (0.0124)$$

$$(4.3.10)$$

$$\ln \gamma^{(c)} = 10.75c + 168.97c^2 - 1153.33c^3 + 3015.750c^4 \quad (0.046)$$

$$(4.3.11)$$

In wt % of solute (w):

$$\ln \gamma^{(w)} = 0.10 + 640\left(1.00 + \frac{10.00w}{1.00 - w}\right)^{10.1} \quad (0.597) \quad (4.3.11a)$$

SUCROSE–0.025 M SODIUM PHOSPHATE BUFFER

$$D \times 10^6 = 5.2659 + 8.2675c \quad (0.4008) \quad (4.3.12)$$

$$\ln \gamma^{(c)} = -6.2787 + 12.1543c^2 + 22.5125c^3 \quad (4.3.13)$$

SUCROSE—DISTILLED WATER

$$D \times 10^6 = 5.2659 - 9.0681c + 6.6518c^2 - 2.4550c^3 \quad (0.1351)$$

$$(4.3.14)$$

$$\ln \gamma^{(c)} = -9.7571c + 26.7647c^2 - 25.5977c^3 + 13.0724c^4 - 3.6839c^5$$

$$(4.3.15)$$

SORBITOL—0.010 M SODIUM PHOSPHATE BUFFER

$$D \times 10^6 = 2.8000 + 13.5397c - 19.7006c^2 \quad (0.6013)$$

$$(4.3.16)$$

$$\ln \gamma^{(c)} = 6.7665c + 1.1449c^2 + 6.6876c^3 + 40.6578c^4 - 47.3264c^5$$

$$(4.3.17)$$

SORBITOL—0.025 M SODIUM PHOSPHATE BUFFER

$$D \times 10^6 = 2.8000 + 11.5405c - 13.1166c^2 \quad (0.9484)$$

$$(4.3.18)$$

$$\ln \gamma^{(c)} = 6.0502c + 1.6321c^2 + 8.1993c^3 + 34.6545c^4 - 31.5098c^5$$

$$(4.3.19)$$

SORBITOL—0.050 M SODIUM PHOSPHATE BUFFER

$$D \times 10^6 = 2.8000 + 6.6653c - 8.9353c^2 \quad (0.4349) \quad (4.3.20)$$

$$\ln \gamma^{(c)} = 4.3090c + 0.6998c^2 + 9.1592c^3 + 20.0149c^4 - 21.4651c^5$$

$$(4.3.21)$$

SORBITOL—0.100 *M* SODIUM PHOSPHATE BUFFER

$$D \times 10^6 = 2.8000 + 7.5500c - 10.3814c^2 \quad (0.5065) \quad (4.3.22)$$

$$\ln \gamma^{(c)} = 4.6250 + 0.7463c^2 + 8.8272c^3 + 22.6718c^4 - 24.9319c^5$$
$$(4.3.23)$$

SORBITOL—DISTILLED WATER

$$D \times 10^6 = 2.8000 + 20.8788c + 64.9780c^2 - 91.9200c^3 \quad (0.7764)$$
$$(4.3.24)$$

$$\ln \gamma^{(c)} = 9.3853c + 33.9806c^2 + 15.1881c^3$$
$$+ 46.8682c^4 + 156.0958c^5 - 184.0151c^6 \quad (4.3.25)$$

In wt % of solute (*w*):

$$\ln \gamma^{(w)} = 0.6773\left(1.000 + \frac{6.7426w}{1.00 - w}\right)^{1.8393} \quad (4.3.25a)$$

METRIZAMIDE—0.005 *M* PHOSPHATE BUFFER AT pH 6.2

$$D \times 10^5 = 0.1725 + 4.4325c \quad (0.0354) \quad (4.3.26)$$

$$\ln \gamma^{(c)} = 27.2092c + 19.8622c^2 + 11.2099c^3 + 92.4010c^4 - 80.0733c^5$$
$$(4.3.27)$$

In wt % of solute (*w*):

$$\ln \gamma^{(w)} = 0.350\left(1.000 + \frac{45.92w}{1.00 - w}\right)^{1.1945} \quad (1.6407) \quad (4.3.28)$$

METRIZAMIDE—1.0 *M* BORATE BUFFER AT pH 8.2

$$D \times 10^5 = 0.1725 + 2.5186c - 5.6888c^2 + 27.1757c^3 - 37.6827c^4 \quad (0.0846)$$
$$(4.3.29)$$

$$\ln \gamma^{(c)} = 16.1163c - 5.0435c^2 + 44.3479c^3 + 50.0149c^4 - 187.8279c^5$$
$$+ 452.8310c^6 - 818.4722c^7 + 425.4621c^8 \quad (4.3.29a)$$

In wt % of solute (*w*):

$$\ln \gamma^{(w)} = 0.3218\left(1.00 + \frac{24.362w}{1.00 - w}\right)^{1.256} \qquad (0.1267)$$

METRIZAMIDE—0.5 M ACETATE BUFFER AT pH 4.6

$$D \times 10^5 = 0.1725 + 2.7920c + 2.7850c^2 - 9.5860c^3 + 9.3180c^4$$
$$- 8.9406c^5 \qquad (0.046) \qquad\qquad (4.3.30)$$

$$\ln \gamma^{(c)} = 17.7007c + 20.7162c^2 - 1.4420c^3 + 52.1049c^4$$
$$- 4.0631c^5 - 187.3444c^6 + 234.0011c^7$$
$$- 202.3248c^8 + 89.7292c^9 \qquad\qquad (4.3.31)$$

In wt % of solute (w):

$$\ln \gamma^{(w)} = 0.141\left(1.00 + \frac{84.33w}{1.00 - w}\right)^{1.150} \qquad (0.1050) \qquad (4.3.32a)$$

METRIZAMIDE—DISTILLED WATER AT pH 7.0

$$D \times 10^5 = 0.1725 + 2.6412c - 4.2837c^2 + 12.3238c^3 - 17.3343c^4 \qquad (0.0370)$$
$$(4.3.33)$$

$$\ln \gamma^{(c)} = 16.8269c - 0.4324c^2 + 19.9460c^3 + 50.9723c^4 - 142.3057c^5$$
$$+ 230.8316c^6 - 374.2156c^7 + 195.7153c^8 \qquad\qquad (4.3.34)$$

In wt % of solute (w):

$$\ln \gamma^{(w)} = 0.1302\left(1.00 + \frac{108.931w}{1.00 - w}\right)^{1.027} \qquad (0.09217)$$
$$(4.3.34a)$$

The numbers in parentheses after the above equations are standard deviations. The agreement of the concentration-dependent diffusivities between experimental data points with calculated values using the polynomials and the empirical formulas of the activity coefficient is generally good. The diffusivities calculated from the van Laar-type activity coefficient formula are slightly lower than the corresponding experimental data points. The thermodynamic term involves a differentiation of the activity coefficient with respect to the concentration. The van Laar-type activity coefficients may have a good fit, but they give a poorer result for the thermodynamic term than that of the polynomial form.

The experiments show that the diffusivities of sucrose in distilled water in the absence of buffer decrease with increase in sucrose concentration but increase with increase in sucrose concentration in a sodium phosphate buffer solution. The diffusivities of sorbitol in distilled water increase rather significantly with increase in sorbitol concentration; however, less variation in diffusivities was observed as the sorbitol concentration varied in sodium phosphate buffer. The diffusivities of Metrizamide in various solutions increase with increase in concentration. The diffusivity is the highest for Metrizamide in 0.005 M phosphate buffer at pH 6.2; diffusivity in 0.5 M acetate buffer at pH 4.6 is next; and diffusivity in 1.0 M borate buffer at pH 8.2 is the lowest among three buffer solutions. Diffusivity of Metrizamide in distilled water above 0.3 M is lower than in buffer solutions. At lower concentration ranges, diffusivities of Metrizamide in distilled water are slightly higher than in 1.0 M borate buffer at pH 8.2, however, which may be due to experimental errors. The experiments also show that the diffusivity of Metrizamide depends highly on the concentration and pH value of buffers in which it is dissolved. This is probably due to the molecular weight, ion charge, and ionization constant, and so on, of buffers.

A trial-and-error procedure was used to obtain a generalization factor ϕ, which could be used to estimate diffusivities of Metrizamide in a new buffer solution from experimental measurements; the diffusivities obtained in various buffer solutions multiplied by the generalization factor will yield the diffusivities of equivalent concentration in distilled water. It was found that if the ratio of a quantity, consisting of formula weight of ion per valence multiplied by its pK value, between water and buffer solution is used as a generalization factor, diffusivities obtained in the experiments can be reduced to that in distilled water at the equivalent concentration within $\pm 12\%$.

Thus, if one knows the properties of a new buffer solution, that is, the formula weight per valence and its pK value, diffusivities of Metrizamide in this buffer solution can be estimated from the value of diffusivity of Metrizamide in distilled water and the generalization factor calculated from those buffer properties. However, the procedure has not been applied to the other gradient materials yet. A further investigation in this direction is warranted.

Formulas for Density and Viscosity of Sucrose Density Gradient

In the application of the zonal centrifuge to separation of biological materials, sucrose solutions are much more often used than any other gradient materials to produce the density-gradient solution. Barber [49] has developed very accurate equations which permit the calculation of the density and viscosity of sucrose solutions as functions of the composition and temperature.

The density and the viscosity of sucrose solutions have been experimentally determined over a range of compositions and temperatures [50–52]. Barber has compiled these data into a table showing densities of sucrose solutions as functions of the weight percent and temperature and used these data to correlate a nine-constant empirical equation expressing the density as a function of the weight fraction of sucrose and the temperature. The density of sucrose solutions is presented in Table 4.6. The form of the empirical equation is given as a quadric interwoven weight fraction and temperature function:

$$\rho_{T,m} = (B_1 + B_2 T + B_3 T^2) + (B_4 + B_5 T + B_6 T^2)Y + (B_7 + B_8 T + B_9 T^2)Y^2$$

$$(4.3.35)$$

where $\rho_{T,m}$ is density of a sucrose solution, T is temperature (°C), Y is weight fraction sucrose, and B_i's are constants; their values are listed in Table 4.7a.

The nine-constant, strictly empirical equation fits the tabulated data covering the concentration range of 0 to 75 wt % sucrose and 0. to 30°C with a maximum deviation of 7 parts in 10,000, and is quite accurate within the fitted range by an extrapolation beyond the fitted range is not recommended. A theoretical equation that is less accurate within the range but which may be extrapolated to 60°C has also been obtained by Barber [49].

The second equation with six adjustable constants has a maximum deviation from the tabulated data covering 0 to 30°C and 0 to 70 wt % sucrose of 4 parts per 1000. It is believed that the accuracy of this equation is the same from 30 to 60°C. No extrapolation beyond the composition limits of the data is required. The form of this equation follows with the constants having the values given in Table 4.7b:

$$\rho_{T,m} = \frac{yM_1 + (1 - y)M_2}{y(C_1 + C_2 T + C_3 T^2) + (1 - y)(A_1 + A_2 T + A_3 T^2)}$$

$$(4.3.36)$$

where A_1, A_2, and A_3 are constants that give the molar volume of water determined from independent data on pure water covering the temperature range of interest; C_1, C_2, and C_3 are constants that give the molar volume of liquid sucrose determined from the sucrose solution data; M_1 and M_2 are the molecular weights of sucrose and water; and y is the mole fraction of sucrose in the solution and is defined as follows:

$$y = \frac{Y/M_1}{Y/M_1 + (1 - Y)/M_2} \qquad (4.3.36a)$$

where Y is the weight fraction of sucrose.

Table 4.6 Absolute Density of Sucrose Solutions

Weight Percent Sucrose in Vacuum	Mole Fraction Sucrose	Density of Solution at Indicated Temperature (g/cc)						
		0°C	10°C	15°C	20°C	25°C	30°C
0	0.00000	1.00000	0.99969	0.99914	0.99823	0.99699	0.99565
5	0.00276	1.02000	1.01976	1.01895	1.01785	1.01654	1.01520
10	0.00582	1.04130	1.04010	1.03929	1.03814	1.03687	1.03527
15	0.00921	1.06300	1.06149	1.06042	1.05916	1.05774	1.05614
20	0.01300	1.08540	1.08368	1.08234	1.08096	1.07940	1.07754
25	0.01726	1.10890	1.10640	1.10506	1.10356	1.10185	1.09999
30	0.02208	1.13280	1.13017	1.12857	1.12698	1.12510	1.12323
35	0.02758	1.15740	1.15473	1.15313	1.15128	1.14940	1.14727
40	0.03393	1.18380	1.18036	1.17849	1.17645	1.17449	1.17210
45	0.04132	1.21030	1.20650	1.20464	1.20254	1.20038	1.19798
50	0.05004	1.23780	1.23398	1.23184	1.22957	1.22732	1.22493
55	0.06049	1.26630	1.26197	1.25984	1.25754	1.25505	1.25266
60	0.07323	1.29550	1.29129	1.28890	1.28646	1.28384	1.28145
65	0.08911	1.32600	1.32140	1.31875	1.31633	1.31369	1.31104
70	0.10946	1.35700	1.35231	1.34965	1.34717	1.34460	1.34168
75	0.13647	1.38930	1.38427	1.38161	1.37897	1.37629	1.37337

Source: Reprinted from Ref. 2.

Table 4.7   Constants for Sucrose Density Calculations

Constant	Value[a]

Constants for Empirical Density Calculations Using (4.3.35)

B_1	1.0003698
B_2	3.9680504×10^{-5}
B_3	$-5.8513271 \times 10^{-6}$
B_4	0.38982371
B_5	$-1.0578919 \times 10^{-3}$
B_6	1.2392833×10^{-5}
B_7	0.17097594
B_8	4.7530081×10^{-4}
B_9	$-8.9239737 \times 10^{-6}$

Constants for Density Function Using (4.3.36)

A_1	18.027525
A_2	4.8318329×10^{-4}
A_3	7.7830857×10^{-5}
M_1	342.30
M_2	18.032
C_1	212.57059
C_2	0.13371672
C_3	$-2.9276449 \times 10^{-4}$

Source: Reprinted from Ref. 2.
[a]Numbers as given are for machine calculations. Use of 5 significant figures for hand calculations is adequate.

The first equation should be used for centrifuge runs at 0 to 30°C and the second one, at 30 to 60°C.

The viscosity data for sucrose solutions chosen for fitting were those tabulated by Swindelle et al. [51]. These data were tabulated as functions of composition at intervals from 20 to 75 wt % sucrose and at 5°C intervals from 0 to 80°C.

In fitting the viscosity data, it was recognized that water is an associated liquid containing clusters of molecules with a high degree of structural order and that sucrose itself is capable of forming hydrogen bonds with the water. Since varying the amounts of sucrose may be expected to affect the structural arrangement of the solution more than varying the temperature, an equation was sought which would accurately describe the change of the viscosity with temperature at constant composition. By use of the "hole" theory of Eyring and associates [53], the Antoine or modified Arrhenius equation so successfully used in vapor pressure treatment was used. The free-volume approach of Doolittle [54] and Miller [55] also leads to similar equations. The Arrhenius form of viscosity formula may be given as

$$\ln \frac{\mu_{T.m}}{\mu_0} = \frac{\Delta H^*}{R(T_A - T_0)} \qquad (4.3.37)$$

where

$$\ln \mu_0 = 2.303A \qquad (4.3.37a)$$

$$\frac{\Delta H^*}{R} = 2.303B \qquad (4.3.37b)$$

$$T_0 = 273.16 - C \qquad (4.3.37c)$$

In these equations, μ_0 is the limiting viscosity, ΔH^* is the heat of activation for viscous flow, and R is the gas constant in units compatible with those of ΔH^*. T_A is the absolute temperature in °K, and T_0 is probably best defined as the temperature in °K at which the free volume of the solution of the given composition becomes zero [56].

Equation 4.3.37 may be arranged to give

$$\log \mu_{T.m} = A + \frac{B}{T + C} \qquad (4.3.38)$$

where $\mu_{T.m}$ is the viscosity of sucrose solution at a given composition and temperature; T is the temperature in °C; and A, B, and C are constants that depend on composition alone. Equation 4.3.38 is known as the Antoine equation, and the constant C is referred to as the Antoine constant, which has the dimensions of temperature in °C. The Antoine constant is defined as a function of the composition over the whole range of compositions by the equation

$$C = G_1 - G_2 \left[1 + \left(\frac{y}{G_3} \right)^2 \right]^{1/2} \qquad (4.3.38a)$$

where y is the mole fraction of sucrose defined in (4.3.36a) and the G_i's are constants given in Table 4.8c.

The constants A and B are defined by a polynomial in the composition of the form

$$A = D_0 + D_1 y + D_2 y^2 + D_3 y^3 + \cdots + D_n y^n \qquad (4.3.38b)$$

where y is the mole fraction defined in (4.3.36a) and the D's are sets of constants defining A (or B) for compositions above and below 48 wt % sucrose. The values of these constants are given in Tables 4.8a and 4.8b.

Table 4.8   Coefficients[a] for Antoine Equation (4.3.38)

Coefficients	Values (wt %)	
	0 to 48[b]	48 to 75

Coefficients for Calculation of the Limiting Viscosity, or A Constant, as a Function of Sucrose Mole Fraction

D_0	2.1169907×10^2	1.3975568×10^2
D_1	1.6077073×10^3	6.6747329×10^3
D_2	1.6911611×10^5	-7.8716105×10^4
D_3	-1.4184371×10^7	9.0967578×10^5
D_4	6.0654775×10^8	-5.5380830×10^6
D_5	$-1.2985834 \times 10^{10}$	1.2451219×10^7
D_6	1.3532907×10^{11}	
D_7	$-5.4970416 \times 10^{11}$	

Coefficients for Calculation for the Activation Energy, or B Constant, as a Function of Sucrose Mole Fraction

D_0	-1.5018327	-1.0803314
D_1	9.4112153	-2.0003484×10^1
D_2	-1.1435741×10^3	4.6066898×10^2
D_3	1.0504137×10^5	-5.9517023×10^3
D_4	-4.6927102×10^6	3.5627216×10^4
D_5	1.0323349×10^8	-7.8542145×10^4
D_6	-1.1028981×10^9	
D_7	4.5921911×10^9	

Coefficients for Calculation of the Antoine Constant C (4.3.38a) as a Function of Sucrose Mole Fraction

G_1		146.06635
G_2		25.251728
G_3		0.070674842

Source: Reprinted from Ref. 2.
[a]Numbers as given are for machine calculations. Use of 5 significant figures for hand calculations is adequate.
[b]Weight fraction.

The standard deviation in using (4.3.38) over the entire range is 2 parts per 1000. No individual value deviates from the tabulated data by more than 3 parts per 1000 between 5 to 40°C and 5 to 40 wt % sucrose. Above 40°C and 40 wt % sucrose, some individual values may differ from tabulated values by as much as 8 parts per 1000, but even here the standard deviation is only 3 parts per 1000 at any given composition. Thus, the set of (4.3.38) gives the viscosity well within the allowed deviation of 1% from the tabulated data.

Properties of some commonly used gradient materials are attached in the appendices. Appendix A presents the tables of density and viscosity of aqueous sucrose solutions as a function of weight fraction concentration calculated by programming (4.3.35) through (4.3.38) for the temperature range between 0 and 30°C at 5°C interval and for weight percentage between 1 and 70% at 1% interval. Densities of CsCl solutions at 25°C as a function of refractive index are also listed in Appendix B. Density, refractive index, and viscosity of aqueous Ficoll solutions as a function of weight percent of Ficoll are given in graphic form in Appendix C.

4 DISPERSION IN DENSITY-GRADIENT CENTRIFUGATIONS

An optimum separation and purification is achieved by having a high resolution in each separated zone or band. Concentration increases as mass fraction of separated material in a band or zone increases, likewise furthering a high resolution in a separated zone. Therefore, a concentration gradient exists between a separated particle zone and outside of this zone in a density-gradient solution. Resolution in centrifugation is due to a complicated phenomenon of interaction of molecular diffusion and sedimentation. Sometimes it is convenient to lump together all the contributing factors, which tend to reduce the concentration difference, and call them the dispersion coefficient.

Dispersion in Dynamic Loading Systems

In the determination of parameters that affect dispersion between sedimenting macromolecules and the sustaining gradient solution, dispersion coefficients of two sizes of polystyrene latex beads (diameter 0.091 ± 0.0058 μm and 0.109 ± 0.0025 μm) and bovine serum albumin (BSA) in sucrose and Ficoll* 9.5 to 10% w/v step gradient solutions and 10 to 25% w/v linear-with-volume gradient solutions have been measured [30] using an Oak Ridge B-XV zonal centrifuge rotor at 20°C at 2500 rpm for various centrifugation times. Both gradient solutions were buffered with 0.1 M NaCl solution when BSA was used as the sample material. Relevant physical properties of gradient materials and samples are summarized in Table 4.9.

Dispersion coefficients were obtained from experimental data of intensity of UV absorbance of macromaterials versus volume fractions using the method of moments. The method is summarized below.

The Lamm equations (4.2.1) describe a material balance for the ith component in a radial sector of a cylindrical rotor; a redistribution of component i has been caused by diffusion and sedimentation within the sector. The process involves diffusion coupling with sedimentation, which is so complex that we have

*Polysucrose, Pharmacia Fine Chemicals, Sweden.

Table 4.9   Relevant Physical Properties of Gradient Materials and Samples Used in Dispersion Experiments

Gradient (20°C)	Density (g/cm^3)	Viscosity (cp)
Sucrose		
9%	1.0339	1.2947
10%	1.0380	1.3371
Ficoll		
9%	1.030	4.8
10%	1.033	5.0

Particle	Diameter (cm)	Sedimentation Coefficient (sec)	Density (g/cm^3)
Polystyrene bead	9.1×10^{-6}	0.230×10^{-10}	1.04
BSA	7.2×10^{-7}	0.430×10^{-12}	1.36

Source: C. T. Rankin, Jr., and L. H. Elrod, "Useful Data for Zonal Centrifugation," The Molecular Anatomy Program, Oak Ridge National Laboratory, January 1970.

called the term D_{ij} dispersion coefficient instead of diffusion coefficient, a term generally used in a sense of molecular diffusion.

The analysis of moments can be used in the determination of dispersion coefficients from experimental measurements. The method is summarized below following Schumaker's derivation [22] for the Lamm equation. If sedimentation and diffusion coefficients are constants, the nth moment of concentration i about the center of rotation is

$$\langle r^n \rangle = \frac{\displaystyle\int_{r_1}^{r_2} r^n c_i r \, dr}{\displaystyle\int_{r_1}^{r_2} c_i r \, dr} \qquad (4.4.1)$$

where the quantities r_2 and r_1 are related to the bandwidth, which is a significant measure of resolution in a separation process. A bandwidth w is defined from an experimental strip chart distribution diagram in such a way that

$$w = \int_{r_1}^{r_2} f(c) dr / h$$

where $f(c)$ is the shape and h is the peak of the distribution diagram. The integral is always equal to the total particle concentration, the theoretical area.

Differentiation of (4.4.1) with respect to times gives

$$\left[\frac{\partial \langle r^n \rangle}{\partial t}\right]_r = \int_{r_1}^{r_2} r^n \left[\frac{\partial c_i}{\partial t}\right]_r r\, dr \Bigg/ \int_{r_1}^{r_2} c_i r^n\, dr$$

$$(4.4.2)$$

Through substitution of (4.4.2) into the Lamm equation, (4.1.1), and integration by parts, a recursion formula between the various moments of the ith component is obtained:

$$\frac{\partial \langle r^n \rangle}{\partial t} = n^2 D \langle r^{n-2} \rangle + nS\omega^2 \langle r^n \rangle \qquad (4.4.3)$$

Integration for the second moment, $n = 2$, yields

$$\frac{\langle r^2 \rangle + 2D/S\omega^2}{\langle r^2 \rangle_0 + 2D/S\omega^2} = \exp[2\omega^2 S(t - t_0)] \qquad (4.4.4)$$

in which subscript $_0$ refers to the reference band without dispersion at time t_0. The reference bandwidth can be calculated from the amount of sample and the geometry of the rotor.

If the sedimentation coefficient S and the angular velocity ω are given, the dispersion coefficient can be obtained from (4.4.4). If the sedimentation coefficient is unknown, (4.4.3) can be integrated for the fourth moment, $n = 4$. Thus,

$$\frac{\langle r^4 \rangle - 2\langle r^2 \rangle^2}{\langle r^4 \rangle_0 - 2\langle r^2 \rangle_0^2} = \exp[4\omega^2 S(t - t_0)] \qquad (4.4.5)$$

The sedimentation and dispersion coefficients may then be obtained by simultaneous solution of (4.4.4) and (4.4.5).

The volume–concentration distribution data from the strip chart were converted into the radius–concentration distribution forms for further processing of the data for both step and linear-with-volume-gradient solutions. The converted results are presented graphically in Figs. 4.10a, 4.10b, 4.10c, and 4.10d for step gradient solutions, and the results for linear-with-volume gradient runs in Figs. 4.11a and 4.11b.

Figure 4.10a shows the effect of sedimentation of polystyrene beads in sucrose. The leading edge of the sample zone, since it is at a greater radius, experiences a larger centrifugal force and the width of the zone base continually expands. The same beads in Ficoll gradient solution are shown in Fig. 4.10b. The distribution curves retain a more symmetrical shape and may be attributed to (a) a larger centrifugal force at the leading edge, which is offset by a fourfold increase in viscosity of Ficoll over sucrose, and (b) the viscosity gradient through the sample zone is steeper in Ficoll than in sucrose, thus offsetting the larger centrifugal force acting on the leading edge. Figures 4.10a and 4.10b show a BAS zone in sucrose and Ficoll gradient solutions. They show the same characteristic distributions as beads in the respective gradient solutions. It is interesting to note that for less than a 1-hr run, the center of concentration distributions has shifted back to the axis instead of moving forward to the rotor edge. This may be attributed to the mixing of loaded sample with the gradient solution and an early rapid dispersion of gradient and overlay and of sample and overlay. This is probably an indication of the great instability that exists during this initial period of centrifugation.

Linear-with-volume gradient solution runs are shown in Figs. 4.11a and 4.11b for beads and a BSA in a sucrose solution, for less than 1-hr runs and for 1- to 3-hr runs, respectively. Instabilities during the initial period of centrifugation were also observed for both cases. After a certain period of centrifugation, the center of concentration distributions moved consistently

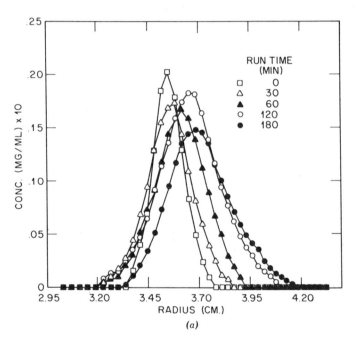

(a)

Fig. 4.10 Macromolecules in step-gradient centrifugation. (a) Polystyrene beads in a step-gradient sucrose solution. (b) Polystyrene beads in a step-gradient Ficoll solution. (c) BSA in a step-gradient sucrose solution. (d) BSA in a step-gradient Ficoll solution. *Source:* Reproduced from Ref. 30 by permission of Marcel Dekker, Inc.

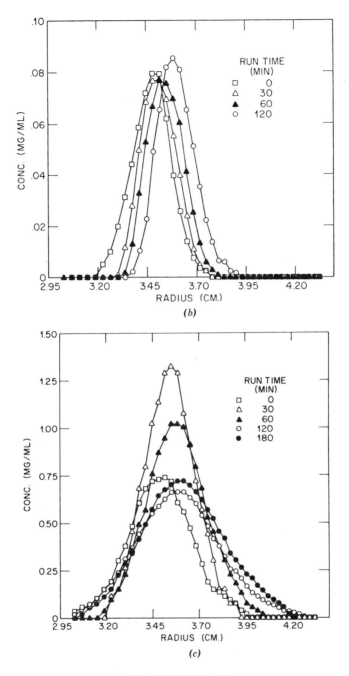

Fig. 4.10 Continued

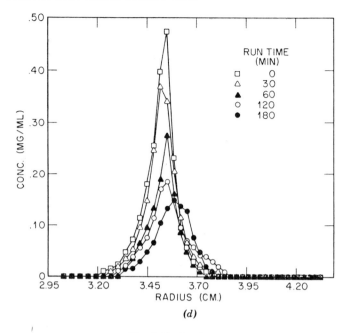

Fig. 4.10 Continued

toward the rotor edge. The zone base broadened with time as the leading edge advanced, while the trailing edge remained in a relatively constant position.

The second moments of each concentration distribution for step-gradient solutions were determined from (4.4.1), based on Figs. 4.10 and 4.11. The integration technique used was Simpson's method with 40 subintervals (5-ml increments through the 120- to 320-ml region of the rotor) for step-gradient solutions, and with eight subintervals (10-ml increments through the 165- to 245-ml region of the rotor) for linear-with-volume gradient solutions. Dispersion coefficients for each centrifugation run were also calculated from (4.4.4). Both results calculated are presented in Table 4.10. Second moments ranged from 0.072 to 0.082, and dispersion coefficients ranged from 1.0×10^{-6} to 1.0×10^{-5} cm^2/sec, and both consistently decreased with centrifugation time. Dispersion coefficients as a function of centrifugation time are also presented in Fig. 4.12. The dispersion coefficient accounts for both eddy and molecular diffusions. From Fig. 4.12, we would like to suggest that the eddy contribution to the dispersion process decreases exponentially until centrifugal work was approximately $\omega^2 t = 8.4 \times 10^8$ sec^{-1}; and after $\omega^2 t > 8.4 \times 10^8$ sec^{-1}, the dispersion coefficient approaches a constant value, a molecular diffusivity. Therefore, the eddy diffusion coefficient may be estimated by subtracting the asymptotic constant from the observed dispersion coefficients.

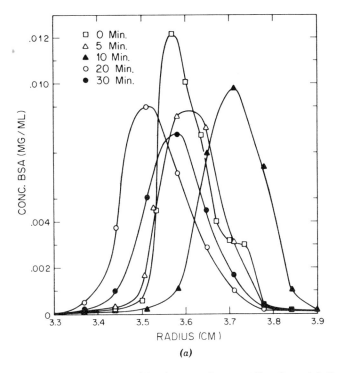

Fig. 4.11   Macromolecules in linear-with-volume gradient centrifugations. (*a*) Polystyrene beads in a sucrose solution. (*b*) BSA in a sucrose solution. *Source:* Reproduced from Ref. 30 by permission of Marcel Dekker, Inc.

In order to make the experimental results general and more useful, dispersion coefficients were correlated to a dimensionless number, the Schmidt number ($Sc = \mu_0/\rho_0 D$), as a function of various dimensionless parameters. Using results from step-gradient solution runs by a trial-and-error method, it is found that

$$Sc = 105[30F^{1.03}(Ta \cdot Q)^{-1} + 19Se^{0.19}\tau]^{1.02} \qquad (4.4.6)$$

fits all the data points within an average deviation of 2.5% and a maximum deviation of 6.3%. The quantities μ_0 and ρ_0 are the viscosity and density of the gradient solution evaluated over the bandwidth values, respectively; D is the dispersion coefficient between sample and a gradient solution, instead of the customary term of the diffusion coefficient; $F = \omega^2 t \rho_0 D_p^2/\mu_0$ is the reduced centrifugal force field strength; D_p is the size of sample dimension; and $Ta = \omega D_p R \rho_0/\mu_0$ is the Taylor number, which is equivalent to the Reynolds number in a rotating system. If $Q < 1$, a sedimentation takes place until $Q = 1$

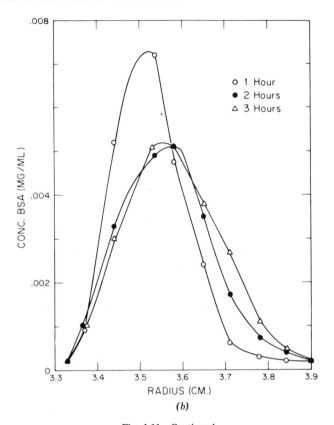

Fig. 4.11 Continued

(isopycnic point). $Se = S\omega^2\rho_0 R^2/\mu_0$ is the reduced sedimentation coefficient as defined in (4.2.22), and S is the local sedimentation coefficient in a gradient solution as defined by Svedberg [10, 11].

The results obtained from the linear-with-volume gradient solution runs were checked against (4.4.6). A good agreement exists. All the data points were within an average of 3.5% deviation and a maximum of 8.9% deviation. They are assumed to be within the range of experimental errors, so that a further correlation to improve the results was not made. Equation 4.4.6 and correlations from both a step-gradient solution and a linear-with-volume gradient solution are presented in Fig. 4.13.

Equation 4.4.6 and Fig. 4.13 may be used to estimate a dispersion coefficient on new biomaterials if sizes of particle and their sedimentation coefficients are available.

Table 4.10 Dispersion Coefficients of Polystyrene Latex Beads and BSA in Sucrose and Ficol Solutions

Second Moments and Dispersion Coefficients from Step-Gradient Runs

Gradient	Sample	Run Time (min)	Second Moment	Dispersion Coefficients (cm²/sec) × 10⁵
Sucrose	Bead	0	0.07913	1.097
		30	0.07897	0.546
		60	0.07668	0.353
		120	0.07456	0.205
		180	0.07227	0.141
		720	0.07250	0.038
Ficoll	Bead	0	0.08275	1.115
		30	0.08106	0.562
		60	0.07946	0.367
		120	0.07709	0.214
Sucrose	BSA	0	0.08145	1.113
		30	0.07933	0.551
		60	0.07744	0.359
		120	0.07671	0.213
		180	0.07557	0.150
Ficoll	BSA	0	0.08093	1.124
		30	0.08042	0.558
		60	0.08022	0.371
		120	0.07885	0.219
		180	0.07799	0.155

Second Moments and Dispersion Coefficients from Linear-with-Volume Sucrose Gradient Runs

Run Time (min)	Second Moment	Dispersion Coefficients (cm²/sec) × 10⁵
0	0.0767	1.065
5	0.0760	0.905
10	0.0727	0.758
20	0.0790	0.658
30	0.0773	0.537
60	0.0799	0.370
120	0.0781	0.217
180	0.0777	0.154

Source: Reprinted from Ref. 30, p. 265, by permission of Marcel Dekker, Inc.

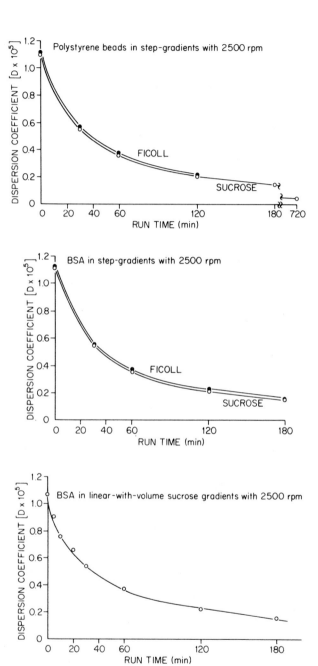

Fig. 4.12 Dispersion coefficients in gradient solutions at various run times. *Source:* Reproduced from Ref. 30 by permission of Marcel Dekker, Inc.

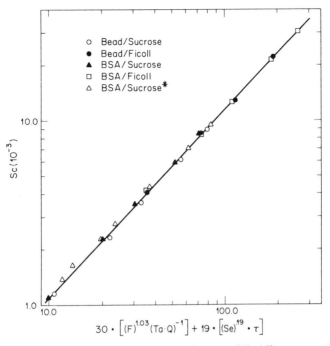

Fig. 4.13 Plot of (4.4.6), $Sc = 105[30F^{1.03}(Ta \cdot Q)^{-1} + 19Se^{0.19}\tau]^{1.02}$, and general correlation of experimental data points. *Runs against linear-with-volume gradients which were not used in the correlation. *Source:* Reproduced from Ref. 30 by permission of Marcel Dekker, Inc.

Dispersion in Density-Gradient Reorienting Systems

In reorienting systems, during the gradient reorientation from rest to a stable orientation in a high centrifugal force field, the shearing force occurring in a liquid confined in a closed cylinder will cause an increase in dispersion of a reorienting gradient system. The dispersion coefficient contributed from reorientation, D_{orien}, may be written as

$$D_{\text{orien}} = \frac{dA}{dt} \quad \text{(area/time)} \qquad (4.4.7)$$

In a given time period, the dispersion due to reorientation shearing forces is

$$\sigma = \int_0^t D_{\text{orien}} \, dt = \int_0^t \frac{dA}{dt} \, dt = A(t) - A(0) \qquad (4.4.8)$$

Investigating the fluid dynamic behavior of reorienting systems, one finds that an isodensity interfacial area will approach a constant value (completely oriented) after a certain rotational speed for a given liquid loading level. A rotational speed, rpm $= (60 \times \omega)/2\pi$, is directly proportional to the angular velocity ω (in $\sec^{-1}$). Thus, (4.4.8) may be rewritten

$$\sigma = \int_0^t \frac{dA}{dt}\, dt = \int_0^{t_c} \frac{dA}{dt}\, dt + \int_{t_c}^t \frac{dA}{dt}\, dt = \int_0^{t_c} \frac{dA}{dt} = A(t_c) - A(0)$$

$$= \int_c^\omega \frac{dA}{d(1/\omega)}\, d\left(\frac{1}{\omega}\right) = \int_0^{\omega_c} \frac{dA}{d\omega}\, d\omega = A(\omega_c) - A(0) \qquad (4.4.8a)$$

where t_c is the time required to reach a completely reoriented paraboloid configuration and ω_c is the angular velocity at which this occurs. From (4.4.8), one may see that $A(\omega_c)$ is a function of rotor configuration and liquid loading level only. Therefore, we conclude that dispersion from a reorienting gradient system is independent of rate of acceleration; it is a constant depending on rotor configuration and liquid loading levels.

Experimental verification of the hypothesis underlying the principle of (4.4.8) and (4.4.8a) was made by Hsu, Brantley, and Breilatt [57] using K-III and J-I rotors, and a sample band broadening due to dispersion was also evaluated for those two rotors. Sample distributions in terms of UV absorbancies after a specified acceleration and deceleration are shown in Figs. 4.14a and 4.14b.

The 435-sec run for the K-III rotor was made by the following procedures: (1) The rotor was filled with 1 to 34% sucrose gradient solution from the bottom of the rotor by pumping. (2) A dense sucrose solution (66%) was pumped into the rotor from the bottom to remove a 300-ml fraction of the gradient solution from the top of the rotor which was stored in a funnel. (3) The dense sucrose solution was pumped into the rotor, and the 300- to 500-ml fraction of the gradient solution from the top of the rotor was discarded. (4) A 50-ml sample of yeast RNA (0.62 mg/ml) in 10% sucrose solution was loaded into the rotor from the top. (5) The 0- to 300-ml fraction of gradient solution was returned through a funnel over the sample layer. (6) The rotor was started from rest to 2000 rpm in 45 sec. (7) Finally, the rotor was brought to rest from 2000 rpm in 390 sec.

In the 465-sec run, the following procedure was used: (1) The 0- to 300-ml fraction of 1 to 35% sucrose gradient solution was pumped into the rotor from the bottom. (2) The 50-ml sample was loaded. (3) The 300- to 500-ml fraction of the gradient solution was discarded from the gradient pumping line. (4) The rotor was then filled with the rest of the sucrose gradient solution. (5) The rotor was started to accelerate to 2000 rpm in 45 sec and from 2000 rpm to a complete stop in 420 sec.

For the 1440-sec run, the gradient solution and sample were loaded into the

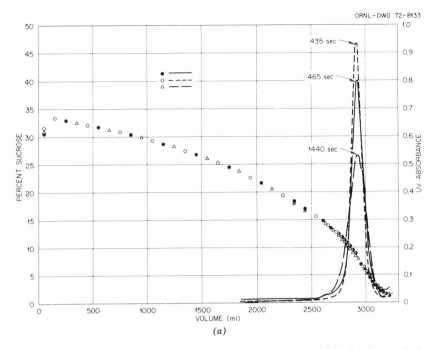

Fig. 4.14   Absorbance curves of yeast RNA in zonal centrifuge runs. (*a*) In K-III rotor. (*b*) In J-I rotor. *Source:* Reproduced from Ref. 57 by permission of Marcel Dekker, Inc.

rotor exactly the same as for the 465-sec run. The rotor was carefully accelerated to 500 rpm in 250 sec, then further accelerated to 2000 rpm in 375 sec, and decelerated to a full stop in 13 min, 35 sec.

In all three runs, the sample was loaded in an exact location between 2870 ml of dense fraction and 300 ml of light fraction of the gradient solution to give the reduced liquid loading level at $0.876 \leq \alpha \leq 0.891$ [where α = (liquid loading height)/(height of rotor)]. In the 465- and 1440-sec runs, the sample was pumped from the bottom; therefore, the sample had been moving along the rotor from the bottom to that loading level.

In Fig. 4.14*b*, the results of experimental runs on the J-I rotor are presented. Runs A, B, and E were made by filling the rotor from the bottom with 400 ml 0 to 34.0% sucrose gradient solution, 10 ml of yeast RNA sample solution (2 ml containing 1.22 mg yeast RNA in 8 ml 10% sucrose solution), and 390 ml 35.5 to 46.4% sucrose gradient solution. In run A, the filled solution was set to rest for 130 sec, and then the rotor was unloaded. In run B, the rotor was set to start from rest to 4000 rpm in 10 sec and decelerated to rest in 120 sec. Run E was made by careful control of acceleration and deceleration of the rotor. The controlling sequences of the rotor were as follows: (1) the initial

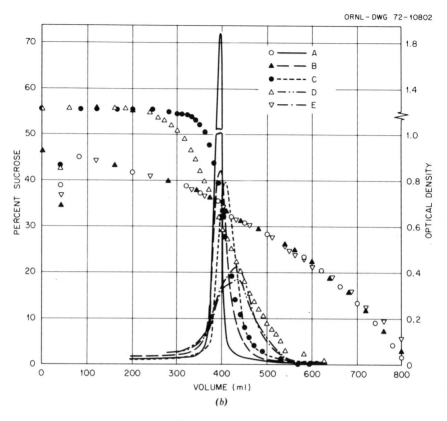

ORNL-DWG 72-10802

Fig. 4.14   Continued

acceleration of the rotor from 0 to 500 rpm in 4 min; (2) acceleration from 500 to 2000 rpm in 7 min; (3) acceleration from 2000 to 4000 rpm in 5 min; (4) deceleration from 4000 to 2000 rpm in 5 min; (5) deceleration from 2000 to 500 rpm in 7 min; and (6) final deceleration from 500 rpm to a complete stop in 4 min. The entire run was completed in 1920 sec.

Runs C and D were made in step-gradient solutions by loading the rotor first with 400 ml water, then adding the 10 ml of the sample, and finally adding the 390 ml of the 55.3% sucrose solution. In run C, the rotor was run exactly as for run B for 120 sec, while run D was made as run E for 1920 sec. In all five runs, the reduced liquid loading levels were at $0.500 \leq \alpha \leq 0.513$.

From Figs. 4.14a and 4.14b, one sees that a sample bandwidth is sharper when the rotor accelerates and decelerates at higher speeds. For both types of rotors, the band spreading is far less for a shorter period of operation time.

In order to gain an insight into band spreading or band dispersion in reorienting gradient systems, the second moment and dispersion coefficient in each experimental run were elevated and are presented in Table 4.11. Dispersion coefficients were obtained from the second moment $\langle x^2 \rangle$ by Einstein's formula [58]:

$$\langle x^2 \rangle = 2Dt \qquad (4.4.9)$$

where D is the dispersion coefficient contributed by molecular diffusion due to the concentration gradient and shearing forces in changes of isodensity area due to the gradient reorientation and other dynamic factors (such as vibration and fluctuation in the rotor speed), and t is the actual measured time. From (4.4.7) and (4.4.8), one also sees that the total dispersion σ defined in (4.4.8) is equal to the second moment defined in (4.4.9). Thus,

$$\sigma = \langle x^2 \rangle \qquad (4.4.10)$$

For evaluation of dispersion constants for K-III and J-I rotors with variation of core radius ($0 < \beta < 1$) and of reduced liquid loading level, α varying from 0.1 to 0.9, a graphic differentiation and integration was performed from Fig. 4.21. The dispersions due to reorienting for K-III and J-I rotors are presented in Figs. 4.15a and 4.15b, respectively; in a unit of the reduced isodense area increased.

The result of run A for the J-I rotor shows the dispersion due to loading and unloading and molecular diffusion of the sample in gradient solution for a 120-sec period. The difference between run A and run B may be interpreted as

Table 4.11 Relative Second Moments and Dispersion Coefficients from K-III and J-I Rotor Runs

Run	Second Moment $\langle x^2 \rangle$ (cm^2)	Dispersion Coefficient $D \times 10^2$ (cm^2/sec)
J-I Rotor: Scaling Factor = 1.983 × 10⁴		
A (120 sec)	16.188	6.226
B (120 sec)	16.234	6.244
C (120 sec)	16.369	6.296
D (1920 sec)	16.503	4.298
E (1920 sec)	16.427	4.278
K-III Rotor: Scaling Factor = 2.696 × 10⁶		
435 sec	42.206	4.851
465 sec	42.256	4.544
1440 sec	42.268	1.467

Source: Reprinted from Ref. 57, p. 74 by permission of Marcel Dekker, Inc.

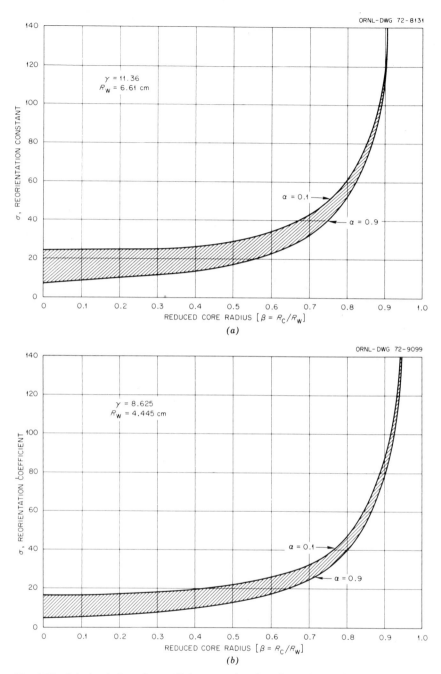

Fig. 4.15 Calculated dispersion coefficients as a function of annular gap (β) in a rotor in reorienting gradient systems. (*a*) For K-III rotor. (*b*) For J-I rotor. *Source:* Reproduced from Ref. 59 by the permission of Marcel Dekker, Inc.

the dispersion due to the gradient reorientation and an increase in molecular diffusion resulting from an increase in interfacial area during the reorientation. The difference in total dispersion between run B and run A. is 0.0467 cm^2.

It is important to point out that the numerical values presented in Table 4.11 are relative values. The optical absorbances or optical densities given in Figs. 4.14a and 4.14b are all relative values which have been scaled to a reference value given as unit value. For a practical estimation of results presented in Table 4.11, one has to obtain a scaling factor for K-III and J-I rotors. In this attempt, we have used the results presented in Figs. 4.21a and 4.21b in conjunction with the value 0.0467 cm^2 just obtained. The values of β, the radius ratio between the core and inside wall of the rotor, for the K-III and J-I rotors are 0.825 and 0.808, respectively. The reduced total dispersion evaluated for run B [α = 0.500 (reduced liquid loading height), β = 0.808 (reduced core radius)] is 43.0. The annular cross-sectional area of the J-I rotor is 21.54 cm^2. Therefore, the estimated dispersion due to reorientation in run B is 926 cm^2. The molecular diffusion coefficient of yeast RNA in sucrose solution is estimated roughly to be 2.6 $\times$ 10^{-7} cm^2/sec. Therefore, the dispersion due to molecular diffusion in a 120-sec run is negligible compared to the magnitude of reorientation dispersion. The scaling factor for the J-I rotor is obtained by the method of proportion, which gives 1.983 $\times$ 10^4. The difference between run E and run B gives dispersion due to the difference in acceleration or deceleration and other dynamic factors causing dispersions. The dispersion arising from those sources totals 0.1927 relative units. This experimental finding is in agreement with the analysis presented in Ref. 59. Therefore, we would like to stress that, in an improvement of resolution for a separation task, a control of rotor acceleration or deceleration does not contribute to the improvement. As demonstrated in (4.4.8), dispersion arising from a gradient reorientation is a constant for a given rotor and for a given loading level. The magnitude of dispersion due to reorientation is much smaller than that due to loading or to the fluctuation in rotor speed or vibration in a rotating system.

A comparison between runs B with C and runs E with C demonstrates that a smooth built-in gradient system offers a better result than a step-gradient system. A diffusion coefficient is proportional to its concentration difference. In a step-gradient system, the concentration difference between a sample zone and the gradient solution is higher than in a built-in gradient system.

In the K-III runs, we attributed the difference in dispersion between a 435-sec run and a 465-sec run to the rate at which the sample moves from the bottom of the rotor to the loading level. Since they were both accelerated in the same way, just the difference in time to bring the rotor to a full stop (which may be categorized as being within experimental error) was probably the cause of the slight difference in dispersion.

With a limited amount of information, we extrapolated runs of 1440 and 465 sec to zero and subtracted the dispersion due to sample moving; we thus determined the dispersion due to loading and unloading in the K-III rotor to be 42.2 reduced units. Then, using Fig. 4.21a, the scaling factor of the K-III rotor was estimated to be 2.7×10^5. Therefore, it is concluded that dispersion due to loading and unloading is the major factor in loss of resolution.

From the foregoing analysis of experimental results, we make the following suggestions: if the sample material is not shear sensitive, one should run the rotor as fast as possible in the start-up to reduce the operating time. If the material is shear sensitive, one should go as slowly as possible up to the critical angular speed and then, after this speed is reached, run the rotor as fast as possible to its full speed to reduce the start-up time. The band broadening due to dispersion in gradient reorienting systems and the resolution losses due to reorientation are much smaller than those due to loading and unloading. The control of rotor acceleration or deceleration is a very difficult task; since the resolution loss due to this factor is negligible, control is unnecessary. Furthermore, at high speeds, the fluctuation of rotor speed is less than that at low speed, and smooth operation is easier to attain.

Improvement of the zonal centrifuge should be directed to an improvement in the loading and unloading method to reduce the dispersion contributed by this operation. Changing of a pumping system to a gravitational loading method by loading a gradient solution from a dense to a light fraction from the top may be worthwhile.

5   FLUID MECHANICAL CONSIDERATIONS

When a cylindrical rotor filled with a density-gradient solution is rotating about its own axis at a steady state at a given angular velocity ω, the fluid in the rotor moves as an element of a rigid body [13]. Thus, the tangential velocity component of the fluid with respect to a space fixed coordinate, v_θ, is ωr and with respect to the rotor wall is zero; the radial velocity component of the fluid, v_r, is zero with respect to both coordinates; and there is no shear stress existing within the fluid. Thus, fluid mechanical consideration of zonal rotors is important only during the acceleration or deceleration periods.

Experimental Investigation on Numbers of Septa in Zonal Rotors

In the development of large preparative zonal centrifuges, one of the ideas is that sedimentation should occur in sector-shape compartments similar to those used in the ultra analytical centrifuge cells to minimize the loss of resolution due to so-called "wall effects" [22, 60, 61]. The general practice has been simply to insert septa into the hollow cylindrical rotor chamber to divide the rotor into compartments which are very nearly sector shaped. The septa serve the additional function of preventing circular flow in the rotor during changes in rotor speed and during radial movement of the sample and gradient as occurs during dynamic loading and unloading.

An experimental investigation to study the effects of septa on the fluid dynamic characteristics during the acceleration and deceleration period has been performed by Pham and Hsu [62]. A Beckman B-XV-size zonal centrifuge rotor was used to investigate two-dimensional transient velocity components v_r and v_θ, shear stress τ_{rr}, and $\tau_{r\theta}$ distribution in the rotor by the technique of birefringence using a NGS-grade milling yellow dye solution as an optical solution. The density–viscosity relationship of 1.2–2.0 wt % milling yellow aqueous solutions is very close to that of sucrose aqueous solutions at the same density. This was the reason that the aqueous milling yellow dye was used as the double refraction liquid to simulate the density sustaining solution in the zonal centrifuge rotors. The investigation was performed with 3, 4, 6, and 12 septa in the rotor with the acceleration and deceleration rates varying from ± 0.4 to ± 3.0 rps per second. The dimensions of the B-XV-type zonal rotor used in the experiment are given in Table 4.12, and a schematic diagram of the equipment setup and assembled rotor are presented in Figs. 4.16a and 4.16b, respectively.

Typical streamline patterns for each rotor configuration with the acceleration rate of $6\pi/\sec^2$ (setting 1/1) at 800 and 1820 rpm are presented in Fig. 4.17. The velocity and shear stress distributions with the same acceleration rate for each rotor configuration are presented in Fig. 4.18.

The results show that all the flow patterns and shear stress distributions exhibit similar characteristics regardless of the rate of acceleration or deceleration, acquire their characteristic profiles almost instantaneously, and exhibit a

Table 4.12 Dimensions of B-XV Type Zonal Rotor Used in Birefringence Studies

	Dimension (cm)
Rotor inside diameter	18.42
Rotor wall thickness	2.03
Height of inside rotor	5.49
Top and bottom thickness	1.91
Core diameter	3.81
Length of septa	7.88
Width of septa	0.25

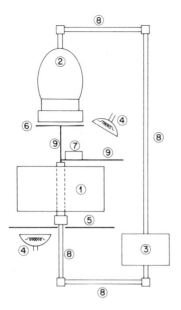

1. Zonal Rotor   6. Analyzer
2. Movie Camera   7. Time Counter, RPM Meter
3. Motor   8. Shafts
4. Strobe Lights   9. Metal Plates
5. Polarizer

(a)

(b)

Fig. 4.16 (a) Schematic diagram of synchronized birefringent patterns recording setup. (b) Partially assembled B-XV-size rotor. *Source:* Reproduced from Ref. 62 by permission of the American Chemical Society.

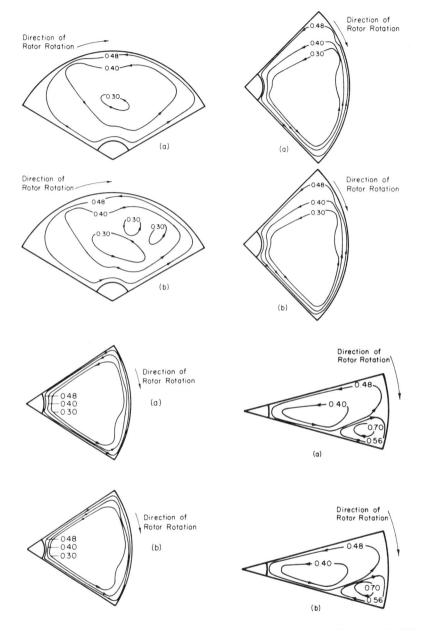

Fig. 4.17 Transient flow patterns for 3-, 4-, 6-, and 12-septra rotors at (*a*) 800 rpm, (*b*) 1800 rpm. *Source:* Reproduced from Ref. 62 by permission of the American Chemical Society.

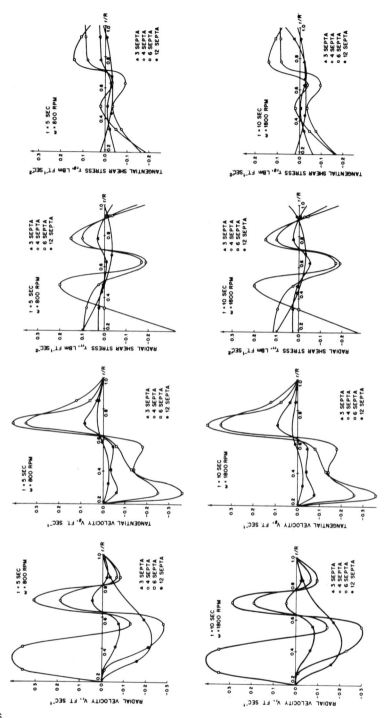

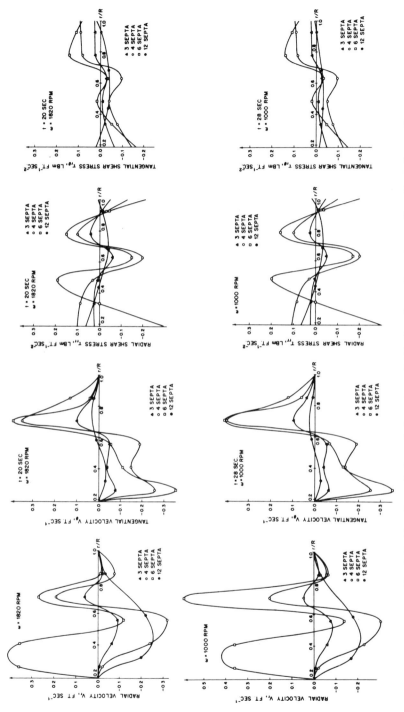

Fig. 4.18 Transient velocity and shear stress distributions for various rotor configurations. *Source:* Reproduced from Ref. 62 by permission of the American Chemical Society.

197

rather constant characteristic value within the same rotor configuration. The numbers of septa in a rotor configuration contribute greatly to the velocity profiles and shear stress distributions. During the transition period, a secondary flow has been developed in a low-streamline region for the 3-septa rotor and in a high-streamline region for the 12-septa rotor. For 4- and 6-septa rotors, the flow patterns appear to be more uniform and no prominent secondary flow develops during the transition periods. The development of secondary flows in a low-streamline region for the 3-septa rotor may be interpreted as a consequence of interactions of the Coriolis force and fluid velocities, causing a local swirling motion during the transition periods; while for the 12-septa rotor configuration at a high-streamline region, the tangential velocity component may interact with the rotor wall and the septa wall to cause a local swirling motion at the corner of the rotor.

The tangential and radial velocities are zero at a normalized radius of about 0.62. This implies that the flow has an axis around this location. The absolute magnitude of the tangential velocity has local maxima at normalized radii of 0.26 and 0.7. The radial velocity changes direction several times and has maximum absolute values at normalized radii of 0.85, 0.70, 0.55, and 0.33 for rotors of 3, 4, 6, and 12 septa, respectively. Also the radial and tangential velocities are highest in magnitude in 6-septum rotors and lowest in 3- and 12-septum rotors.

Shear stresses, unlike velocity components, have local maxima at a normalized radius equal to 0.62. The radial shear stress approaches zero at normalized radii of 0.5, 0.7, 0.91, except for the case of a 6-septum rotor. It is also zero at the normalized radius of 0.33, where the radial velocity is maximum. The tangential shear stresses are small at the normalized radii of 0.5 and 0.7. Both radial and tangential shear stresses are found to be the highest at the walls of the rotor. It is also noticed that high shear stresses correspond to low velocity components.

Figure 4.19 presents the variations of the maximum velocity components and the maximum shear stress distributions with respect to the number of septa in the zonal rotor. These quantities are approximately equal in 3- and 12-septum rotors. The maximum velocities and the maximum shear stresses developed are the highest in a 6-septum rotor; the next highest are found in a 4-septum rotor.

From these experimental results, we suggest the following configurations for the separation task: (1) Rotors with 3- or 12-septa configurations are recommended for separation of long-chain particles, where low shear stresses are the criterion for preventing damage to the particles. (2) A rotor with 4-septa configuration is recommended for separation of spherical particles, in which the shear stresses are not too critical during the starting period of operation. This configuration does not induce secondary flows, and remixing is minimized

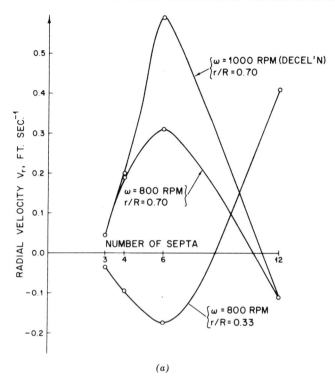

(a)

Fig. 4.19   Variations of the maximum velocity components and the maximum shear stress distributions with respect to the number of septa in B-XV-size rotor. (a) Variation of maximum radial velocity with number of septa. (b) Variation of local maximum radial shear stress with number of septa. (c) Variation of local maximum tangential velocity with number of septa. (d) Variation of local maximum tangential shear stress with number of septa. *Source:* Reproduced from Ref. 62 by permission of the American Chemical Society.

during acceleration for the reorienting period. A 6-septa rotor has the same characteristics as this 4-septa rotor, but in it shear stresses are higher. A 4-septa rotor is also easier to balance and to construct than a 6-septa rotor.

Changes of Interfacial Area of Reorienting Isodensity Layers

The qualitative evaluation of the shearing forces between each isodense layer of liquid confined in a closed cylindrical rotor during the transition from rest to a stable orientation has been made [13, 59] by evaluating the changes of surfaces at various levels as a function of the rotational speeds.

Deformations occurring at the various levels may be best understood by describing the changes which occur in layers originally at the top, middle, and bottom of the rotor. The fluid originally against the upper rotor cap becomes

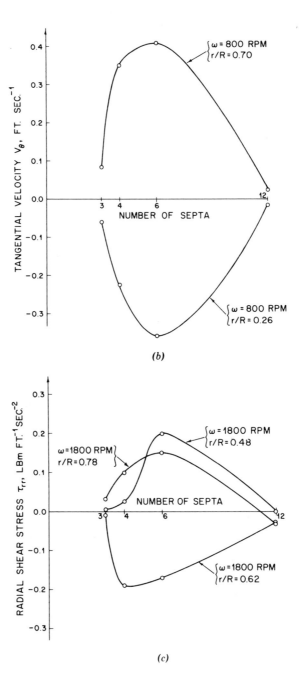

(b)

(c)

Fig. 4.19 Continued

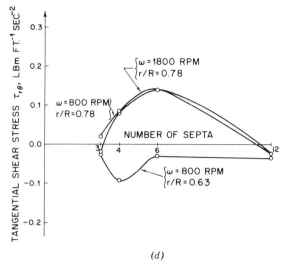

(d)

Fig. 4.19   Continued

squeezed into a small paraboloid of revolution during acceleration (Figs. 4.20*b* and 4.20*c*) and then occupies the center of the rotor at high speed (Fig. 4.20*d*). A zone in the middle of the rotor at rest (Fig. 4.20*a*) increases in area during acceleration and then decreases in area slightly as the vertical position is approached. The greatest area changes, therefore, occur in the zone near the top at rest but near the center and the edge at high speed.

The reoriented gradient before the particles have sedimented appreciably is shown at the left and after sedimentation at the right of Fig. 4.20*d*. The distribution during deceleration is shown in Fig. 4.20*e*, with the distribution at rest shown in Fig. 4.20*f*. The separated zones are recovered by draining the gradient out the bottom of the rotor or by displacing it out the top.

A cylindrical rotor system with inside rotor wall radius R_w, outer core radius R_c, and height H was considered. The rotor is filled, at rest, with density gradient and sample layer as shown schematically in Fig. 4.20*a*. Then the rotor is set to accelerate to a given rotational speed. During acceleration, each isodense surface becomes part of a paraboloid of revolution and can go through a series of configurations. There are four types of paraboloid configurations which will occur, depending on the liquid loading level and change in angular velocity. They are shown in Fig. 4.20*b* as types 1, 2, 3, and 4:

Type 1:

$$Z_c > 0 \qquad r_H = R_w \qquad Z_w < H$$

Type 2:

$$Z_c > 0 \qquad r_H < R_w \qquad Z_w = H$$

Type 3:

$$Z_c < 0 \qquad r_H = R_w \qquad Z_w < H$$

Type 4:

$$Z_c < 0 \qquad r_H < R_w \qquad Z_w = H$$

where Z_c and Z_w are the heights at the intersection of the paraboloid with the rotor outer core and the inside rotor wall, respectively.

The sample layer sediments through the reoriented gradient while the rotor is at speed. The separated particles band in respective isodensity zones. Figures 4.20c and 4.20d are schematic diagrams of the reorienting gradient rotor system in acceleration and the rotor at speed, before the particles have sedimented appreciably and after sedimentation banding in respective isodensity zones. The distribution during deceleration is shown in Fig. 4.20e, with the distribution at rest shown in Fig. 4.20f. The separated zones are recovered by draining the gradient out the bottom of the rotor or by displacing it out the top.

The equation describing the parabolic surface of revolution is well known and is given by Bird et al. [2] as

$$Z = \frac{\omega^2 r^2}{2g} + Z_0 \qquad (4.5.1)$$

where Z is a vertical axial coordinate, r is a radial coordinate as shown in Fig. 4.20e, ω is angular velocity, g is the gravitational force, and Z_0 is the minimum of Z in the paraboloid, which depends on the angular velocity and the loading level of liquid. For a rotor with core, Z_0 is a hypothetical point which lies inside the core and changes from positive to negative with increase in angular velocity ω or decrease in loading level. The equation describing the paraboloid interfacial area can be obtained from (4.5.1) to give

$$S_i = \int_l^u r\left[1 + \left(\frac{dr}{dZ}\right)^2\right]^{1/2} dZ$$

$$= \frac{2\pi g}{\omega^2} \int_l^u \left[\frac{2\omega^2}{g}(Z - Z_0) + 1\right]^{1/2} dZ \qquad (4.5.2)$$

The quantities l and u are the lower and upper integration limits. These values depend on the paraboloid configuration and must be constrained by

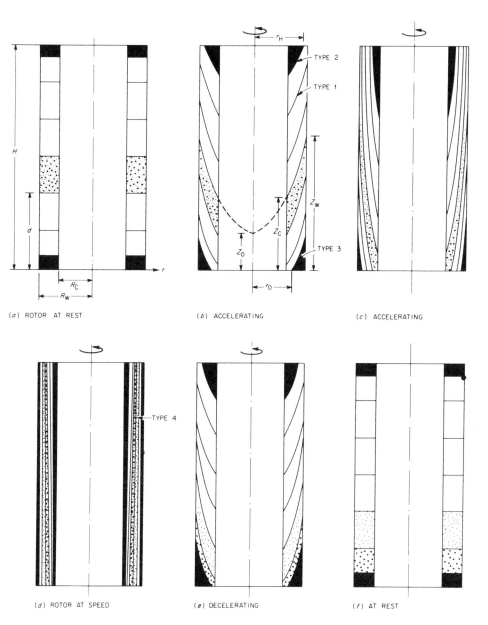

Fig. 4.20 Schematic diagram of reorienting gradient rotors with core. *Source:* Reproduced from Ref. 59 by permission of Marcel Dekker, Inc.

the volume of liquid loaded. The constrained liquid volumes in the paraboloid for configuration types 1 and 2 are

$$V = \pi(R_w{}^2 - R_c{}^2)\alpha H = \pi(R_w{}^2 - R_c{}^2)Z_c + \pi \int_{Z_c}^{Z_w} (R_w{}^2 - r^2)\, dZ \quad (4.5.3a)$$

and

$$V = \pi(R_w{}^2 - R_c{}^2)\alpha H = \pi(R_w{}^2 - R_c{}^2)Z_c + \pi \int_{Z_c}^{H} (R_w{}^2 - r^2)\, dZ \quad (4.5.3b)$$

respectively. For types 3 and 4, they are

$$V = \pi(R_w{}^2 - R_c{}^2)\alpha H = \pi \int_{0}^{Z_w} (R_w{}^2 - r^2)\, dZ \quad (4.5.3c)$$

and

$$V = \pi(R_w{}^2 - R_c{}^2)\alpha H = \pi \int_{0}^{H} (R_w{}^2 - r^2)\, dZ \quad (4.5.3d)$$

respectively. The quantity $\alpha = d/H$ in (4.5.3) is the liquid loading level as shown in Fig. 4.20a. One may visualize from Fig. 4.20b that if $u = H$, the meaningful limit is r_H, the radius of an isodense paraboloid at the upper wall of a rotor; if $l = 0$, the meaningful limit is r_0, the radius of an isodense paraboloid at the bottom wall of a rotor. Thus, it is more convenient to express (4.5.2) in terms of a radial variable than in terms of a height variable.

If we define the following reduced variables

$$\alpha = \frac{d}{H} \qquad \beta = \frac{R_c}{R_w} \qquad \gamma = \frac{H}{R_w} \qquad (4.5.4a, b, c)$$

$$\rho_H = \frac{r_H}{R_w} \qquad \rho_0 = \frac{r_0}{R_w} \qquad \zeta_c = \frac{Z_c}{R_w} \qquad \zeta_0 = \frac{Z_0}{R_w} \qquad (4.5.4d, e, f, g)$$

$$\Omega = \frac{\omega^2 R_w}{g} \qquad A_i = \frac{S_i}{\pi(R_w{}^2 - R_c{}^2)} \qquad (4.5.4h, i)$$

we obtain the reduced isodensity surface area for each type of configuration together with its constrained specifications as follows:

TYPE 1: $0 < \zeta_c < \gamma, \rho_H = 1$

$$A_1 = \frac{2}{3\Omega^2(1 - \beta^2)} [(1 + \Omega^2)^{3/2} - (1 + \Omega^2\beta^2)^{3/2}] \quad (4.5.5)$$

with constraints

$$\zeta_c = \alpha\gamma - \frac{\Omega}{4}(1 - \beta^2) > 0 \quad (4.5.5a)$$

and

$$\rho_H = \left[\frac{2(1 - \alpha)}{2} + \frac{1 + \gamma^2}{2}\right]^{1/2} > 1 \qquad (4.5.5b)$$

It is seen that for this type, $u = Z_w$, and the term $Z - Z_0$ at Z_w can be obtained from (4.5.1) to give

$$Z_w - Z_0 = \frac{\omega^2 R_w{}^2}{2g} \qquad (4.5.5c)$$

TYPE 2: $0 < \zeta_c < \gamma, \beta < \rho_H < 1$

$$A_2 = \frac{2}{3\Omega^2(1 - \beta^2)} \left[(1 + \Omega^2\rho_H{}^2)^{3/2} - (1 + \Omega^2\beta^2)^{3/2}\right] \qquad (4.5.6)$$

with constraints

$$\zeta_c = \gamma - [\Omega\gamma(1 - \alpha)(1 - \beta^2)]^{1/2} > 0 \qquad (4.5.6a)$$

$$\rho_H = \beta\left\{1 + \frac{2}{\beta^2}\left[\frac{\gamma}{\Omega}(1 - \alpha)(1 - \beta^2)\right]^{1/2}\right\}^{1/2} < 1 \qquad (4.5.6b)$$

TYPE 3: $\zeta_c < 0, \rho_H = 1$

$$A_3 = \frac{2}{3\Omega^2(1 - \beta^2)} \left[(1 + \Omega^2)^{3/2} - (1 + \Omega^2\rho_0{}^2)^{3/2}\right] \qquad (4.5.7)$$

with constraints

$$\zeta_c = -\frac{\Omega}{2}\left\{1 - \beta^2 - 2\left[\frac{\alpha\gamma}{\Omega}(1 - \beta^2)\right]^{1/2}\right\} < 0 \qquad (4.5.7a)$$

$$\rho_H = \left\{\frac{2\gamma}{\Omega} + 1 - 2\left[\frac{\alpha\gamma}{\Omega}(1 - \beta^2)\right]^{1/2}\right\}^{1/2} > 1 \qquad (4.5.7b)$$

$$\beta < \rho_0 = \left\{1 - \left[2\frac{\alpha\gamma}{\Omega}(1 - \beta^2)\right]^{1/2}\right\}^{1/2} < 1 \qquad (4.5.7c)$$

TYPE 4: $\zeta_c < 0, \beta < \rho_H < 1$

$$A_4 = \frac{2}{3\Omega^2(1 - \beta^2)} \left[(1 + \Omega^2\rho_H{}^2)^{3/2} - (1 + \Omega^2\rho_0{}^2)^{3/2}\right] \qquad (4.5.8)$$

with constraints

$$\zeta_c = \frac{1}{2}\{\Omega[1 - (1 - \alpha)(1 - \beta^2)] - \gamma\} < 0 \qquad (4.5.8a)$$

$$\rho_0 < \rho_H = \left[\frac{\gamma}{\Omega} + 1 - (1 - \beta^2)\alpha\right]^{1/2} < 1 \qquad (4.5.8b)$$

$$\beta < \rho_0 = \left[1 - (1 - \beta^2)\alpha - \frac{\gamma}{\Omega}\right]^{1/2} < \rho_H \qquad (4.5.8c)$$

The reduced paraboloid interfacial area A_i is a function of speeds of revolution and the loading levels of liquid. By use of (4.5.5) through (4.5.8), the reduced paraboloid interfacial area was calculated for K-111 and J-1 rotors; the results are presented in Figs. 4.21a and 4.21b. The dimensions of these rotors are listed in Table 4.13.

From Figs. 4.21a and 4.21b, it is found that the variation of the reduced paraboloid interfacial area as a function of both speeds of revolution and the loading levels of liquid exhibits the same general profiles as the rotor without core [13]. Therefore, the variation with respect to γ $[= H/R_w]$ for rotors with core is expected to be the same as rotors without core. Hence, the variation of the paraboloid interfacial area with respect to γ has not been performed.

6   EXPERIMENTAL DESIGN IN ZONAL CENTRIFUGES

A variety of methods have been published for the isolation or the purification of bioparticles from cells, tissues, and suspensions using zonal centrifuge rotors. Many of these have been empirically devised and apply to a specific particle from one source of material. Isolation procedures have not generally been viewed as a broader problem, namely, the fractionation of cells, tissues, and culture media. The fractionation involved ranges from the separation of different cell types—large-scale vaccine purification (particles roughly 1 to 3 μ) to separation of very small subcellular components resolvable with the electron microscope (approximately 10 to 30 Å) to the separation of different organic compounds and simple inorganic salts. If the particles have physical properties which are unique (for example, size and density unlike those of other particles), the procedures used to isolate or purify these particles should exploit the unique physical properties. It is desirable, therefore, to define the pertinent physical properties not only of the particle to be isolated but also of all other particles found in the source material. A rational approach to particle isolation can then be planned. The analytical developments described in the

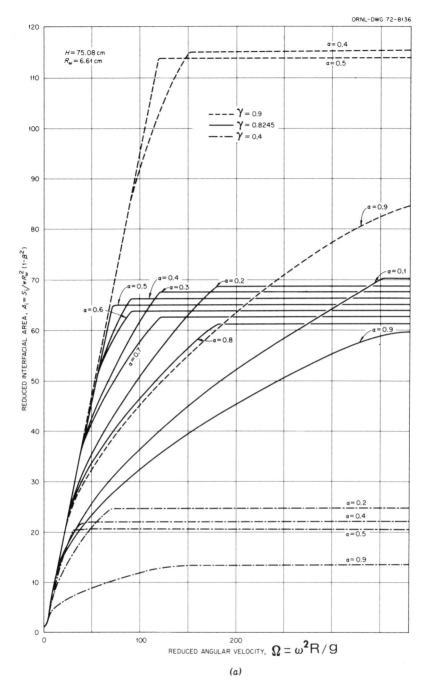

Fig. 4.21 Variation of reduced interfacial area with respect to speeds of revolution for rotors with core. (*a*) For K-III. (*b*) For J-I. *Source:* Reproduced from Ref. 59 by permission of Marcel Dekker, Inc.

207

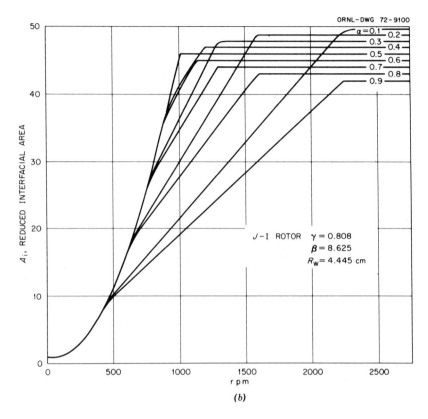

ORNL-DWG 72-9100

J-I ROTOR $\gamma = 0.808$
$\beta = 8.625$
$R_w = 4.445$ cm

A_i, REDUCED INTERFACIAL AREA

r p m

(b)

Fig. 4.21 Continued

Table 4.13 Dimensions of Rotors

	Dimension (cm)	
	K-111	J-1
Height H	75.08	38.338
Rotor diameter R_w	13.22	8.890
Core diameter R_c	10.90	7.184

previous section will serve as a basis for the rational approach to an experimental design.

Preparative ultracentrifuges were first developed largely for virus isolation [63, 64]. Separation methods based on sedimentation rate were most often used in early studies, whereas separation methods based on particle density came into general use in the 1960's [24, 65]. Ideally, one would like to separate particles on the basis of size and density independently. Although separations based on buoyant density can be made easily by the sedimentation equilibrium method (isopycnic banding), separations based on size cannot be made independent of density in centrifuges. Instead, sedimentation-rate (sedimentation-velocity) separations are made depending on (a) the square of the particle diameter, (b) the difference in density of the particle and the suspending medium (the local gradient density), (c) the local gradient viscosity, (d) the strength of centrifugal force $\omega^2 r$, and (e) the particle shape, and so on, as can be seen from (4.2.3). If a given virus or a given particle differs in sedimentation rate and/or buoyant density from other particulates, then rate and isopycnic separations can be used to achieve a clean separation. The sedimentation coefficients and buoyant densities of a variety of subcellular particles and viruses are given by Anderson et al. [66] as "the virus window." The values indicated in their sedimentation coefficients are corrected to water at 20°C, and the buoyant densities were measured by isopycnic banding in cesium chloride density gradients. The experimental problem is, therefore, to develop a method for particle isolation in which separations are made on the basis of sedimentation rate S and buoyant density ρ and then to see whether the results will be in concord with the virus window, as reproduced in Fig. 4.22. When these two separation procedures were employed sequentially, the combined procedure was termed for convenience an "S-ρ separation" method by Anderson et al. [67]. The choice of order depends largely on which separation can be most conveniently done on a large number of fractions. Since large-volume rate separations may be easily performed and are less time consuming, the rate separation is usually performed first. Pelleting appears to inactivate or at least reduce the titer of a number of biological particles. It is desirable in such instances to use a concentration method in which pelleting is avoided. For liter volumes of bioparticle harvest solution, this may be done by allowing the bioparticle-rich stream to flow over a liquid density gradient imprisoned in a rotor in such a way that the bioparticles sediment out of the stream and band in the gradient solution (continuous sample flow with isopycnic banding). Reimer et al. [68] have demonstrated, by using rate-zonal centrifugation followed by isopycnic banding, that the commercial influenza virus vaccine could be purified an additional tenfold and that the protective antigens could be separated from most of the pyrogens usually found in such partially purified material.

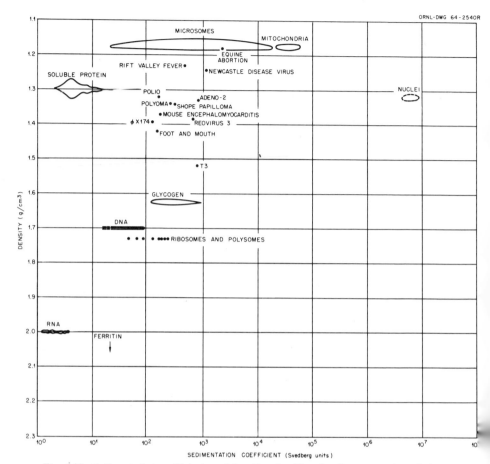

Fig. 4.22 Sedimentation coefficients and banding densities of cell components and viruses. *Source:* Reproduced from Ref. 66. Courtesy of Oak Ridge National Laboratory.

For small volumes of bioparticle harvest solution, isopycnic banding separation can be economically carried out in the standard Spinco angle-head rotors [69, 70]. Methods for setting up the gradient depend on the sample size, the gradient solute used, the particle size, and the steepness of the gradient as expressed in (4.2.32).

The particles separated in preparative centrifuges may have sedimentation coefficients which range over six orders of magnitude, 4 to 10^6 S [66]. While zonal centrifuges now available allow high-resolution separation to be made over a very wide range of particle sizes, no single zonal centrifuge experiment can achieve separations over more than a part of this range. Angle-head centrifugation is therefore often a necessary prelude to high-resolution zonal separations, especially where a minor component is to be concentrated and resolved. Methods for optimizing preparative separations are therefore of continuing interest.

It is suggested that one should consult the available literature before one runs his specific separations. Applications of centrifugation for various bioparticle separation are now so numerous that it is almost impossible to trace all the activities. A brief survey of various bioparticle separation or purification procedures is presented in Appendix D.

If one has to modify the reported procedure for adaption into his separation, the basis of procedure modification should be the principle of dynamic similarity, which is briefly presented for the case of zonal centrifugation.

Dynamic Similarity

Two-dimensional $S-\rho$ separations are most conveniently done on a laboratory scale using the B-IV-, B-XIV-, or B-XV-type rotors for the rate separations and angle-head centrifuge tubes for the isopycnic separations and in a large production scale using two K-series rotor in series. One should refer to the listed literature for separations of a specific bioparticles—gradients, centrifugation time, and speed, and so on. If the rotor available for the separation is different from that used in the literature or there are differences in the characteristics of bioparticles, the conditions applied to the reported separation should be modified according to the rule of dynamic similarity.

Moving systems are considered to be dynamically similar (a) if they are geometrically similar, and (b) if the forces acting in one system are in the same ratio to each other as similar forces in the second system. One might also say that moving systems are dynamically similar if the dimensionless parameters obtained in a dimensional analysis of the systems are the same for both. This is essentially the same as conditions (a) and (b).

The number of dimensionless quantities is always less than the total number of physical quantities, and this fact enables one to correlate experimental data more easily, which has been demonstrated in the correlation given in (4.4.6). The dimensionless quantities obtained in the analytical development section may be used to adjust the experimental conditions if one's conditions or facilities differ from those reported literatures. The useful dimensionless quantities are listed below:

$$\mathbf{F} = \left[\frac{\omega^2 t \, \rho_0 D_p{}^2}{\mu_0}\right] \qquad \text{the reduced centrifugal force field strength}$$

$$\mathbf{N} = \left[\frac{\rho_0 \omega R D_p}{\mu_0}\right]^2 \qquad \text{the equivalent Reynolds number in a rotating system, (4.2.14c)}$$

$$\mathbf{Se} = \left[\frac{S\omega^2 R^2 \rho_0}{\mu_0} \right] \qquad \text{the reduced observed sedimentation coefficient, (4.2.22)}$$

$$\mathbf{Ta} = \left[\frac{\omega R D_p \rho_0}{\mu_0} \right] \qquad \text{the Taylor number}$$

$$\tau = \left[\frac{\mu_0 t}{\rho_0 R^2} \right] \qquad \text{the reduced time (4.2.14e)}$$

Th reference physical characteristics of the density gradient are based on density and viscosity, at the light end of the gradient and the gradient shape is to be exactly the same in the reduced form. If one uses different gradient salts, the characteristics of a density-gradient solution by viscosity and density profiles will be completely different from those in the literatures. For this case, it is suggested that one must first find viscosity and density profiles of a gradient solution in a rotor by curve fitting to determine λ_i and ϵ_i in (4.2.10). If the diameter D_p and density ρ_p of particles are known, one can use the mathematical results, (4.2.19), to predict where and when the particles will be in a given gradient solution in a rotor. If one is interested in a rate sedimentation, (4.2.21) together with λ_i and ϵ_i, information on the gradient solution can be used to determine when and where the centrifugal separation can be maximum.

If information on the gradient solution and on experimental runs for new bioparticle is available, one can use the analytical results. Equations 4.2.19 through 4.2.22, obtained to analyze the separated particle's characteristics such as diameter, density, and sedimentation coefficient. Thus, the analysis will serve as a method of analytical zonal centrifugation.

SYMBOLS

A	Characteristic constant $\{ = [(1/Q) - 1](N/18) \}$ as defined in (4.2.16)
A	Constant
A_i	Reduced paraboloid interfacial area of type i [$= S_i/[\pi(R_w^2 - R_c^2)]$] as defined in (4.5.4i)
B_i	Constant ($i = 1, 2, \ldots, n$)
B	Reduced parameter [$= (\beta_i - \alpha)$] as defined in (4.2.41h)
c_i	Concentration of component i in molarity
C	Reduced parameter [$= (1 - \alpha)/\zeta_c^{2/3}$] as defined in (4.2.41i)

$c_1{}', c_2{}', c_3{}', c_4{}'$	Constants as defined in Ref. 14
d	Height of liquid loading level
D	Dispersion coefficient
D_i	Constant ($i = 1, 2, \ldots, n$)
D_{ij}	Multicomponent diffusivity
D_p	Particle diameter
f	Fractional cleanout of continuous flow of centrifuge defined as [influx $-$ outflux]/influx
f_e	Fractional elution
F	Reduced centrifugal force ($[= \omega^2 t \rho_0 D_p{}^2 / \mu_0]$)
h	Peak of concentration distribution diagram
H^*	Heat of activation
H	Height of centrifugal rotor
k	Proportionality constant
k_i	Parameter defined as $S_i{}^0(H_2O, 20°C)\mu(H_2O, 20°C)$; $i = 1, 2$
l	Lower limit of integration
M	Parameter ($= C/B$) as defined in (4.2.41f)
M_i	Molecular weight of ith species
n	Index for moment
N	Integers
$\mathbf{N}$	Reduced parameter [$= (\rho_0 \omega R D_p / \mu_0)^2$] as defined (4.2.14c)
P	Reduced particle diameter [$= (D_p/R)^2$] as defined in (4.2.14a)
Q	Volumetric flow rate
Q	Density ratio ($= \rho_0/\rho_p$) as defined in (4.2.14b)
r	Radial coordinate distance
r_i	Radial position of component i
r_0	Radius of isodense paraboloid at bottom wall for rotor with core
r_H	Radius of isodense paraboloid at top wall for rotor with core
R	Radius of rotor
R_c	Outside radius of core
R_w	Inside radius of rotor
S_i	Svedberg's sedimentation coefficient of component i as defined in (4.2.2)

S_i	Paraboloid interfacial area of type i
Se	Dimensionless observed sedimentation coefficient $(=S\omega^2 R^2 \rho_0/\mu_0)$ as defined in (4.2.22)
Sc	Schmidt number $(= \mu/\rho D)$
$S(x)$	Step function as defined in (4.2.54)
t	Time
t_f	Final time
T	Temperature
T_{rr}	Reduced shear stress $(= \tau_{rr}/2S\omega^2\mu_0)$ as defined in (4.2.23)
Ta	Taylor number $(= \omega RD_P\rho_0/\mu_0)$
u	Upper limit of integration
$\bar{v}_i$	Partial specific volume of component i
V	Volume
$\mathbf{V}_r^*$	Reduced sedimentation velocity as defined in (4.2.21)
$\mathbf{v}_{ri}$	Radial velocity component of ith species
$\mathbf{v}_\theta$	Tangential velocity component
w	Bandwidth
x	Variable in step function as defined in (4.2.54)
y	Mole fraction
Y	Weight fraction
z	Vertical axis coordinate
Z_c	Height of isodense paraboloid at core
Z_w	Height of isodense paraboloid at rotor wall for rotor with core
z_0	Minimum of the height of isodense paraboloid

Greek Symbols

α	Wave number
α	Reduced height of loading level for rotor $(= d/H)$ as defined in (4.5.4a)
β	Ratio of core radius to rotor wall $(= R_c/R_w)$ as defined in (4.5.4b)
β	Parameter $(= \rho_i/\rho_0)$ as defined in (4.2.41d)
γ	Ratio of rotor height to radius $(= H/R_w)$ as defined in (4.5.4c)
γ	Activity coefficient
γ	Growth rate constant

σ	The second moment
δ	Parameter $[= \mu(H_2O, 20°C)/\mu_0]$ as defined in (4.2.41e)
ϵ_i, ϵ_i'	Characteristic coefficients for density gradient variation on density as defined in (4.2.10) and (4.2.14g)
ζ	Reduced radius variable $(= r/R)$ as defined in (4.2.14d)
ζ_c	Reduced height of isodensed paraboloid at core $(= Z_c/R_w)$ as defined in (4.5.4f)
ζ_0	Reduced minimum height of isodensed paraboloid $(= Z_0/R_w)$ as defined in (4.5.4g)
λ	Characteristic constant as defined in (4.2.5c)
λ_i, λ_i'	Characteristic coefficients for density-gradient variation on viscosity as defined in (4.2.10) and (4.2.14f)
μ	Dynamic viscosity
μ_0	Dynamic viscosity of gradient solution at the light end
ν_0	Kinematic viscosity at minimum end
ξ_0	Reduced inital position
$\rho(r)$	Density of gradient solution
ρ_i	Density of ith component
ρ_0	Density of gradient solution at the light end
$\mu(r)$	Viscosity of gradient solution
ρ_H	Reduced radius of isodense paraboloid at the top wall $(= r_H/R_w)$ as defined in (4.5.4d)
ρ_0	Reduced radius of isodense paraboloid at the bottom wall $(= r_0/R_w)$ as defined in (4.5.4e)
τ	Reduced time $(= \mu_0 t/\rho_0 R^2)$ as defined in (4.2.14e)
τ_{rr}	Shear stress as expressed in (4.2.23a)
ψ	Stream function
ω	Angular velocity
Ω	Reduced angular velocity $(= \omega^2 R_w/g)$ as defined in (4.5.4h)

Subscripts

f	Final state
i	Initial position
G	Density gradient
T	Property evaluated at temperature T
0	Property evaluated at minimum position
H	Quantity evaluated at top wall for rotor with core

w	Quantity evaluated at wall for rotor with core
P	Paraboloid
p	Particle
r	Radial direction
θ	Tangential direction
0, 1, 2, 3, . . .	Indices for the degree of approximation
< >	Average quantity

Superscripts

$'$	Perturbed quantity
$-$	Unperturbed quantity
$\char"005E$	Complex quantity

References

1. N. G. Anderson, *Q. Rev. Biophys.,* **1** (3), 217 (1968).
2. N. G. Anderson, Ed., *The Development of Zonal Centrifuges and Ancillary System for Tissue Fractionation and Analysis. Natl. Cancer Inst. Monogr. 21,* U.S. Government Printing Office, Washington, D.C., 1966.
3. H. P. Barringer, "The Design of Zonal Centrifuges," in *The Development of Zonal Centrifuges. Natl. Cancer Inst. Monogr. 21,* N. G. Anderson, Ed., U. S. Government Printing Office, Washington, D.C., 1966, pp. 77–111.
4. O. M. Griffith, *Technique of Preparative Zonal and Continuous Flow Ultracentrifugation,* Beckman Instruments, Inc., April 1975.
5. H. Svensson, L. Hagdohl, and K. D. Lerner, *Sci. Tools,* **4** (1), 1 (1957).
6. M. K. Brakke, *Arch. Biochem. Biophys.,* **107**, 388 (1964).
7. A. S. Berman, "Theory of Centrifugation", in *The Development of Zonal Centrifuges. Natl. Cancer Inst. Monogr. 21,* N. G. Anderson, Ed., U.S. Government Printing Office, Washington, D.C., 1966, pp. 41–73.
8. S. P. Spragg and C. T. Rnakin, Jr., *Biochem. Biophys. Acta,* **141**, 164 (1967).
9. N. G. Anderson, H. P. Barringer, J. W. Amburgey, Jr., G. B. Cline, C. E. Nunley, and A. S. Berman, "Continuous-Flow Centrifugation Combined with Isopycnic Banding: Rotors B-VIII and B-IX," in *The Development of Zonal Centrifuges, Natl. Cancer Inst. Monogr. 21,* N. G. Anderson, Ed., U.S. Government Printing Office, Washington, D.C., 1966, pp. 199–216.
10. T. Svedberg, *Phys. Chem.,* **127**, 51 (1927).
11. T. Svedberg and K. O. Pederson, *The Ultracentrifuge,* Clarendon Press, Oxford, 1940.
12. J. P. Breillatt, W. K. Sartory, and J. N. Brantley, Separation Method Design by Simulation Techniques: CsCl Gradients in the K-III Rotor, ORNL-TM 3185, Oak Ridge National Laboratory, May 1971.

13. H. W. Hsu and N. G. Anderson, *Biophys. J.,* **9**, 173 (1969).
14. H. W. Hsu, *Separat. Sci.,* **6** (5), 699 (1971).
15. J. F. Thomson and E. T. Mikuta, *Arch. Biochem. Biophys.,* **51**, 487 (1954).
16. J. Rosenblum and E. C. Cox, *Biopolymers,* **4**, 747 (1966).
17. C. deDuve, J. Berthet, and H. Beaufay, *Progr. Biophys. Biochem.,* **9**, 325 (1959).
18. T. G. Pretlow, C. W. Boone, R. I. Schrager, and G. H. Weiss, *Anal. Biochem.,* **29**, 230 (1969).
19. N. G. Anderson, *Exp. Cell Res.,* **9**, 445 (1955).
20. M. K. Brakke, *Arch. Biochem. Biophys.,* **55**, 175 (1955).
21. P. V. Nason, N. V. Shumaker, H. B. Halsall, and J. Schweder, *Biopolymers,* **1**, 241 (1969).
22. V. N. Schumaker, "Zonal Centrifugation," in *Advances in Biological and Medical Physics,* C. A. Tobias and J. H. Lawrence, Eds., Academic, New York, 1967, pp. 245-339.
23. W. K. Sartory, *Biopolymers,* **7**, 251 (1969).
24. M. K. Brakker, *Adv. Virus Res.,* **7**, 193 (1960).
25. J. A. T. P. Meuwissen and K. P. N. Heirwegh, *Biochem. Biophys. Res. Comm.,* **41**, 675 (1970).
26. H. W. Hsu, *Math. Biosci.,* **13**, 361 (1972).
27. H. Fujita and L. J. Gosting, *J. Phys. Chem.,* **64**, 1256 (1960).
28. H. W. Hsu, *Math Biosci.,* **23**, 179 (1975).
29. L. S. Pontryagin, V. G. Boltyanskii, R. V. Gamkrelidge, and E. F. Mishchenko, *The Mathematical Theory of Optimal Processes,* translated by K. N. Trirogoff, L. W. Neustadt, Ed., Interscience, New York, 1962.
30. R. K. Genung and H. W. Hsu, *Separ. Sci.,* **7** (3), 249 (1972).
31. W. K. Sartory, *Separat. Sci.,* **5** (2), 137 (1970).
32. E. J. Barber, P. F. Shorten, L. L. McCauley, and J. F. Simmons, *Oak Ridge National Laboratory Report 3978* (Special), July 1, 1965, to June 30, 1966, p. 81.
33. E. M. Boyd, D. R. Thomas, and B. F. Horton, *Clin. Chim. Acta,* **16**, 333 (1967).
34. J. D. Biggers and R. A. McFeely, *Nature,* **199**, 718 (1963).
35. C. V. Lusena and F. Depocas, *Can. J. Biochem.,* **44**, 497 (1966).
36. B. S. McEwen, V. G. Allfrey, and A. E. Mirsky, *J. Biol. Chem.,* **238**, 758 (1963).
37. M. K. Johnson, *Biochem. J.,* **77**, 610 (1960).
38. D. H. Brown and C. A. Price, *Plant Physiol.,* **47**, 217 (1971).
39. S. I. Honda, T. Hongladarom, and G. G. Laties, *J. Exp. Botany,* **17**, 460 (1966).
40. A. P. Mathias and C. A. Wynther, *FEBS Lett.,* **33**, 18 (1973).
41. M. Aas, *Proc. 9th Intl. Congress of Biochem.,* Stockholm, 1973 (Abstract).
42. B. M. Mullock and R. H. Hilton, *Biochem. Soc. Trans.,* **1**, 27 (1973).
43. D. M. McDonald and H. W. Hsu, *Separ. Sci.,* **7** (5), 491 (1972).
44. L. R. Bell and H. W. Hsu, *Separ. Sci.,* **9** (5), 401 (1974).
45. V. S. Shah and H. W. Hsu, *Separ. Sci.,* **10** (6), 787 (1975).
46. W. H. Lu and H. W. Hsu, *Separ. Sci.,* **15** (6), 1393 (1980).
47. M. K. Brakke, *Science,* **148**, 387 (1965).
48. L. J. Gosting, "Measurement and Interpretation of Diffusion Coefficients of Proteins," in *Advances Protein Chemistry,* Vol. II, M. L. Anson, K. Bailey, and John T. Edsall, Ed., Academic, New York, 1956, p. 430.

49. E. J. Barber, "Calculation of Density and Viscosity of Sucrose Solutions as a Function of Concentration and Temperature," in *The Development of Zonal Centrifuges. Natl. Cancer Inst. Monogr. 21,* (N. G. Anderson, Ed., U.S. Government Printing Office, Washington, D.C., (1966), pp. 219-239.

50. C. deDuve, J. Berthet, and H. Beaufay, "Gradient Centrifugation of Cell Particles. Theory and Application," in *Progress in Biophysics and Biophysical Chemistry,* Vol. 9, J. A. V. Butler and B. Katz, Eds., Pergamon, London, 1959, Chap. 7.

51. J. F. Swindell, C. F. Snyder, R. C. Hardy, and P. E. Golden, *Viscosities of Sucrose Solutions at Various Temperatures; Tables of Recalculated Values. Supplement to National Bureau of Standards Circular 440,* U.S. Government Printing Office, Washington, D.C., 1947.

52. C. F. Snyder and L. D. Hammond, *Weights per United States Gallon and Weights Per Cubic Foot of Sugar Solution, National Bureau of Standards Circular C-457,* U. S. Government Printing Office, Washington, D.C., 1947.

53. S. Glasstone, K. J. Laidler, and H. Eyring, *The Theory of Rate Processes,* McGraw-Hill, New York, 1951, Chap. 9.

54. A. K. Doolittle, *The Technology of Solvents and Plasticizers,* Wiley, New York, 1954, Chap. 13.

55. A. A. Miller, *J. Chem. Phys.,* **38,** 1568 (1963).

56. A. A. Miller, *J. Phys. Chem.,* **67,** 1031 (1963).

57. H. W. Hsu, J. N. Brantley, and J. P. Breillatt, *Separ. Sci.,* **14** (1), 69 (1979).

58. A. Einstein, *Ann. Phys.,* **17,** 439 (1905).

59. H. W. Hsu, *Separ. Sci.,* **8** (3), 537 (1973).

60. N. G. Anderson, *Bull. Am. Phys. Soc.,* **1** (Ser. II), 267 (1956).

61. C. DeDuve, J. Berthet, and H. Beaufay, *Progr. Biophys.,* **9,** 325 (1959).

62. H. D. Pham and H. W. Hsu, *Ind. Eng. Chem., Process Des. Develop.,* **11,** 556 (1972).

63. J. H. Bauer and E. G. Pickels, *Exp. Med.,* **64,** 503 (1936).

64. W. M. Stanley and R. W. G. Wyckoff, *Science,* **85,** 181 (1937).

65. J. Vinograd and J. E. Hearst, *Fortschr. Chem. Org. Naturstoffe,* **20,** 372 (1962).

66. N. G. Anderson, W. W. Harris, A. A. Barber, C. T. Rankin, and E. L. Candler, in *The Development of Zonal Centrifuges. Natl. Cancer Inst. Monogr. 21,* N. G. Anderson, Ed., U.S. Government Printing Office, Washington, D.C., 1966, p. 253.

67. N. G. Anderson and G. B. Cline, "New Centrifugal Methods for Virus Isolation," in *Methods in Virology,* Vol. 2, H. Koprowski and K. Maramorosch, Eds., Academic, New York, 1967, p. 137.

68. C. B. Reimer, R. S. Baker, R. M. vanFrank, T. E. Newln, G. B. Cline, and N. G. Anderson, *Am. Soc. Microbiol.,* **1** (6), 1207 (1967).

69. W. D. Fisher, G. B. Cline, and N. G. Anderson, *Physiologist,* **6,** 179 (1963).

70. W. D. Fisher, G. B. Cline, and N. G. Anderson, *Anal. Biochem.,* **9,** 477 (1964).

Chapter V

ANALYTICAL LIQUID CENTRIFUGES

In laboratories, the analytical ultracentrifuge has made a major contribution by helping investigators characterize biological materials in terms of molecular properties such as molecular weight, sedimentation coefficient, buoyant densities, diffusion coefficients, and so on. Data from these measurements are also employed in determining molecular size and shape, frictional coefficients, virial coefficients, compressibility, net solvation, and partial specific volumes of molecules. The same techniques are used equally well with synthetic polymers and inorganic molecules. An analytical ultracentrifugation usually is conducted on a small scale using 1 ml or less solution, and the progress of sedimentation is followed by means of an optical system. The objective is to make accurate measurements of sedimentation rate, banding density, molecular weight, and diffusivity, or of anomalies associated with attempts to measure these. The individual fractions are not normally recovered at the end of the run.

Analytical ultracentrifugation is generally carried out on pure, or nearly pure, samples. It is usually advisable to have pure samples for molecular determinations because the presence of impurities may result in macromolecular interactions, apart from complicating the mathematics involved in the calculation. The analytical ultracentrifuge is a very sophisticated piece of equipment. It has an accurate temperature control and advanced optical systems which make it a very precise analytical instrument.

Another type of centrifugal analysis is exemplified by the mechanized analytical systems developed by Anderson, Scott, and co-workers at Oak Ridge National Laboratory [1–12] in the form of a centrifugal photometer. This versatile mechanized analytical system is known as the General Medical Sciences-Atomic Energy Commission or GeMSAEC Fast Analyzer, or the Centrifugal Fast Analyzer (CFA). Since the original development of this analyzer, there has been a sequence of instrumental advances. Following the first, relatively cumbersome systems, a miniaturized CFA was developed and fabricated [13, 14]. This instrument was compact, weighing 11.4 kg and requiring only 0.0283 m^3 of space. However, to utilize its full capabilities, the miniaturized analytical module required the data acquisition analysis

219

capabilities of a separate, relatively large minicomputer system. Although the analytical module itself was light and compact, the entire CFA system was immobile because of the attendant computer. A recent development was a portable CFA which incorporated a microprocessor within the instrumental package enabling the complete instrument (analytical module, data acquisition, and analysis module) to be contained within a compact (47 × 28 × 30.5 cm) package weighing approximately 25 kg. Further improvements have added even more versatility, and optional battery-powered operation enables it to be used in small mobile clinics, at a patient's bedside, or in remote areas. Its capability for monitoring chemical reactions within 100 ms after the mixing of sample reagents is of special interest in the study of fast kinetics.

Recent rapid technical progress and development now permit performing analytical measurements with the density-gradient centrifugation used in preparatory work. A new method of obtaining quantitative information on physical adsorption and chemical interaction using the density-gradient isopycnic banding centrifugation method has been demonstrated [15, 16]. Subtractive measurements of density changes in complexes formed by the interaction between antigen and antibody immobilized on polystyrene latex beads has allowed calculation of the mass immobilized and of the interacting mass of antigen under the specified experimental conditions. From those data, various thermodynamic quantities characteristic of the system, such as adsorption isotherms, interaction equilibrium constants, the free energy, enthapy, and entropy for the interaction can be obtained.

In this chapter, basic analytical techniques used in the analytical ultracentrifuge are discussed together with fundamental components such as cells, rotors and optical systems. A method of using carriers to characterize macromolecules in preparative centrifuges by the density-gradient equilibrium method is described with examples, and the recent development of the centrifugal fast analyzer is described.

1   ANALYTICAL ULTRACENTRIFUGE

By the analytical ultracentrifuge, the "Svedberg ultracentrifuge," is meant a centrifuge with an optical system for obtaining quantitative data on the sedimentation of solute molecules. Since Professor Svedberg built the first ultracentrifuge in 1923 and used the sedimentation data to determine molecular weights, the analytical ultracentrifuge now has come into widespread use. It is the fundamental instrument for determining molecular properties of natural polymers—proteins, polysaccharides, viruses, nucleic acids, and cell particulates—as well as synthetic polymers and inorganic molecules.

A major contribution to the rapid growth of ultracentrifugation was the development by Spinco Division of Beckman Instruments, Inc., in 1947 of a commercial instrument that contained all the equipment and controls necessary for precise ultracentrifugal measurements. Through developments by ultracentrifuge owners and the Spinco research and engineering departments, the instrument has kept pace with modern advances in technology and engineering. Measuring and Scientific Equipment Ltd. (MSE) in England and Hitachi Manufacturing Co. Ltd. in Japan also produce analytical ultracentrifuges.

The analytical ultracentrifuge subjects a fluid sample to a carefully controlled centrifugal force field and records the behavior of solute molecules or suspended particles by photographic or electronic means. From such measurements, various properties of the molecules or particles can be calculated. The Beckman Model E Analytical Centrifuge can drive rotors at speeds of 68,000 rpm and can generate centrifugal forces to 372,200 g. The instrument features refrigeration and vacuum systems, automatic temperature control, electronic speed control, and schlieren and interference optical systems. Major accessories include an ultraviolet absorption optical system, a monochromator, a high-intensity light source, a high temperature system, and a photoelectric scanning system [17].

Rotors and Cells

The basic design of the analytical rotor has not changed appreciably since the time of Svedberg and co-workers [18]. Figure 5.1a shows a typical commercial Spinco two-place analytical rotor. Four- and six-place rotors shown in Figs. 5.1b and 5.1c are now available for multisample analysis. Most analytical rotors are made of aluminum, but several of the aluminum rotors have titanium counterparts. The basic advantage of titanium is that it can be taken to higher speeds to generate stronger centrifugal fields. However, titanium rotors are heavier than comparable aluminum rotors, which means that they require more time for acceleration and deceleration.

Most rotors accept cell assemblies holding centerpieces up to 12 mm in thickness; a few are extra-thick, with corresponding deeper cell holes for the longer assemblies that hold 18- and 30-mm centerpieces. Rotors with deeper cell holes are indicated when low solute concentrations are to be used, since the thicker centerpiece provides a longer light path through the samples. When a full complement of cells is run in two- and four-hole rotors, the outer reference image is provided by a small hole in the outer edge of the rotor. When a counterbalance is used instead, it provides both outer and inner reference images, by having two slits in it at a fixed radial distance to allow light to pass through to give reference lines.

The heart of the analytical cell is the centerpiece, which contains the sample and is sealed, top and bottom, by windows fitted into window holders. The

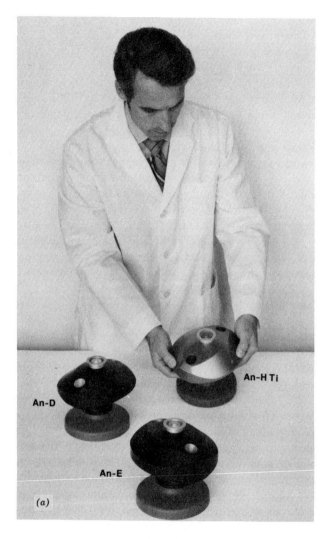

Fig. 5.1   Analytical centrifuge rotors. (*a*) Two-place analytical rotors. (*b*) Four-place analytical rotors. (*c*) Six-place analytical rotors. Illustration courtesy of Beckman Instrument Inc., Spinco Division.

centerpiece, windows, and holders are positioned in a hollow tube, the cell housing, which fits into the rotor hole. A typical Beckman analytical cell and various centerpieces are shown in Fig. 5.2. All centerpieces can be classified as either single sector or double sector. The usual procedure with double-sector centerpieces is to run sample and solvent in one sector and solvent only in the other. The choice of optical system or use of the photoelectric scanner will

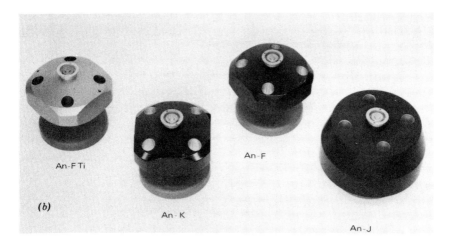

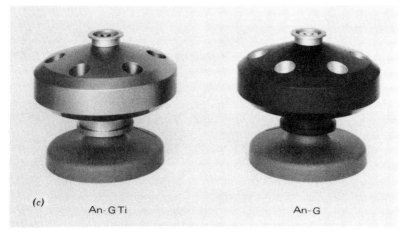

Fig. 5.1 Continued

determine which basic type of centerpiece should be used. Originally, the sector angle was $4°$, but now $3°$, $2°$, and even $2\frac{1}{2}°$ sector angles are available. A smaller angle means that less solution is required, though more care must be taken in alignment. The cell centerpiece is usually made of either aluminum or epoxy resin, epoxy resin being better for corrosive liquids although it does tend to chip easily. The materials vary in chemical resistance, temperature tolerance, and strength. Centrifugation also affects the strength of the centerpiece. Windows may be plane or wedge-shaped to deflect the image on the photographic plate; the windows are usually made of either quartz or sapphire. For interference optics, special window holders are required to allow formation of fringes.

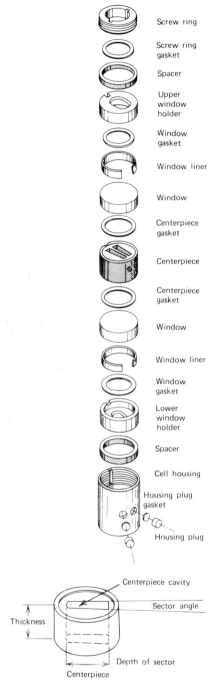

Screw ring

Screw ring
gasket

Spacer

Upper
window
holder

Window
gasket

Window liner

Window

Centerpiece
gasket

Centerpiece

Centerpiece
gasket

Window

Window liner

Window
gasket

Lower
window
holder

Spacer

Cell housing

Housing plug
gasket

Housing plug

Centerpiece cavity

Sector angle

Thickness

Depth of sector

Centerpiece

Fig. 5.2 Typical Beckman analytical cell and various centerpieces. (*a*) Centerpiece and its related parts.

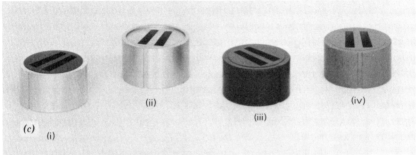

Fig. 5.2 (b) Single-sector and double-sector centerpieces. (c) (i) Charcoal-filled Epon, (ii) aluminum-filled Epon, (iii) Kel-F, (iv) aluminum. (d) 1°, 4°, and 2½° sectors. Illustration courtesy of Beckman Instrument Inc., Spinco Division.

When two or more cells are run in a rotor, it is possible to compare results obtained under virtually identical conditions, thus minimizing the effects of minor differences in temperature and rotor speed between successive runs. The variety of cells, cell components, counterbalances, and rotors that are available gives rise to some confusion in the selection of the best combination for a particular experiment. A component selection guide recommended by Chervenka of Beckman Co. [18] is presented in the following. Centerpiece selection is based upon maximum allowable speed, chemical resistance, configuration, and thickness required. Double-sector cells are required for use with interference optics and are recommended for routine use with schlieren optics because of the baseline obtained with these cells. Selection of centerpiece material and thickness depends upon the properties of the solution under study. The aluminum alloy centerpieces, which were standard for many years, have been replaced largely by plastic ones in order to avoid the possible contamination of samples with aluminum ion. In certain instances—high-temperature runs and the use of special solvents, for example—aluminum centerpieces may still be required. Aluminum-filled Epon* centerpieces are the best choice, in general, for aqueous solutions. If absolutely no contact with aluminum is required, then charcoal-filled Epon or Kel-F† pieces should be used. The first choice of centerpiece thickness (equal to the effective light path through the assembled cell) is the standard 12-mm thickness. Since the areas under schlieren peaks, shifts of interference fringes obtained refractometric optics, and the absorbances obtained with absorption optics are all proportional to the product of light path and solution concentration, usable concentration ranges can be extended considerably by choice of centerpiece thickness. Thus, the area under a schlieren peak obtained with a particular solution in a centerpiece 30 mm thick will be 2.5 times that obtained with the same solution in a 12-mm centerpiece.

Quartz windows are satisfactory for most applications, with two principal exceptions. Sapphire windows are recommended for all experiments using the photoelectric scanner and are absolutely required for interference work at 40,000 rpm or higher. Routine use of sapphire windows has an advantage in that interference fringes can be photographed with a cell containing schlieren or scanner window holders.

Wedge windows are of great convenience in multisample runs. In any optical system, however, determinations requiring highly accurate results should be run only in cells with plane windows and centerpieces.

The choice of counterbalance does not present any particular problem. Interference counterbalances can be used for either interference or schlieren op-

*Registered trademark of Shell Chemical Corp.
†Registered trademark of 3-M Company.

tics, and so should be used any time there is any possibility that fringes will be photographed during the run. Routine use of interference cells and counterbalances for schlieren runs leaves the operator the option of photographing fringes, if desired, with the minor disadvantage that longer exposure times are required. A special counterbalance is provided for use with the photoelectric scanner.

The choice of a rotor depends upon the maximum speed to be used, the number and size of cells to be run, and the temperature of operation.

Optical Systems

Analytical ultracentrifugation requires that the behavior of a sample be observed during a centrifugal run. This is made possible by using cells fitted with quartz or sapphire windows contained in the rotor in such a way that light may pass through the solution in the cell while it is in motion. Quartz windows are used for ultraviolet absorption and schlieren methods but are rather more susceptible to distortion at high speeds (where the pressure in the cell may be of the order of 300 atm) than sapphire, the use of which is usually necessary to obtain good interference patterns.

The distance from the center of the rotor to the inside edge of the reference hole is 57 mm. An interference counterbalance has two pairs of reference holes—two holes at each end of the sector opening. It is essential that this inner sector reference hole be aligned with the outer notch to make sure that the walls are pointing radially, otherwise a wall effect will occur. Figure 5.3 shows a set of typical measurements for the radial alignment of the walls of a cell sector.

In general, three types of optical system are now in common use. One is based on absorption, and two—the schlieren and the interference methods—are based on refraction of light.

Ultraviolet Absorption

In principle, this system is simple. Its basic components are a source of monochromatic radiation, obtained from a monochromator or a mercury vapor lamp (fitted with appropriate filters to isolate particular spectral lines), the rotating centrifuge cell, a system of lenses, and a photographic film or plate. The use of ultraviolet radiation superseded that of visible light as interest developed in the study of biological macromolecules, which, in general, absorb strongly in the ultraviolet region.

The concentration of the macrospecies at various levels in the cell will determine the amount of radiation absorbed, and hence the lack of image on the photographic record. Thus, if a calibration curve in terms of photographic density and concentration is prepared, the distribution of the macrospecies in the cell at any time during the centrifugal run may be determined by means of

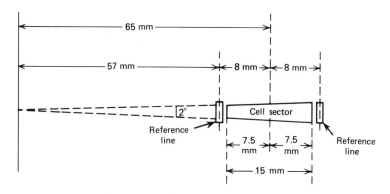

Fig. 5.3 Radial distance of reference holes in an analytical rotor.

a densitometer trace of the photograph. Care must be taken with exposure and development times in order to obtain reproducible and meaningful results.

One disadvantage of this system is that the position of the boundary or the distribution of the macrospecies at any particular time cannot be seen until after the photographs have been developed, unless the concentration is sufficiently high for a schlieren trace to be observed through the eyepiece, whereupon the optical system may be adjusted to take an ultraviolet absorption photograph when required. This problem has been overcome by the development of photoelectric scanner systems in which the film has been replaced by a photomultiplier tube. In front of this tube moves a mask with a slit in it so that the radiation transmitted by successive portions of the solution in the centrifuge is registered and translated into a pen recorder trace of absorbance against distance. A typical trace from an ultraviolet scanner is shown Fig. 5.4*a*. A densitometer trace of a photographic record of such a sample would have a very similar appearance.

Rayleigh Interference

Rayleigh interference patterns are formed as a result of the change in refractive index associated with the concentration gradient set up in the centrifuge cell. The optical system is arranged in such a way that monochromatic radiation from the source is rendered parallel by a collimating lens and passes through two fine slits, whereupon constructive and destructive interference occurs as a result of the two rays traversing different path lengths when brought to a focus on the viewing screen or film. This produces alternate regions of light and dark (where constructive and destructive interference respectively occur) which are known as fringes. The distance between the fringes, known as the spacing, is proportional to the wavelength of the radiation used, and the number of fringes formed depends upon the distance between the slits. Thus, a series of parallel lines is registered on the viewing screen or film.

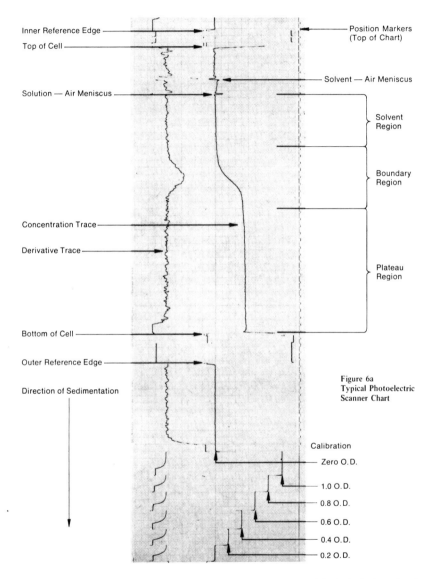

Inner Reference Edge

Top of Cell

Solution — Air Meniscus

Concentration Trace

Derivative Trace

Bottom of Cell

Outer Reference Edge

Direction of Sedimentation

Position Markers
(Top of Chart)

Solvent — Air Meniscus

Solvent
Region

Boundary
Region

Plateau
Region

Figure 6a
Typical Photoelectric
Scanner Chart

Calibration

Zero O.D.

1.0 O.D.

0.8 O.D.

0.6 O.D.

0.4 O.D.

0.2 O.D.

Fig. 5.4 Various optical patterns in analytical ultracentrifugation. (*a*) Typical photoelectric scanner chart (*above*). (*b*) Typical interference fringe pattern. (*c*) Typical Schlieren pattern. Illustration courtesy of Beckman Instruments Inc., Spinco Divisions.

229

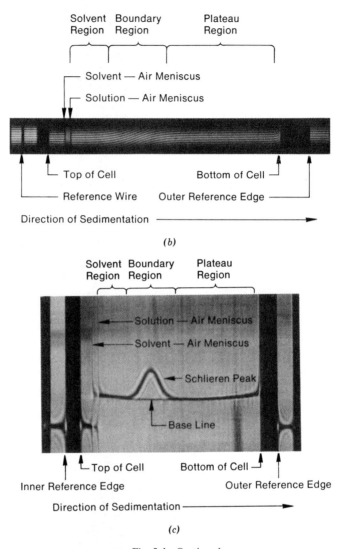

Solvent   Boundary   Plateau
Region   Region   Region

Solvent — Air Meniscus

Solution — Air Meniscus

Top of Cell   Bottom of Cell

Reference Wire   Outer Reference Edge

Direction of Sedimentation

(b)

Solvent Boundary   Plateau
Region   Region   Region

Solution — Air Meniscus

Solvent — Air Meniscus

Schlieren Peak

Base Line

Top of Cell   Bottom of Cell

Inner Reference Edge   Outer Reference Edge

Direction of Sedimentation

(c)

Fig. 5.4 Continued

If a double-sector cell containing solvent only in each sector is now placed in the system and the spacing between the slits is such that radiation passing through each will pass respectively through each sector of the cell, there is no change in fringe pattern since each beam is subjected to the same obstacle. If the solvent in one sector is replaced by a solution (of different refractive index), the recorded image is still apparently unchanged; but if one of the fringes could be marked, it would be seen that they are all displaced in a direction perpendicular to their length by a distance proportional to the difference in the refractive index between the solvent and the solution, which is, in turn, proportional to the concentration of the latter. Thus, the concentration gradient formed by the sedimentation of the macrospecies in one of the sectors results in a similar gradient in the refractive index which is mirrored by the fringe pattern (see Fig. 5.4b).

The Schlieren System

The schlieren optical system also depends upon differences in refractive index, but in this case the trace obtained represents the variation of the refractive index gradient (dn/dr) with radial distance (r). Thus, the point of maximum gradient indicates the center of the boundary. A typical sedimenting boundary and the relationship between the schlieren peak and the concentration of the macrospecies in the cell are shown in Fig. 5.4c.

The selection of the optical system is governed by availability; but if all types are on hand, the ultraviolet absorption method in conjunction with a scanner would in general be the best choice. This system will yield results rapidly, and the sample may be examined at any desired time.

The Rayleigh interference technique is rather less sensitive, and the schlieren method requires a considerably higher concentration of the macrospecies, although the latter does have an advantage in that boundaries are easily observed, and two boundaries very close together are perhaps more readily recognized than from the interference pattern.

Some of the considerations which aid in the selection of an optical system given by Chervenka [18] are presented in Table 5.1.

2 BASIC PRINCIPLES OF ULTRACENTRIFUGAL ANALYSIS

The basic principle of ultracentrifugal analysis is exactly the same as that used in the liquid preparatory centrifuges. The analytical ultracentrifuge is used to produce high centrifugal forces in order to measure the movement or redistribution of sedimenting particles. Centrifugal investigations of macromolecules may be carried out under conditions of thermodynamic equilibrium or conditions of transport, which are generally known as the sedimentation-equilibrium method and the sedimentation-velocity method, respectively. As we shall see later, the final equilibrium state will depend on the speed of rotation. At relatively low speed, macromolecules will, at equilibrium, still be distributed over the entire cell, with concentration increasing gradually from top to bottom of the cell. At high speed, however, the equilibrium concentration gradient would become much greater, so that virtually all the macromolecules will eventually become packed into a small

Table 5.1. Comparison of Optical Systems Used in Analytical Ultracentrifuge

	Schlieren	Interference	Absorption	
			Photographic UV	Photoelectric Scanner
Most important applications	Sedimentation velocity Studies of heterogeneity Approach-to-equilibrium	Sedimentation equilibrium Concentration determination	Sedimentation velocity (dilute solutions) Equilibrium banding	Sedimentation equilibrium (dilute solutions) Sedimentation velocity (dilute solutions) Equilibrium banding
Other usual applications	Sedimentation equilibrium Concentration determination	Diffusion studies Sedimentation velocity		Small molecule binding Differential sedimentation
Usual concentration ranges (12 mm cells)	1.0 to 10 mg/ml	0.5 to 5 mg/ml	Proteins—0.1 to 1.0 mg/ml Nucleic acids—0.01 to 0.1 mg/ml	Proteins—0.05 to 1.0 mg/ml Nucleic acids—0.01 to 0.1 mg/ml
Special advantages	Direct viewing Direct determination of concentration gradients Heterogeneity visualized Variable sensitivity (phase plate angle)	Direct viewing Direct determination of relative concentrations High accuracy	Discrimination among solutes Applicable to very dilute solutions	Baseline provided Equivalent to direct viewing Discrimination among solutes Applicable to very dilute solutions
Particular disadvantages	Salts may interfere Integration required to obtain concentration Patterns sometimes difficult to read accurately	Differentiation required to obtain concentration gradients Sensitive to cell distortion	No direct viewing Inconvenient Sensitive to oil and dirt on optical components	Sensitive to oil and dirt on optical components

Source: Reproduced from Ref. 18 by the permission of Beckman Instruments, Inc., Spinco Division.

region near the bottom of the cell. The sedimentation-velocity method will be primarily concerned with high-speed operation, and one will be interested in the velocity with which macromolecules move to the bottom of the cell. The approach-to-equilibrium method studies the manner in which the equilibrium state is approached during a sedimentation-velocity run in slow-speed operation of the ultracentrifuge by investigating solute distributions at various times.

A brief introduction to the basic theory of ultracentrifugal analysis is presented here to supplement the discussion. For more comprehensive treatment of this subject, one should refer to the review article by Williams, Van Holde, Baldwin, and Fujita [19] and books by Fujita [20, 21] and Williams [22].

Sedimentation-Equilibrium Method [22, 23]

The working equations for the interpretation of sedimentation-equilibrium experiments are the criteria for equilibrium given in (4.2.5) through (4.2.7). Here again, classical thermodynamics provides the description of the sedimentation equilibrium. The system is assumed to be made up of a continuous sequence of phases having fixed volumes and infinitesimal depth in the direction of the ultracentrifugal field. It is further assumed that the density of the solution and the partial specific volumes of the chemical components are independent of pressure and composition. For organic high polymers, it is assumed that the macromolecular components are made up of units that differ only in molecular chain length, and it can be reasonably assumed that the partial specific volumes are the same for all such components.

The chemical potential of each macromolecular component will be a function of the pressure and of the concentration of all components. Thus, a set of $(\nu - 1)$ equations is required to describe the equilibrium. The set in explicit form from (4.2.7) is

$$M_i[1 - \bar{v}_i(r)\rho(r)]\omega^2 r = \sum_{\kappa=1}^{\nu-1}\left[\frac{\partial \mu_i}{\partial c_\kappa(r)}\right]_{T,P,c_{j\neq\kappa}}\frac{dc_\kappa(r)}{dr}$$

$$(i = 1, 2, 3, \ldots, \nu - 1) \qquad (5.2.1)$$

Making use of the definition of the chemical potential of component i defined as $(\mu_i = \mu_i^0(T,P) + RT \ln \gamma_i^{(c)}c_i)$, one has

$$\frac{M_i\omega^2 r}{RT}[1 - \bar{v}_i(r)\rho(r)] = \frac{1}{c_i(r)}\frac{dc_i(r)}{dr}$$

$$+ \sum_{\kappa=1}^{\nu-1}\left[\frac{\partial \ln \gamma_i^{(c)}}{\partial c_\kappa(r)}\right]_{T,P,c_{j\neq\kappa}}\frac{\partial c_\kappa(r)}{dr} \qquad (5.2.2)$$

The (r) refers to the quantities at cell position r at equilibrium. If the solutions are sufficiently dilute, the solution density ρ may be replaced by the density of the solvent, ρ_0. Then, with constant partial specific volumes for all solute components, (5.2.2) becomes

$$\frac{M_i(1 - \bar{v}_i\rho_0)\omega^2 r c_i}{RT} = \frac{dc_i}{dr} + c_i \sum_{\kappa=1}^{\nu-1} \left(\frac{\partial \ln \gamma_i^{(c)}}{\partial c_\kappa}\right)_{T,P,c_{j \neq \kappa}} \frac{dc_\kappa}{dr} \qquad (5.2.3)$$

In molecular weight determination of many protein solutions, the solutions can be treated as neutral molecules of incompressible binary solutions. Many protein solutions appear to exhibit ideal behavior, and for such systems the activity coefficient equals unity. An integrated form of (5.2.3) for binary ideal solutions yields (the subscript 1 is omitted for binary systems),

$$M = \frac{2RT}{(1 - \bar{v}\rho)\omega^2} \frac{d \ln c}{d(r^2)} \qquad (5.2.4)$$

Equation 5.2.4 was used in early sedimentation equilibrium work for determining molecular weight. In these cases, ln c is plotted against (r^2), and the slope of the resulting straight line gives a measure of molecular weight. Plots which are convex upward provide an index of polydispersity, and those which are concave downward are an index of nonideality. For nonideal solutions, an integrated form of (5.2.3) for binary system is

$$M^{(\text{app})} = M\left[1 + c\frac{\partial \ln \gamma^{(c)}}{\partial c}\right]^{-1} = \frac{RT}{(1 - \bar{v}\rho)\omega^2} \frac{1}{rc} \frac{dc}{dr} \qquad (5.2.5)$$

in which $M^{(\text{app})}$ is an apparent molecular weight which approaches M as c approaches zero. If a suitable experimental arrangement is made to measure c as a function of r, for example, by light absorption, then the quantity $(1/rc)(dc/dr)$ can be measured at various points in the sedimentation cell. Several values of $M^{(\text{app})}$ are thus obtained at a given concentration, and the determination of such values at several concentrations permits extrapolation to obtain M. Archibald [24] suggested examining data on the basis of (5.2.5) by plotting $(1/rc)(dc/dr)$ versus r. This plot provides a precise index of the attainment of equilibrium and also gives a reliable index of nonideality, polydispersity, and molecular weight.

Another critical method for handling data was contributed by Van Holde and Baldwin [25, 26]. It is well known that the determination of c as a function of r over the short distance of 1 cm or so which is available in a sedimentation cell is not easily performed with high precision. In the majority of sedimentation equilibrium measurements, use has been made of the schlieren optical system or similar techniques which measure refractive index gradient dn/dr. This quantity is proportional to the concentration gradient dc/dr, that is, these measurements determine $K(dc/dr)$ versus r, where K is a constant. Equation 5.2.5 may be rearranged as

$$\frac{M^{(app)}(1 - \bar{v}\rho)\omega^2}{RT}[(c - c_a) + c_a] = \frac{1}{r}\frac{dc}{dr} \tag{5.2.6}$$

Although c cannot be evaluated from refractive index gradient data, it is possible to evaluate c relative to its value c_a at some arbitrary reference point, for example, at the meniscus, if desired. $M^{(app)}$ is obtained from the slope of a plot of $K(1/r)(dc/dr)$ versus $K(c - c_a)$. With ideal solutions, this plot provides a straight line, and the slope of the line is used to calculate molecular weight. Actual values for c are not required with this method, since only the slope is necessary for the calculation.

Because most macromolecules of biological interest are hydrated in solution, the effect of solvation in the interpretation of sedimentation equilibrium experiments must be considered [27, 28, 29]. Goldberg [30] pointed out that the thermodynamic description of the system permits a choice of components as either a dry substance or solvated macromolecules. If a fixed amount of solvent is bound to the anhydrous solute to form a complex, the chemical potential of the complex can be written in terms of the chemical potentials of the solute and the solvent in equilibrium with the solute. Differentiation of this equation with respect to pressure gives directly the partial specific volume of the complex in terms of the partial specific volumes of the anhydrous solute and solvent and the number of grams of solvent bound to the solute. And as Goldberg has shown, the sedimentation equilibrium experiments give $M(1 - \bar{v}\rho)$ for either the complex or the anhydrous solute. Since $\bar{v}$ for the complex is rarely measurable, it is preferable to use the terms for the anhydrous material even if the solute molecules are hydrated. It should be noted that $M(1 - \bar{v}\rho)$ for the anhydrous solute is equal to $M(1 - \bar{v}\rho)$ for the complex at infinite dilution.

However, for systems which contain a third component such as sucrose or an electrolyte, selective association of either of the components of the solvent with the macromolecules may lead to serious error [29, 31]. If the solute does not combine with either of the two components of the solvent or interacts with no preference for either of the components, then the solvent can be considered as a one-component solvent. It should be noted that for protein studies made in dilute salt solutions, the salt solution is considered a one-component solvent. However, when the concentration of the added third component is large enough to affect markedly the density of the solution, such an approximation is not permissible. In such cases, the possibilities of preferential interaction between the macromolecules and one of the solvent components must be excluded; otherwise, corrections for such interactions are required.

First, one considers that there is just a single additional low-molecular-weight component (component 2). Then one obtains from (5.2.1)

$$\frac{RT}{c_1} \frac{dc_1}{dr} \left[1 + c_1 \left(\frac{\partial \ln \gamma_1^{(c)}}{\partial c_1} \right)_{T.P.c_2} \right]$$

$$= M_1(1 - \bar{v}_1\rho)\omega^2 r - RT\left(\frac{\partial \mu_1}{\partial c_2}\right)_{T.P.c_1} \frac{dc_2}{dr} \qquad (5.2.7a)$$

$$\frac{RT}{c_2} \frac{dc_2}{dr} \left[1 + c_2 \left(\frac{\partial \ln \gamma_2^{(c)}}{\partial c_2} \right)_{T.P.c_1} \right]$$

$$= M_2(1 - \bar{v}_2\rho)\omega^2 r - RT\left(\frac{\partial \mu_2}{\partial c_1}\right)_{T.P.c_2} \frac{dc_1}{dr} \qquad (5.2.7b)$$

where

$$\left(\frac{\partial \mu_i}{\partial c_\kappa}\right)_{T.P.c_{j\neq\kappa}} = \frac{\partial \ln \gamma_i^{(c)}}{\partial c_\kappa}$$

The difficulty in these equations lies in the last term of the equation for c_2, which does not become zero as c_2 approaches zero. Thus, if $M_1^{(app)}$ is defined by (5.2.5), it will not reduce the M_2 when extrapolated to $c_2 = 0$. The sedimentation equilibrium method will thus be generally adaptable to systems of this kind only if $\partial \ln \gamma_1^{(c)}/\partial c_2 = 0$, so that the last term of (5.2.7a) vanishes, or if $M_2(1 - \bar{v}_2\rho) = 0$, so that dc_2/dr vanishes when c_1 and dc_1/dr become zero. In this respect, osmotic pressure is a simpler tool than sedimentation equilibrium, for a second low-molecular-weight component will always distribute itself equally on both sides of the membrane in an osmometer when c_1 approaches zero.

The presence of several macromolecular components is of greater interest. This situation is only likely to arise when all the macromolecular components are of the same chemical nature (e.g., a heterogeneous polymer mixture) so that the partial specific volume and the refractive index increment become the same for each component. If it is possible to find for such a system a solvent in which all solutes behave ideally (e.g., a theta solvent in the case of a synthetic polymer mixture), the activity coefficient terms vanish, and one may write for each component from (5.2.5)

$$\frac{RT}{c_i} \frac{dc_i}{dr} = M_i(1 - \bar{v}_1\rho)\omega^2 r \qquad (5.2.8)$$

where $\bar{v}_1$ is the partial specific volume common to each solute component. Moreover, because of the equality of the refractive index gradient for each component, the optical systems which measure this quality will now measure $K(dc_1/dr)$, $K\langle c_1 \rangle$, and so forth, where

$$\langle c_1 \rangle = \frac{\int_{r_m}^{r_b} c_1 r \, dr}{\int_{r_m}^{r_b} r \, dr} \tag{5.2.9a}$$

$$c_1 = \sum_i c_i \tag{5.2.9b}$$

where r_m and r_b are the radial distances from the rotational axis to the meniscus and the bottom of the cell, respectively. Using (5.2.9a) and (5.2.9b) and integrating (5.2.8) for each component, one has

$$\frac{2RT(c_{ib} - c_{im})}{\omega^2(1 - \bar{v}_1\rho)} = \langle c_i \rangle M_i(r_b^2 - r_m^2) \tag{5.2.10}$$

where $\langle c_i \rangle$ is the initial concentration and c_{ib} and c_{im} are the equilibrium concentrations at r_b and r_m for each component. Summing over all components and dividing both sides by $\sum_i \langle c_i \rangle$, one has

$$\frac{2RT}{\omega^2(1 - \bar{v}_1\rho)} \frac{\sum_i (c_{ib} - c_{im})}{\sum_i \langle c_i \rangle} = \frac{\sum_i \langle c_i \rangle M_i}{\sum_i \langle c_i \rangle} (r_b^2 - r_m^2) \tag{5.2.11}$$

or, with the definition of the weight-average molecular weight $\langle M_w \rangle$, based on the concentrations $\langle c_i \rangle$ of the original polymer sample defined as

$$\langle M_w \rangle = \frac{\sum_i \langle c_i \rangle M_i}{\sum_i \langle c_i \rangle} \tag{5.2.12}$$

together with (5.2.9b), (5.2.11) may be written as

$$\frac{2RT}{\omega^2(1 - \bar{v}_1\rho)} \frac{(c_{1b} - c_{1m})}{c_1} = \langle M_w \rangle (r_b^2 - r_m^2) \tag{5.2.13}$$

The concentration terms on the left-hand side of (5.2.13) are those which are readily obtained from refractive index gradients.

When (5.2.8) is rearranged and summed over all components, one has

$$\frac{1}{r} \sum_i \frac{dc_i}{dr} = \frac{(1 - \bar{v}_1\rho)\omega^2}{RT} \frac{\sum\limits_i M_i c_i}{\sum\limits_i c_i} \sum_i c_i \tag{5.2.14}$$

There is a formal correspondence with (5.2.6), in that one may plot $K(1/r)$ (dc_1/dr) versus $K(c_1 - c_{1a})$ and obtain a slope equal to a weight-average molecular weight, which is based on the equilibrium concentrations at the point r and is clearly one that will have a different value at each point. Thus, the desired molecular weight of the original sample, $\sum_i \langle c_i \rangle M_i / \sum_i \langle c_i \rangle$, is not directly obtainable.

A relation in terms of $\langle c_i \rangle$ can, however, be obtained from the straight line joining the terminal points of such a plot, that is, from evaluation of

$$\left[\frac{1}{r_b} \sum_i \left(\frac{dc_i}{dr} \right)_{r_b} - \frac{1}{r_m} \sum_i \left(\frac{dc_i}{dr} \right)_{r_m} \right] \Big/ [\sum_i c_{ib} - \sum_i c_{im}] \tag{5.2.15}$$

which is one of the quantities obtained directly from refractive index gradient data. Using (5.2.8) for each dc_i/dr and then summing, one obtains for the numerator of (5.1.15)

$$\left[\frac{1}{r_b} \sum_i \left(\frac{dc_i}{dr} \right)_{r_b} - \frac{1}{r_m} \sum_i \left(\frac{dc_i}{dr} \right)_{r_m} \right]$$
$$= \frac{(1 - \bar{v}\rho)\omega^2}{RT} \sum_i M_i(c_{ib} - c_{im}) \tag{5.2.16}$$

From (5.2.10), one gets each $(c_{ib} - c_{im})$ term as

$$c_{ib} - c_{im} = \frac{1}{2} \frac{(1 - \bar{v}_2\rho)\omega^2}{RT} (r_b^2 - r_m^2)\langle c_i \rangle M_i \tag{5.2.17}$$

Substituting (5.1.17) into (5.1.16), one obtains

$$\left[\frac{1}{r_b} \sum_i \left(\frac{dc_i}{dr} \right)_{r_b} - \frac{1}{r_m} \sum_i \left(\frac{dc_i}{dr} \right)_{r_m} \right]$$
$$= \frac{1}{2} \left[\frac{(1 - \bar{v}\rho)\omega^2}{RT} \right]^2 (r_b^2 - r_m^2) \sum_i M_i^2 \langle c_i \rangle \tag{5.2.18}$$

Similarly, the denominator of (5.2.15) may be obtained as

$$\sum_i (c_{ib} - c_{im}) = \frac{1}{2} \frac{(1 - \bar{v}_1 \rho)\omega^2}{RT} (r_b^2 - r_m^2) \sum_i M_i \langle c_i \rangle \qquad (5.2.19)$$

The experimental quantity of (5.2.16) is thus clearly proportional to $\sum_i M_i^2 \langle c_i \rangle / \sum_i M_i \langle c_i \rangle$, which is equal to the "Z-average" defined by [32]

$$\langle M_z \rangle = \frac{\displaystyle\int_0^\infty M^3 N \, dM}{\displaystyle\int_0^\infty M^2 N \, dM} \qquad (5.2.20)$$

in which N is number of molecules, thus $c_i = N_i M_i / V$.

The fact that $\langle M_w \rangle$ and $\langle M_z \rangle$ can be obtained from the same experiments is of great importance, for the difference between these quantities affords a measure of the extent of heterogeneity of a polymer sample. Wales, Williams, and co-workers [33–36] have shown that it is possible to obtain higher averages $\langle M_{z+1} \rangle$, and so on, from the data, that it is often possible to evaluate $\langle M_n \rangle$ also, and that a distribution function for molecular weight can be computed if the mathematical analysis is carried to the higher-order average.

The theory of sedimentation equilibrium of solutions containing ionizing solutes was early treated by Tiselius [37] and was subsequently developed by Pedersen [38], Lamm [39], Johnson et al. [40, 41] and Williams et al. [19]. These latter investigators concerned themselves particularly with the system that contains, in addition to water, a dilute macromolecular electrolyte and an excess of a supporting low-molecular-weight electrolyte. It is the general practice in the physicochemical studies of proteins and of long-chain electrolytes (polyelectrolytes) to add a simple electrolyte in order to repress the electrostatic effects of macroions which produce appreciable deviations from nonelectrolyte systems. True nonideality results from the electrostatic interaction between the charges on macroions and those of other macroions and small ions in its vicinity.

The present status of the theory is not very satisfactory. This is mainly due to the fact that all the previous investigators have neglected taking into account the thermodynamic nonideality of the system. It is very unlikely that, in electrolyte solutions where long-range electrostatic forces act between solute ions, the mean activity coefficient of each electrolyte component is independent of concentration. The solution of a macromolecular electrolyte in the presence of a supporting electrolyte forms a three-component system.

Therefore, the complication that arises when the thermodynamic nonideality is taken into account is primarily of the type we have encountered in systems of three nonelectrolyte components. This is the solvation effect. No theory has yet been formulated to show how the solvation modifies the existing equations for the molecular weight determination of macromolecular electrolytes. Besides the solvation problem, the existing theories contain various difficulties which mainly arise from the fact that the degree of ionization of the macroion appears as another unknown parameter in addition to its molecular weight [20].

The theoretical treatment of electrolyte solutions given by Lamm [39] in 1944 is presented in the following [19]: A ternary solution which contains a polyelectrolyte PX_z and a supporting low-molecular-weight electrolyte BX in pure water is considered. It is assumed that these electrolytes ionize completely according to the schemes

$$PX_z \rightarrow P^{z+} + zX^- \qquad (5.2.21a)$$

$$BX \rightarrow B^+ + X^- \qquad (5.2.21b)$$

where z is the number of ionizable sites on one polyelectrolyte molecule and B^+ and X^- represent, respectively, a univalent cation and a univalent anion. For simplicity, we assume that the solution is incompressible and that the mean activity coefficient on the molality scale of each electrolyte component is equal to unity. The latter assumption is not very realistic and may be removed if desired, but it does not affect the essential points of the results derived below.

We label the neutral PX_z as component 1 and neutral BX as component 2. The sedimentation equilibrium equations for these components are given by (5.2.1), which may be written

$$RT\left(\frac{d \ln a_1}{dr}\right) = 2A_1 r \qquad (5.2.22a)$$

$$RT\left(\frac{d \ln a_2}{dr}\right) = 2A_2 r \qquad (5.2.22b)$$

in which A_i is defined by

$$A_i = \frac{M_i}{2}(1 - \bar{v}_i\rho)\omega^2 \qquad (5.2.23)$$

and a_i is the activity of component i. According to electrochemistry, one may write

$$a_1 = (\gamma_\pm)_1^{z+1}(m_{P^{z+}})(m_{X^-})^z \tag{5.2.24a}$$

$$a_2 = (\gamma_\pm)_2^{2}(m_{B^+})(m_{X^-}) \tag{5.2.24b}$$

where $(\gamma_\pm)_i$ is the mean activity coefficient on the molality scale of component i, and $m_{P^{z+}}$, m_{B^+}, and m_{X^-} are the molalities of ionic species P^{z+}, B^+, and X^-, respectively. Under the assumption that the mean activity coefficient of each electrolyte component is unity, substitution of (5.2.24a and b) into (5.2.22a and b) yields

$$RT\left(\frac{d \ln m_1}{dr} + z\,\frac{d \ln m_{X^-}}{dr}\right) = 2A_1 r \tag{5.2.25a}$$

$$RT\left(\frac{d \ln m_2}{dr} + \frac{d \ln m_{X^-}}{dr}\right) = 2A_2 r \tag{5.2.25b}$$

where account has been taken of the fact that $m_{P^{z+}}$ and m_{B^+} are equal to the molalities of the neutral components m_1 and m_2, respectively. The condition of electric neutrality is

$$zm_{P^{z+}} + m_{B^+} = m_{X^-} \tag{5.2.26a}$$

or

$$zm_1 + m_2 = m_{X^-} \tag{5.2.26b}$$

Equations 5.2.25a and 5.2.25b may be solved simultaneously together with the aid of (5.2.26) to obtain expressions for dm_1/dr and dm_2/dr. For practical purposes it is convenient to use concentrations expressed in molarity. The relationship between the molality concentration (m_i) and the molarity concentration (c_i) is

$$m_i = \frac{(1000/M_i)c_i\bar{v}_0}{1 - \bar{v}_1 c_1 - \bar{v}_2 c_2}$$

If the solution is dilute in both polyelectrolytes and supporting low-molecular-weight electrolyte, the results obtained are

$$\frac{d \ln c_1}{d(r^2)} = \frac{2A_1\left(\dfrac{z}{2}\,\dfrac{M_2}{M_1}\,\dfrac{c_1}{c_2} + 1\right) - A_2 z}{RT\left[2 + z\,\dfrac{M_2}{M_1}\,\dfrac{c_1}{c_2}\,(1 + z)\right]} \tag{5.2.27a}$$

$$\frac{dc_2}{d(r^2)} = \frac{A_2\left(\dfrac{M_1}{M_2}\dfrac{c_2}{c_1} + z + z^2\right) - A_1 z}{RT\left[2 + z\,\dfrac{M_2}{M_1}\dfrac{c_1}{c_2}(1+z)\right]}\,\frac{M_2}{M_1}\,c_1 \qquad (5.2.27b)$$

From (5.2.27a), it follows that if c_2 is not zero, then

$$\lim_{c_1 \to 0}\frac{d\ln c_1}{d(r^2)} = \frac{(1 - \bar{v}_1\rho^0)\omega^2}{2RT}\left[M_1 - \left(\frac{z}{2}\right)M_2\left(\frac{1 - \bar{v}_2\rho^0}{1 - \bar{v}_1\rho^0}\right)\right] \qquad (5.2.28)$$

where ρ^0 is the density of the solution in the absence of the polyelectrolyte ($c_1 \to 0$).

Equation 5.2.28 indicates that if redistribution of the low-molecular-weight electrolyte does not occur (so that ρ^0 may be treated as constant), a plot of $\ln c_1$ versus r^2 becomes linear, as is the case with a very dilute binary solution of a nonelectrolyte solute, but its slope gives an apparent molecular weight $M_1^{(app)}$ and which differs from the true molecular weight M_1 of the polyelectrolyte component by the amount $-(zM_2/2)(1 - \bar{v}_2\rho^0)/(1 - \bar{v}_1\rho^0)$. The difference between $M_1^{(app)}$ and M_1 is usually called the *residual charge effect*, or the *secondary charge effect*. The difference from the true molecular weight may, however, become small. For typical inorganic salts (e.g., NaCl, KCl) in water, $\bar{v}_1 \sim 0.3$ and $M_2 \sim 65$. Thus, the second term of (5.1.28) becomes (with $\bar{v}_2 \sim 0.75$, typical of proteins in water) approximately 90Z. If Z is fairly small (say, two charges per 10,000), the error lies within 2% of M_1. In the case of a protein, which can be made essentially electrically neutral by proper adjustment of pH, it should be possible to come even closer to M_1.

The term $(z/2)(M_2/M_1)(c_1/c_2)$ in (5.2.27a) is usually referred to as the *primary charge effect*. It depends on the concentration of the polyelectrolyte relative to that of the supporting electrolyte. This effect can therefore be eliminated either by diluting the solution to zero polyelectrolyte concentration or by increasing the concentration of the supporting electrolyte. It is important to comprehend that such an operation can be of effective use only for elimination of the primary charge effect.

If one treats all components in the solution as nonelectrolytes and also assumes that the solution is incompressible, then the basic equations describing the sedimentation equilibrium are given by (5.2.1):

$$2A_1 r = \left(\frac{\partial \mu_1}{\partial m_1}\right)_{T.P.m_2}\frac{dm_1}{dr} + \left(\frac{\partial \mu_1}{\partial m_2}\right)_{T.P.m_1}\frac{dm_2}{dr} \qquad (5.2.29a)$$

$$2A_2 r = \left(\frac{\partial \mu_2}{\partial m_1}\right)_{T.P.m_2}\frac{dm_1}{dr} + \left(\frac{\partial \mu_2}{\partial m_2}\right)_{T.P.m_1}\frac{dm_2}{dr} \qquad (5.2.29b)$$

And using the identity below due to Maxwell relations, one has

$$\left(\frac{\partial \mu_1}{\partial m_2}\right)_{T,P,m_1} = \left(\frac{\partial \mu_2}{\partial m_1}\right)_{T,P,m_2} \tag{5.2.30}$$

A dimensionless factor Γ, usually called the "binding coefficient," is defined by

$$\Gamma = - \frac{(\partial \mu_2/\partial m_1)_{P,m_2}}{(\partial \mu_2/\partial m_2)_{P,m_1}} \tag{5.2.31a}$$

It can be shown easily that this expression may be rewritten as

$$\Gamma = \left(\frac{\partial m_2}{\partial m_1}\right)_{P,\mu_2} \tag{5.2.31b}$$

Making use of (5.2.31a) and solving (5.2.29a) with (5.2.30) and (5.2.29b) for dm_1/dr and dm_2/dr, one obtains

$$\frac{dm_1}{dr} = \frac{2r(A_1 + \Gamma A_2)}{(\partial \mu_1/\partial m_1)_{P,m_2} - \Gamma^2(\partial \mu_2/\partial m_2)_{P,m_1}} \tag{5.2.32a}$$

$$\frac{dm_2}{dr} = \frac{2r\{\Gamma A_1 + A_2[(\partial \mu_1/\partial m_1)_{P,m_2}/(\partial \mu_2/\partial m_2)_{P,m_1}]\}}{(\partial \mu_1/\partial m_1)_{P,m_2} - \Gamma^2(\partial \mu_2/\partial m_2)_{P,m_1}} \tag{5.2.32b}$$

Using the chemical potential $\mu_i = \mu_i^0(T, P) + RT \ln \gamma_i^{(m)} m_i$ for solutes ($i = 1, 2$) and the limiting property

$$\lim_{\substack{m_1 \to 0 \\ m_2 \to 0}} \gamma_i^{(m)}(m_i) = 1 \qquad (i = 1, 2) \tag{5.2.33}$$

one may rewrite (5.2.32a) in molarity concentration for a dilute solution. If c_2 is nonzero,

$$\lim_{c_1 \to 0} \frac{d \ln c_1}{d(r^2)} = \frac{(1 - \bar{v}_1 \rho^0)\omega^2}{RT} \left[M_1 + \Gamma^0 M_2 \left(\frac{1 - \bar{v}_2 \rho^0}{1 - \bar{v}_1 \rho^0}\right) \right] \tag{5.2.34}$$

where

$$\Gamma^0 = \lim_{c_1 \to 0} \left(\frac{\partial m_2}{\partial m_1}\right)_{T,P,\mu_2}$$

Thus, a plot of $\ln c_1$ versus r^2 also follows a straight line for the computation of the quantity in the bracket. Since there exists no experimental means of estimating Γ^0, it is not possible to proceed further.

Comparison of (5.2.34) and (5.2.28) shows that the ternary solution under consideration has a binding coefficient Γ^0 equal to $-z/2$. This implies that the residual charge effect may be regarded as being due to an apparent binding of the added supporting electrolyte to the polyelectrolyte solute. Close examination of the derivation given above reveals that this apparent binding has its root not in a special thermodynamic interaction between the two electrolytes but in the requirement that the condition of electric neutrality be satisfied in any volume element of the solution.

It is concluded then that the sedimentation equilibrium method for a two-component system, as outlined earlier, may be applied to a macromolecular electrolyte such as a protein in the presence of an inorganic salt, provided that the salt concentration is moderately high and provided that z is low, that is, in the case of proteins, provided that the pH is close to that of the isoelectric point.

Sedimentation-Velocity Method

The application of a high centrifugal force causes macromolecules to sediment outward, that is, toward the bottom of a cell. In a single-solute-component, noninteracting system, this transport of molecules produces both a region in the cell containing only solvent and a region in the cell where solute concentration is uniform. This region of uniform concentration is called the plateau region. Between the supernatant and the plateau region is a transition zone, called the boundary, in which concentration varies with distance from the axis of rotation. Since the movement of the boundary is actually a measure of the movement of solute molecules in the plateau region, the sedimentation-velocity method is generally used to study the rate of this boundary movement. Thus, the sedimentation-velocity method is also known as rate centrifugation and is basically a transport method; the parameter used to describe the transport of particles is the "sedimentation coefficient." It is simply a measure of velocity of sedimentation particles per unit centrifugal force field, defined by Svedberg [42, 43] as equation 4.2.2, in which the term dr/dt really represents the time rate of displacement of a moving boundary and $\omega^2 r$ is the measure of centrifugal field strength. Thus, the sedimentation coefficient of component i, S_i, is a typical mobility [22], in centimeters per second per unit field, and has the dimensions of seconds.

For theoretical discussions of solutions that contain more than one solute component, it is convenient to classify an almost infinite variety of such solutions into two categories usually referred to as *paucidisperse* and *polydisperse*. A solution is called paucidisperse if it contains a small number of molecularly homogeneous macromolecular solutes, regardless of whether the solvent consists of a single pure liquid, whether it contains a few low-molecular-weight solutes, or whether it is a mixture of two or more liquid compounds. If the number of such macromolecular solutes is indefinitely large, the solution is called polydisperse. This mode of classification is a mere convention, because there is a continuous spectrum between these two extremes.

In general, biochemical preparations are considered to be molecularly homogeneous or nearly so. Hence, a mixture of several of them is typically paucidisperse. Proteins often tend to aggregate with each other to form oligomers or dissociate into subunits. Therefore, it is to be noted that a protein solution cannot always be treated as composed of a single solute even though the dissolved protein preparation is considered to be pure. Biochemists and molecular biologists are primarily interested in utilizing the ultracentrifuge for separating paucidisperse macromolecular solutions into component solutes and estimating the concentrations and sedimentation rates of the separated components.

For polydisperse solutions, the methods used in paucidisperse systems become ineffective, and one has to give up the method of separating and characterizing individual solute components from observed boundary curves. Solutions of synthetic (homo) polymers are the best known examples of polydisperse systems. Any sample of this class of macromolecular substance is a mixture of a large number of long-chain molecules which are chemically indistinguishable but heterogeneous with respect to physical characteristics such as molecular weight, microstructure, and so forth. Hence when dissolved in a solvent, such a sample gives a great number of polymer solutes differing, in the simplest case, only in molecular weight. The ultracentrifuge is primarily used in the determination of distributions of molecular weights in polymeric substances.

The theory of sedimentation processes in two-component systems is not only essential for a rational interpretation and analysis of ultracentrifugal measurements on binary solutions but also a prerequisite for the study of similar problems on multicomponent solutions.

The sedimentation velocity method provides a measure of the transport of solute materials across a surface of the cell. The equation governing this transport of solute for binary system is the binary Lamm equation, which can be written from (4.2.1) in terms of convenient molarity concentration:

$$\frac{\partial c}{\partial t} = -\frac{1}{r}\left[\frac{\partial}{\partial r}\left(S\omega^2 r^2 c - Dr\frac{\partial c}{\partial r}\right)\right] \qquad (5.2.35)$$

The exact solution of (5.2.35), being a formidable infinite series in terms of integrals which can be computed only by numerical integration, was obtained by Archibald [44] in 1942. A number of approximate solutions have been obtained, all of which are reviewed by Williams et al. [19]. The simplest is the solution which applies whenever sedimentation is rapid in comparison with diffusion by assuming $D = 0$. Experimental data exhibit the general trend that the sedimentation coefficient increases and the diffusion coefficient decreases as the molecular size of solute becomes greater. Accordingly, the results from (5.2.35) with no diffusion term are of great significance in describing the limiting behavior of very high-molecular-weight substances such as DNA and viruses. For this case, (5.2.35) reduces to

$$\frac{\partial c}{\partial t} = -\frac{1}{r}\frac{\partial}{\partial r}[r^2\omega^2 S(c)c] \qquad (5.2.36)$$

Then, (5.2.36) can be rewritten in the form

$$\frac{\partial c}{\partial t} + r\omega^2\frac{d[S(c)c]}{dc}\frac{\partial c}{\partial r} + 2\omega^2 S(c)c = 0 \qquad (5.2.37)$$

which is Lagrange's linear partial differential equation [45] and can be solved by the method of characteristics or the subsidary equations for Lagrange's equation. It is found to be

$$\frac{dt}{1} = \frac{dr}{r\omega^2\dfrac{d[S(c)c]}{dc}} = -\frac{dc}{2\omega^2 S(c)c} \qquad (5.2.38)$$

Any arbitrary functional relation $\phi[u(c, t, r), v(c, t, r)] = 0$ satisfies the linear partial differential equation (5.2.37), provided that $u = a$ and $v = b$ are independent integrals of the subsidiary equations (5.2.38). The solution $\phi[u, v] = 0$, with ϕ arbitrary, is called the general solution of (5.2.37) in which a and b are arbitrary constants [45].

If one defines $\bar{S}(c) = S(c)/S_0$ as the reduced sedimentation coefficient, two independent subsidiary equations of (5.2.38) are

$$u(c, t, r) = 2S_0\omega^2 t + \int^c \frac{dc}{\bar{S}(c)c} = a \qquad (5.2.39a)$$

$$v(c, t, r) = r^2\bar{S}(c)c = b \qquad (5.2.39b)$$

in which S_0 is the value of S at $c = 0$. The most convenient form of the general solution to (5.2.37), $\phi[u, v] = 0$, for the development which follows is [20, 21]

$$\exp\left[\int^c \frac{dc}{\bar{S}(c)c}\right] = G\left\{r^2\bar{S}(c)c\exp\left[-2S_0\omega^2 t\right.\right.$$

$$\left.\left. - \int^c \frac{dc}{\bar{S}(c)c}\right]\right\}\exp(-2S_0\omega^2 t) \qquad (5.2.40)$$

where G stands for an arbitrary function of its argument. The form of G is to be determined by the initial and boundary conditions to (5.2.36).

The simplest case for the complete solution is to assume the sedimentation coefficient to be independent of concentration, that is, $S = S_0 = $ constant. Thus, $\bar{S}(c) = 1$. Therefore, (5.2.40) for the case reduces to

$$c = G\{r^2 \exp(-2S_0\omega^2 t)\}\exp(-2S_0\omega^2 t) \tag{5.2.41}$$

If one uses a conventional cell, filling the cell cavity uniformly with a given concentration c_0, the initial condition may be written as

$$t = 0, \quad c = c_a \quad \text{for} \quad r_m < r < r_b \tag{5.2.42}$$

It is assumed that the solute does not sediment before the rotor reaches its given full angular velocity ω and that time count starts when the angular velocity reaches its given constant value. Thus, r_m and r_b denote the radial position of the air–liquid meniscus and the cell bottom, respectively.

Substituting (5.2.42) into (5.2.41) yields

$$c_0 = G(r^2) \quad r_m < r < r_b \tag{5.2.43}$$

Hence, if one lets

$$G(y) = c_0 \quad r_m{}^2 < y < r_b{}^2 \tag{5.2.44}$$

it follows from (5.2.41) for $t \geq 0$ that

$$c = c_0 \exp(-2S_0\omega^2 t) \quad r_m \exp(S_0\omega^2 t) < r < r_b \tag{5.2.45}$$

because the physically meaningful region of r is $r_m \leq r \leq r_b$, $r_b \exp(S_0\omega^2 t)$ being replaced by r_b in (5.2.45). When $D = 0$, the boundary conditions are

$$c = 0 \quad \text{at} \quad r = r_m \text{ and } r_b \quad \text{for} \quad t \geq 0 \tag{5.2.46}$$

Because (5.2.45) obviously does not satisfy the condition at $r = r_b$, it has to be satisfied with the fulfillment of the condition at $r = r_m$. Substitution of this condition into (5.2.41) yields

$$G[r_m{}^2 \exp(-2S_0\omega^2 t)] = 0 \quad \text{for } t \geq 0 \tag{5.2.47}$$

Hence,

$$G(y) = 0 \quad 0 < y < r_m{}^2 \tag{5.2.48}$$

Therefore, it follows from (5.2.41) that

$$c = 0 \quad 0 < r < r_m \exp(S_0\omega^2 t) \quad \text{for } t \geq 0 \tag{5.2.49a}$$

$$c = c_0 \exp(-2S_0\omega^2 t) \quad r_m \exp(S_0\omega^2 t) < r < r_b \quad \text{for } t \geq 0 \tag{5.2.49b}$$

This indicates that the concentration distribution is represented by a step function and that the step point, denoted by $r^*(t)$, moves toward the cell bottom with time, in accordance with the equation

$$r^*(t) = r_m \exp(S_0 \omega^2 t) \quad \text{or} \quad \ln r^* = \ln r_m + S_0 \omega^2 t \qquad (5.2.50)$$

This indicates that a plot of $\ln r^*$ versus $\omega^2 t$ is a straight line with a slope S_0, passing $\ln r_m$ at $\omega^2 t = 0$. Thus, when $D = 0$ and $S = S_0 = $ constant, it is possible to evaluate S_0 by measuring the position of the solution–solvent separation boundary, that is, $r^*(t)$ as a function of time. Differentiation of (5.2.50) gives the concentration gradient distribution along the cell which for the case is characterized by an infinitely steep line located at the moving boundary $r^*(t)$.

The sedimentation velocity method is basically a transport method. The above discussion can also proceed from momentum equations instead of continuity equations, if $D = 0$. For a binary and uniform density solvent system, an equation of motion may be written from (4.2.3), neglecting the subscript 1 for the solute component, as

$$\frac{d\mathbf{v}_r}{dt} + \frac{18\mu}{\rho D_p{}^2} \mathbf{v}_r = \frac{\omega^2 r}{\rho} (\rho - \rho_0) \qquad (i = 0, 1) \qquad (5.2.51)$$

where $\mathbf{v}_r = dr/dt$, the rate of solute mobility. Since $\mathbf{v}_r$ is very small compared with other quantities such as μ (viscosity), ρ, D_p, and ω, and so on, one may neglect the acceleration term to approximate (5.2.51). Then, one has

$$\frac{18\mu}{\rho D_p{}^2} \frac{dr}{dt} = \frac{\omega^2 r}{\rho} (\rho - \rho_0) \qquad (5.2.52)$$

The movement of a boundary between the supernatant and the plateau region provides a measure of the migration rate of individual molecules in the plateau region. Therefore, the boundary condition may be written for (5.2.52) as

$$t = 0 \qquad r = r_m \qquad (5.2.53a)$$

$$t > 0 \qquad r = r^* \qquad (5.2.53b)$$

After integration with the boundary conditions, the solution for the solute position in the cell becomes

$$\ln r^* = \ln r_m + \frac{D_p{}^2 \rho}{18\mu} \left(1 - \frac{\rho_0}{\rho}\right) \omega^2 t \qquad (5.2.54)$$

Thus, a plot of $\ln r^*$ versus $\omega^2 t$ is a straight line with a slope $(D_p{}^2 \rho / 18\mu)(1 - \rho_0/\rho)$, which is equivalent to S_0 in (5.2.50).

In a real case, a sedimentation coefficient of a solute is concentration dependent. To correct this shortcoming, two forms of concentration-dependent sedimentation coefficient, namely, $S = S_0(1 - kc)$ and $S = S_0/(1 + kc)$, $k > 0$, have been used by Fujita [20, 21] to obtain the solution of (5.2.40) for the $D = 0$ case. If momentum equations are used instead of Lamm's continuity equation, the analyses are equivalent to treatment of the problem of particle sedimentation in concentration-dependent viscosity as given in (5.2.52).

To obtain an experimental expression which will allow r^* to be determined, one notes, from a statistical consideration and a material balance of solute in the cell during the course of an ultracentrifugation, the boundary position or boundaries of any shape ($D \neq 0$) which can be derived in terms of the second moment of the concentration distribution curve depending on the optical system used. It follows that for the initial condition given by (5.2.42),

$$\frac{1}{2}c_0(r_b^2 - r_m^2) = \int_{r_m}^{r_b} cr\,dr \tag{5.2.55}$$

Equation 5.2.55 is integrated by parts to give

$$\frac{1}{2}c_0(r_b^2 - r_m^2) = \frac{1}{2}[r_b^2 c_b - r_m^2 c_m] - \frac{1}{2}\int_{r_m}^{r_b} r^2\left(\frac{\partial c}{\partial r}\right) dr \tag{5.2.56}$$

On the other hand, there is the following identity:

$$c_b - c_m = \int_{r_m}^{r_b} \left(\frac{\partial c}{\partial r}\right) dr \tag{5.2.57}$$

where c_b and c_m are the concentration at the cell bottom and meniscus, respectively. If, as is usually the case, the sedimentation boundary has completely separated away from the meniscus before any measurements are made, c_m may be set equal to zero. Furthermore, the solute concentration at the radius position beyond the plateau region is zero; thus, one may replace $r_b^2 c_b$ by $r_p^2 c_p$ or c_b by c_p, so that

$$\langle r^2 \rangle = \frac{\displaystyle\int_{r_m}^{r_p} r^2\left(\frac{\partial c}{\partial r}\right) dr}{\displaystyle\int_{r_m}^{r_p} \left(\frac{\partial c}{\partial r}\right) dr} \tag{5.2.58a}$$

or

$$\langle r^2 \rangle = r_p^2 - \frac{2}{c_p}\int_{r_m}^{r_p} cr^2\,dr \tag{5.2.58b}$$

where r_m and r_p are the distance from the axis of rotation to the meniscus and to a surface in the plateau region, respectively. If the schlieren optical system is used, $\partial c/\partial r$ in (5.2.58a) can be obtained directly from experiments. If boundaries are symmetrical in shape and relatively sharp, the second moment of the gradient curves becomes almost identical to the maximum ordinate r^* of the curve.

In most instances, therefore, the movement of the maximum ordinate can be used as an accurate indicator of individual particle movement in the plateau region. This provides the basis for the common practice of measuring sedimentation coefficients from the maximum ordinate boundary positions. On the other hand, when a boundary is asymmetric or broad, the second moment of the gradient curve must be used to determine the correct average sedimentation coefficient. This evaluation of the second moment, often quite complicated, provides a weight-average sedimentation coefficient for all species migrating in the solution. For the absorption and interference optical systems, which produce curves of concentration or refractive index versus distance rather than concentration gradient curves versus distance, (5.2.58b) is used for the second moment. The quantity $\langle r \rangle$ obtained from (5.2.58a) and (5.2.58b) may be called the equivalent boundary position. The movement of $\langle r \rangle$ is an indication of the movement of the individual solute molecules in the plateau region rather than in the boundary region.

Another relation which is readily obtained is "the radial dilution law" [46], which gives c_p as a function of $\langle r \rangle$. Combining (5.2.36) and (4.2.2) and integrating between $t = 0$ (at which time $\langle r \rangle = r_m$) and $t = t$, one has

$$\frac{c_p}{c_0} = \left[\frac{r_m}{\langle r \rangle} \right]^2 \tag{5.2.59}$$

Further details of the experimental determination of sedimentation coefficients are described in the detailed treaties of Schachman [47]. Rigorous discussions of the complete solution of the Lamm equation are given by Williams et al. [19, 22] and Fujita [20, 21].

Approach-to-Equilibrium Method

The approach-to-equilibrium method, often called the "Archibald method," is, in many respects, directly related to sedimentation equilibrium or may be termed a pseudoequilibrium centrifugation. The major drawback to the sedimentation-equilibrium method is the fact that a long time is required to reach equilibrium. The necessity to keep an ultracentrifuge in operation at constant speed and temperature for such a long time is a distinct disadvantage. The approach-to-equilibrium method preserves several advantages of the equilibrium method while eliminating the excessive time required and has proved to be of great value in the study of many enzymes and proteins.

Attainment of the pseudoequilibrium conditions is achieved by counterbalancing the two types of transport phenomena, as given in (4.2.9). On the other hand, the thermodynamics for sedimentation equilibrium given in (5.2.1) must also hold. If one defines $\mu_{i\kappa} = (\partial\mu_i/\partial c_\kappa)_{T,P,c_{j\neq\kappa}}$ and $\mu_{i\kappa}$ and Γ_κ as determinants of all $\mu_{i\kappa}$ in (5.2.1) in such a manner that the term (grad c_k) in (5.2.1) can be expressed in terms of $|\mu_{i\kappa}|$ and $|\Gamma_\kappa|$, then one obtains these two determinants as

$$|\mu_{i,\kappa}| = \begin{vmatrix} \mu_{1,1} & \mu_{1,2} & \cdots & \mu_{1,\kappa} & \cdots & \mu_{1,\nu-1} \\ \mu_{2,1} & \mu_{2,2} & \cdots & \mu_{3,\kappa} & \cdots & \mu_{2,\nu-1} \\ \vdots & \vdots & & \vdots & & \vdots \\ \mu_{\nu-1,1} & \mu_{\nu-1,2} & \cdots & \mu_{\nu-1,\kappa} & \cdots & \mu_{\nu-1,\nu-1} \end{vmatrix} \qquad (5.2.60a)$$

$$|\Gamma_\kappa| = \begin{vmatrix} \mu_{1,1} & \mu_{1,2} & \cdots & M_1(1-\bar{v}_1\rho) & \mu_{1,\kappa+1} & \cdots & \mu_{1,\nu-1} \\ \mu_{2,1} & \mu_{2,2} & \cdots & M_2(1-\bar{v}_2\rho) & \mu_{2,\kappa+1} & \cdots & \mu_{2,\nu-1} \\ \vdots & \vdots & & \vdots & \vdots & & \vdots \\ \mu_{\nu-1,1} & \mu_{\nu-1,2} & \cdots & M_{\nu-1}(1-\bar{v}_{\nu-1}\rho) & \mu_{\nu-1,\kappa+1} & \cdots & \mu_{\nu-1,\nu-1} \end{vmatrix}$$

$$(5.2.60b)$$

Substitution of (5.2.60a) and (5.2.60b) into (5.2.1) gives

$$\text{grad } c_k = \frac{|\Gamma_\kappa|}{|\mu_{i\kappa}|} \qquad (5.2.61)$$

The relation between the sedimentation coefficient and the diffusion coefficients in a pseudoequilibrium system is then obtained by substituting (5.2.61) into (4.2.9), which yields

$$S_i = (c_i|\mu_{i\kappa}|)^{-1} \sum_{\kappa=1}^{\nu-1} |\Gamma_\kappa| D_{i\kappa} \qquad (i = 1, 2, \ldots, \nu-1) \qquad (5.2.62)$$

This general relationship between the measurable sedimentation coefficients S_i and the measurable diffusion coefficients $D_{i\kappa}$ contains only measurable quantities. It should be noted that all quantities refer to the same pressure. Thus, the sedimentation coefficients measured at the high pressures in an ultracentrifuge must be converted to atmospheric pressure, since diffusion coefficients and the other quantities $\mu_{i\kappa}$, $\bar{v}_i$, and ρ in (5.2.62) are usually determined at atmospheric pressure.

The most important use of sedimentation during the past 40 years has been in the determination of molecular weights by the combined measurement of sedimentation and diffusion coefficients. If the sedimentation coefficient S and the diffusion coefficient D are measured in solutions of identical composition for a binary system, or for systems which behave like binary systems, (4.2.7) and (4.2.9) may be combined together to yield

$$\frac{S}{D} = \frac{M(1 - \bar{v}_1 \rho)}{RT \left[1 + \dfrac{\partial \ln \gamma^{(c)}}{\partial \ln c} \right]} \tag{5.2.63}$$

Equation 5.2.63 shows that S/D is a thermodynamic quantity resembling the apparent molecular weight determinable from sedimentation equilibrium, (5.2.5). As Archibald [24] was the first to point out, however, (5.2.63) must also apply at any time during the approach to equilibrium at any cross section of the cell across which no flow occurs. The true molecular weight can be determined by extrapolation of S/D to zero solute concentration, and values of $\gamma^{(c)}$ can be evaluated from the concentration dependence.

An alternative procedure for determining molecular weight is to extrapolate S and D to zero concentration separately, giving extrapolated values S^0 and D^0 from which M is evaluated as

$$M = \frac{S^0 RT}{D^0 (1 - \bar{v}_1 \rho_0)} \tag{5.2.64}$$

where ρ_0 is the density of the solvent.

Proteins and viruses are the most common macromolecular substances which are homogeneous with respect to molecular weights, and it is for these substances that the sedimentation-diffusion method, or approach-to-equilibrium method, has been primarily used.

For ideal binary systems, from (4.2.9), the following equation can also be obtained:

$$\frac{S}{D} = \frac{(\partial c/\partial r)_m}{\omega^2 r_m c_m} = \frac{(\partial c/\partial r)_b}{\omega^2 r_b c_b} \tag{5.2.65}$$

The subscripts m and b denote that the quantities are to be evaluated at the meniscus and the bottom of the cell, respectively. In obtaining (5.2.65), it is assumed that the solution is ideal so that S and D remain constant throughout the experiment.

It is clear that at the cell extremities both the concentration and the concentration gradient vary continuously during a run until equilibrium is reached. However, (5.2.65) does not describe how these terms change with time. It merely states that for a homogeneous substance the ratio of the concentration gradient to concentration remains constant at a given speed and that the term $(\partial c/\partial r)/\omega^2 r$ remains constant even if speed changes.

If use is made of (2.8.6a) and (2.8.6b) for a binary system, one obtains the so-called "Svedberg equation" for binary systems [20, 21], yielding

$$M = \frac{RTS}{D(1 - \bar{v}_1\rho)} \qquad (5.2.66)$$

If (5.2.65) is combined with the Svedberg equation (5.2.66), the following expressions are obtained for the molecular weight M_m and M_b at the top and bottom of the cell, respectively:

$$M_m = \frac{RT}{\omega^2(1 - \bar{v}_1\rho)} \frac{(\partial c/\partial r_m)}{c_m r_m} \qquad (5.2.67a)$$

$$M_b = \frac{RT}{\omega^2(1 - \bar{v}_1\rho)} \frac{(\partial c/\partial r_b)}{c_b r_b} \qquad (5.2.67b)$$

Separate equations are used for the top and the bottom of the cell to emphasize the fact that, for polydisperse systems, the apparent molecular weight at these positions may differ and may vary with time. Equations 5.2.67a and 5.2.67b are identical to (5.2.9a) and (5.2.9b).

The discussion so far presented has been concerned only with systems containing a single sedimenting solute. Equations 5.2.67a and 5.2.67b may, however, be readily applied to a mixture of sedimenting components. A weight-average molecular weight is obtained, as the derivation below shows [23]. The quantity determined at the meniscus, after extrapolation to zero concentration, is

$$\frac{1}{r_m}\left[\frac{1}{c}\left(\frac{\partial c}{\partial r}\right)\right]_{r=r_m} = \frac{1}{r_m}\left[\frac{1}{\Sigma_i c_i}\Sigma\left(\frac{\partial c_i}{\partial r}\right)\right]_{r=r_m}$$

where c_i is the concentration of the ith component, all concentrations being expressed in grams per cubic centimeter. For the simple case in which $\bar{v}_1$ is the same for all components, for example, a heterogeneous polymer preparation,

$$\frac{1}{r_m}\left[\frac{1}{c_i}\left(\frac{\partial c_i}{\partial r}\right)\right]_{r=r_m} = M_i \frac{(1 - \bar{v}_1\rho)\omega^2}{RT}$$

so that

$$\frac{1}{r_m}\left[\frac{1}{c}\left(\frac{\partial c}{\partial r}\right)\right]_{r=r_m} = \frac{(1-\bar{v}_1\rho)\omega^2}{RT}\left(\frac{\Sigma_i c_i M_i}{\Sigma_i c_i}\right)_{r=r_m} = \langle M_w \rangle_{r=r_m}\frac{(1-\bar{v}_1\rho)\omega^2}{RT}$$

The Archibald method thus measures $\langle M_w \rangle_m$ for the mixture at the meniscus at the time of measurement or, by the corresponding equation at r_b, the value of $\langle M_w \rangle_b$ at the bottom of the cell. These values will be different. However, both may be evaluated at various times, and extrapolation to $t = 0$ will yield $\langle M_w \rangle$ for the original mixture.

Nonideal systems can also be treated by the approach-to-equilibrium method. Since there is no solute flow through the extremities of the solution column, there must be a balance of driving forces such that at each instant the gradient of the total potential vanishes. This leads to an equation relating the apparent molecular weight $M^{(app)}$, as obtained experimentally, to the activity coefficient $\gamma^{(c)}$ of the solute:

$$M^{(app)} = \frac{RT(dc/dr)}{(1-\bar{v}_1\rho)\omega^2 rc} = \frac{M}{\left[1 + \left(\dfrac{d\ln\gamma^{(c)}}{d\ln c}\right)\right]} \tag{5.2.68}$$

This equation applies to either the top or bottom of the cell. The true molecular weight is obtained by plotting $1/M^{(app)}$ versus concentration and extrapolation to $c = 0$, where the activity coefficient term disappears. Rearranging (5.2.68), one also obtains an expression for the activity coefficient $\gamma^{(c)}$ of the solute at a given concentration, which gives

$$\ln\gamma^{(c)} = \int_0^c \left[\frac{M}{M^{(app)}} - 1\right] d\ln c \tag{5.2.69}$$

The approach-to-equilibrium method was used to study interacting systems by Adams et al. [48–53]. Such systems include the association of monomers to form higher molecular aggregates or the association of two different species to form a third. For an ideal solution, $\langle M_w \rangle$ represents the weight average of the various species present. If the rates of association and dissociation are very rapid, thermodynamic equilibrium is maintained at all levels in the cell. The composition of the mixture present at the meniscus at any given instant is dependent upon the various equilibrium constants that may be involved.

Sedimentation Equilibrium in a Density-Gradient

If a solution of low-molecular-weight solute is placed in a centrifugal field of sufficient strength, an appreciable concentration gradient will result. This will lead to a gradient of density whose density increases with radial distance from the axis of rotation. For example, if 7.7 molal CsCl in water is centrifuged at 45,000 rpm in a cell placed 65 mm from the center of rotation, a density gradient of 0.12 $g/(cm^3)(cm)$ will be established at equilibrium; the density will range from 1.64 to 1.76 g/cm^3 if the liquid column is 1 cm high [23].

Sedimentation equilibrium in a density gradient is a recent addition to various centrifugation techniques first employed by Meselson, Stahl, and Vinograd [54] in 1957. In sedimentation in a density-gradient, the density gradient serves another purpose. The density of the solution containing a macromolecular solute is increased to a value very near that of the solvated macromolecule by adding a dense third component of low molecular weight. If a cell containing this solution is rotated at high speeds for a sufficiently long period of time, an equilibrium state will be achieved. The distribution of the third component is described by (5.2.7b). The change in the concentration of this component produces a density gradient. Instead of the macromolecule being found as a thin pellet at the bottom of the cell, it is distributed as a narrow band at a position in the cell where the density of the solution is the same as that of the solvated macromolecule.

If one assumes that the density of the solution is at some point r_0 equal to the reciprocal of the partial specific volume of some macromolecular substance which has been added to the solution, then no net force will act on the macromolecular particles at r_0. At $r < r_0$, ρ will be less than $1/\bar{v}_2$ (component 2 is the macromolecular solute), $\rho\bar{v}_2 < 1$, and the particles will sediment toward the bottom of the cell. At $r > r_0$, on the other hand, $\rho\bar{v}_2 > 1$, and the particles will rise or flow back. At equilibrium, all macromolecules will be distributed in a narrow band about r_0.

The theory for sedimentation equilibrium in a density-gradient is complicated for three reasons: (1) The third component is present in such high concentrations that nonideality considerations become very important. (2) The third component is almost always an electrolyte, thus making it necessary to consider the effect of an electric field on a charged macromolecule. (3) The experiments are performed at such high speeds that the effect of pressure must be taken into account. No theory yet advanced considers all three effects at the same time. The most important of the three factors, nonideality, has been considered by Fujita [20, 21], and results are given in (5.2.7a) and (5.2.7b). Pressure effects have been discussed by Vinograd and Hearst [55]. In order to obtain an expression giving the distribution of the macromolecule, (5.2.32a) must be integrated. It is convenient to measure the radial parameter, not from the center of rotation, but from the center of mass of the macromolecular particle at r_0, that is, $r = r_0 + (r - r_0)$. Thus, ρ and Γ may be written in a Taylor series as

$$\rho = \rho_0 + \left(\frac{d\rho}{dr}\right)_0 (r - r_0) + \cdots$$

$$= \frac{1}{\bar{v}_2} + \left(\frac{d\rho}{dr}\right)_0 (r - r_0) + \cdots \tag{5.2.70a}$$

$$\Gamma = \Gamma_0 + \left(\frac{d\Gamma}{dr}\right)_0 (r - r_0) + \cdots \tag{5.2.70b}$$

Neglecting the higher-order terms of the Taylor series, (5.2.34a) may be rewritten as

$$\frac{d \ln c}{d(r - r_0)} = - \frac{M_2^{(\mathrm{app})} \omega^2 \bar{v}_2 (d\rho/dr)}{RT} [r_0 + (r - r_0)](r - r_0) \tag{5.2.71}$$

If the higher-order terms of $r - r_0$ are neglected on integration, one obtains

$$\frac{c_2}{(c_2)_{r=r_0}} = e^{-(r-r_0)^2/2\sigma^2} \tag{5.2.72}$$

where

$$\sigma^2 = \frac{RT}{\bar{v}_2 (d\rho/dr) \omega^2 r_0 M_2^{(\mathrm{app})}} \tag{5.2.73}$$

Equation 5.2.72 represents a Gaussian distribution about $r = r_0$. Equation 5.2.73 shows that $M_2^{(\mathrm{app})}$ can be evaluated from σ. The value of $\bar{v}_2$ is obtained directly as $1/r_0$, and $d\rho/dr$ is determinable in a number of ways, most of which are based on the supposition that its value is not appreciably affected by the presence of the macromolecules.

The density gradient method does not possess an advantage over ordinary sedimentation equilibrium in the determination of molecular weight, especially as any slight heterogeneity among the sedimenting molecules will result in broadening of the distribution which cannot be readily distinguished from broadening due to a large value of σ. The method is, however, a powerful one for distinguishing between macromolecules with markedly different values of $\bar{v}_2$. Of particular importance is the fact that a mixture of macromolecules characterized by different values of $\bar{v}_2$, that is, by different effective densities, will in the absence of strong interactions separate into discrete bands, each located at a value of r at which $\rho = (\bar{v}_i)^{-1}$.

3 ANALYSES BY PREPARATIVE DENSITY-GRADIENT EQUILIBRIUM CENTRIFUGATION

The preparative centrifuge is a simpler instrument and does not have an optical system. However, recent rapid technical progress and development permit the preparative centrifuge not only to completely separate several or all of the components in a mixture, but also to perform analytical measurements. This versatile technique is density-gradient centrifugation. Isopycnic banding, or sedimentation-equilibrium in a density-gradient solution, can also be used to measure the buoyant densities of various substances. A technique using latex beads as a carrier to detect a very small amount of biological materials either in immunological or enzymatic reactions has become widely adapted as a procedure in biological research work, because of its simplicity, high sensitivity, and stability in maintaining the reactivity of immobilized biological materials. Using the density-gradient isopycnic banding technique with a carrier method, one can measure the buoyant density changes either in the immobilization or in the reacted masses from simple hydrostatic relationships to obtain quantitative information on the characteristics of the equilibrium systems.

Since the report by Singer and Plotz [56] on the use of synthetic polymer latex beads for serological purposes to detect rheumatoid factors, a number of articles have appeared describing this carrier method of detecting small amounts of biological activity. The notable ones are those by Anderson and Breillatt [57], Bloomfield et al. [58], Dezelic et al. [59, 60], Duboczy and White [61], Oreskes and Singer [62] and Singer et al. [63–66]. The majority of these works have dealt with the carrier reactions in serology. However, very little has been reported on the quantitative aspect of the carrier method.

In the following, an application of the density-gradient isopycnic banding technique with a carrier method to obtain quantitative information on microquantities of biological active mass is demonstrated. The procedures consist, in general, of coating the carrier particles with a suitable antigen or antibody (immobilized antigen or antibody on the carrier particles) and using these coated (or immobilized) particles to react with specific antibody or antigen in serums or other fluids. The immobilized masses or the reacted masses are calculated from the difference in banding densities between immobilized or reacted particles (complexes) and simple carrier particles.

Methods and Procedures

The experiments are divided into two phases. The first phase consists of the coating of latex beads with antigen (IgG). The second phase consists of the interaction of antigen-coated latex beads with specific antibody (anti-IgG IgG). The method used in both phases of experiments are isopycnic banding of respective hybrid particles in a density gradient centrifugation to observe their density changes.

In a typical example [15, 16], four commercially available IgG's were used, namely, bovine, dog, rabbit, and sheep immunoglobulins. Each of the IgG's was incubated in solution with latex beads at 24°C for about 30 min. Samples of the equilibrium incubation mixture were layered on sucrose density gradients and isopycnically banded. Then, using the band-sampling apparatus and the refractometer, the density of the immobilized IgG–latex bead complex was determined by measuring the gradient density at the banding position of the latex complex.

The pH values were chosen so that the IgG molecules would carry a positive net surface charge (pH 4.6), a relatively neutral surface charge (pH 6.2), and a negative net surface charge (pH 8.2). The normal NaCl concentration in serum is 0.85% w/v. NaCl concentrations of 0.1, 0.3, 1.0, and 1.5% w/v were used in this study. The procedures of an immobilization run have appeared elsewhere [15, 67]. A schematic summary of an immobilization of IgG on latex beads is given in Fig. 5.5a. The experimental procedures of the second phase, the antigen-antibody interaction are also given in Refs. 16 and 67. A schematic outline of the experimental procedure is presented in Fig. 5.5b.

Results and Analysis

The data taken in the experimental runs were buoyant density of hybrid particles formed by immobilization of immunoglobulin (IgG) on the surface of the latex beads and the buoyant densities of complexes formed between immobilized immunoglobulins as antigen and their specific antibodies resulting from interaction, which were measured for bovine IgG and anti-(bovine IgG) IgG produced in the rabbit at 24°C, 1.0% w/v NaCl with pH 8.2 and 4.6, and for rabbit IgG and anti-(rabbit IgG) IgG produced in the goat at the same condition with pH 8.2. The amounts of immobilized IgG and the complexes formed by interaction were calculated by the simple hydrostatic relationship [57]

$$\rho_H = \frac{v_c \rho_c + v_L \rho_L}{v_c + v_L} \tag{5.3.1}$$

where v and ρ are volume and density, and subscripts H, c, and L refer to hybrid, carrier, and load, respectively. The mass of the loads can be calculated by

$$M_L = \frac{v_c(\rho_H - \rho_c)}{1 - \rho_H/\rho_L} \tag{5.3.2}$$

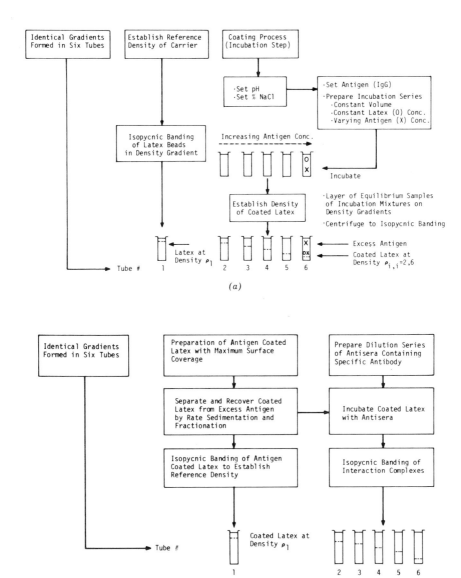

Fig. 5.5 Schematic summary of immobilization and interaction on polystyrene latex beads. (*a*) Immobilization of immunoglobulins. *Source:* Reproduced from Ref. 15 by permission of John Wiley. (*b*) Interaction between antibody and antigen immobilized on latex beads. *Source:* Reproduced from Ref. 16 by permission of Academic Press.

259

Equation 5.3.2 was obtained by rearranging (5.3.1) and using $v_L = M_L/\rho_L$. For the latex beads used, an average diameter of 0.109 μm and a density of 1.050 g/ml, and for a density of 1.350 g/ml for IgG, the mass of IgG immobilized on a latex bead becomes

$$m_L = 9.153 \times 10^{-16} \frac{\rho_H - 1.050}{1.350 - \rho_H} \qquad (5.3.3)$$

The molecular weight of IgG is estimated to range between 156,000 and 161,000; thus, the mass of an IgG molecule may range from 2.56×10^{-19} to 2.67×10^{-19} g. Using an average weight, the number of IgG molecules coated on a single bead, n_L, can be obtained from (5.3.3) as

$$n_L = 3645 \frac{\rho_H - 1.050}{1.350 - \rho_H} \qquad (5.3.4)$$

The number of latex beads present during the coating process was estimated from the weight-by-volume percentage of stock latex solution which was accurately measured gravimetrically. The total mass of load (IgG) immobilized can then be calculated from (5.3.3). This quantity of immobilized IgG was substracted from the total quantity of IgG incubated to give the quantity of soluble IgG in solution. The soluble IgG was in equilibrium with the IgG immobilized on the solid phase in solution.

In summary, a direct measurement of ρ_H in (5.3.2) gives the solid-phase concentration of antigen, $\bar{C}_s$ (μg antigen/mg latex), at equilibrium, and a mass balance was used to obtain the liquid-phase concentration of antigen, $\bar{C}_L$ (μg antigen/ml). An experimental verification of the mass balance was possible spectrophotometrically only at high concentrations of incubated IgG when significant fractions of the IgG were immobilized (see also Ref. 56). In practice, the spectrophotometric verification was not generally possible due to the microquantities involved.

The amount of specific antibody (anti-IgG IgG) which had reacted with the immobilized antigen, including a small amount of antibody immobilized on the surface of vacant sites of polystyrene latex beads, was calculated by the simple hydrostatic relationship given in (5.3.1). For the second-phase experiments, subscript H refers to the antigen–antibody complex formed by the interaction, c refers to the immobilized antigen on the latex beads [(IgG)–(beads)] as a carrier, and L refers to the load (anti-IgG IgG). The mass of the specific antibody which has reacted with the immobilized antigen on the latex beads can be evaluated in a similar manner by (5.3.2), in which the density of the specific antibody (anti-IgG IgG), ρ_L, is 1.35 g/ml for immunoglobulins and densities of the immobilized antigen–beads complex, ρ_c, are measured in the first phase of immobilization runs. These values are listed in Table 5.2.

The number of IgG molecules as antibody reacting with antigen, n_L, can be obtained from (5.3.4). This quantity of reacted antibody anti-IgG IgG was substracted from the total quantity of anti-body layered on the immobilized IgG as antigen to give the quantity of soluble IgG in solution. The soluble antibody was in equilibrium with the antibody–antigen complex in solution. The equilibrium mass of antibody present in solution and unreacted was determined by mass balance.

The number of antibody molecules interacting with immobilized antigen was determined from the total specific antibody mass per complex, using 2.56×10^{-19} g as the mass of antibody molecule. Thus, knowing the number of antigen molecules on the latex surface, the equilibrium stoichiometry between the bound antigen–antibody complexes and the antibody in solution was established. The data, that is, the estimated total numbers of bound and free antibody in the system, are listed in Table 5.3.

To test the specificity of the interaction, latex beads coated with bovine IgG were incubated with antisheep IgG produced in rabbits and the resulting equilibrium complexes banded isopycnically. The resulting data are presented in Table 5.4. The nonspecific interaction was demonstrated by small density increases, approximately 0.003 g/ml, as shown in Table 5.4, compared to the increases found with the specific antisera for equivalent antibody (Ab) concentrations, of approximately 0.030 g/ml. The small density increases resulting from the nonspecific interaction may be attributed to nonspecific binding of protein components in serum smaller than IgG and known to adsorb to a latex surface, probably albumin [65].

Adsorption Isotherms

The mass of immunoglobulin immobilized on a unit mass of latex beads was characterized with an adsorption isotherm. Adsorption isotherms between bovine IgG and latex beads at 24°C and pH 8.2 with NaCl concentrations of 0.1, 0.3, 1.0 and 1.5% w/v were determined. The results are presented in Fig. 5.6. Except for NaCl concentrations exceeding 1.0% w/v, the binding isotherms show monolayer adsorptions. Adsorptions isotherms for the four immunoglobulins (bovine, dog, rabbit, and sheep IgG) at 24°C, pH 8.2, and 1.0% w/v NaCl and for bovine IgG at 24°C with 1.0% w/v NaCl using three different buffer solutions were also made. Based on the monolayer adsorption characteristic, these results were correlated in double reciprocal forms and are presented in Figs. 5.6a and 5.6b, respectively.

As the data indicate, there is a breakpoint in each isotherm, thus indicating a departure from the assumptions of the classical Langmuir isotherm [68]. For example, if the immobilized IgG carries a surface charge, the

Table 5.2 Properties of Immobilized Immunoglobulins as Antigen

Antigen Run No.	Conc. of IgG Incubated (µg/ml)	Wt % Sucrose at Band Center	Sucrose Density at Band Center	Grams IgG Per Bead ($\times 10^{16}$)	Molecules of IgG per Bead	$\bar{C}_s$ (µg IgG per mg latex)	$\bar{C}_L$ (µg IgG per ml soln.)
Bovine IgG-10	22.6	19.9	1.0788	1.0103	387	137	15.7
	70.6	21.3	1.0852	1.2646	484	171	62
	116.3	22.6	1.0912	1.5145	580	205	106
	137.4	22.8	1.0918	1.5402	590	208	127
	306.4	22.8	1.0918	1.5402	590	208	296
	385.9	23.1	1.0934	1.6091	616	217	375
Bovine IgG-20	530.8	23.1	1.0932	1.6004	613	216	520
	575.4	22.8	1.0918	1.5402	590	208	565
	1104.9	23.2	1.0935	1.6134	618	218	1094
	1925.5	23.6	1.0955	1.7008	652	230	1914
	2011.4	23.5	1.0952	1.6876	647	228	2000
	171.3	22.7	1.0914	1.5230	584	206	161
Bovine IgG-25	15.3	18.15	1.0712	0.7234	277	98	10.4
	78.7	21.3	1.0850	1.2565	481	170	70.2
Rabbit IgG-50	1672.5	25.65	1.1050	2.1357	818	289	1658
	809.3	24.90	1.1014	1.9670	754	266	796
	296.7	22.95	1.0926	1.5745	603	213	286
	104.1	21.8	1.0874	1.3549	519	182	95
	42.7	20.6	1.0819	1.1320	434	153	35
	19.4	20.1	1.0797	1.0453	400	141	12.3

first bound IgG molecule will exert a repulsive force on the approach of the second IgG molecule to the same neighborhood. Therefore, since the Langmuir assumption of noninteraction would fail in this case, the binding free energy change has to be modified. The adsorption coefficient K is the equilibrium coefficient between $\overline{C}_s$ and $\overline{C}_L$ at a given temperature. The free-energy change ΔG and the adsorption coefficient K are given by

$$\Delta G = -RT \ln K = RT \ln \frac{\overline{C}_s}{\overline{C}_L} \qquad (5.3.5)$$

If there is a repulsive interaction in the adsorption, the free-energy change of the above adsorption without intermolecular interaction has to be modified to

$$\Delta G' = \Delta G + \Delta G_I = -RT \ln K - RT \ln K_I = -RT \ln K' \qquad (5.3.6)$$

in which ΔG_I and K_I are the free energy change and the adsorption coefficient when there is significant repulsive interaction between immobilized molecules. The value of ΔG_I depends on the concentration of molecules involved in charge interactions. As can be seen from Fig. 5.6, the number of such molecules depends on surface coverage.

Therefore, an additional energy ΔG_I associated with immobilization when there are repulsive interactions can be defined as

$$\Delta G_I = \Gamma \cdot \xi \qquad (5.3.7)$$

where Γ is the energy required to overcome repulsive forces during the establishment of maximum immobilization and ξ is the fraction of maximum immobilization obtained. These parameters can be defined as

$$\Gamma = n_{max} F_z \qquad (5.3.8a)$$

$$\xi = \frac{\overline{C}_s}{(\overline{C}_s)_{max}} \qquad (5.3.8b)$$

where n_{max} is the number of molecules associated with $(\overline{C}_s)_{max}$ and F_z is the repulsive force, dependent on the molecular surface charge z, by an average immobilized molecule in the n_{max} population.

Combining (5.3.6) and (5.3.7), one obtains

$$K_I = \exp\left(\frac{-\Gamma\xi}{RT}\right) \qquad (5.3.9)$$

Table 5.3 Results of Reaction Between Antibody and Antigen Immobilized on Latex Beads (Each Sample Contained 10 μg Coated Latex in Total Volume of 0.2 ml, 1.0% w/v NaCl, 2 hr Incubation at 24°C)

Run No.	Conc. of Ab Incubated (μg/ml)	Wt % Sucrose at Isopycnic Band Center	Density of Complex ρ (g/ml)	Total No. of Ab Molecules Bound per Total No. of Ag Molecules in System[a]	No. of Free Ab Molecules in System ($\times 10^{-13}$)
Bovine IgG and Anti-(Bovine IgG) at pH 8.2					
200	0[b1]	20.1	1.0801	—	—
210	0	22.1	1.0891	—	—
220	0	22.0	1.0882	—	—
200	1.8	23.8	1.0968	55.0	0.064
210	1.8	24.2	1.0989	88.6	0.017
220	1.8	23.2	1.0940	12.9	0.123
200	18	30.1	1.1262	559.3	0.62
210	18	31.8	1.1344	609	0.55
200	180	33.0	1.1403	851	12.9
210	180	34.6	1.1481	1031	12.6
220	180	33.5	1.1427	905	12.6
200	900	36.9	1.1596	1321	68.6
210	900	36.4	1.1571	1257	68.5
200	1800	38.8	1.1692	1501	138.4
210	1800	37.6	1.1631	1421	139.0
220	1800	37.0	1.1601	1334	139.0

Bovine IgG and Anti-(Bovine IgG) IgG at pH 4.6

230	0^{b_2}				
230		20.0	1.0789	—	—
230	2.15	23.5	1.0947	34.3	0.12
230	21.50	29.0	1.1202	457.0	1.04
230	215.00	31.8	1.1336	719.0	15.80
230	1075.00	33.5	1.1420	902.0	82.00
230	2150.00	34.4	1.1468	816.0	167.00

Rabbit IgG and Anti-(Rabbit IgG) at pH 8.2

240	0^{b_3}				
240		22.15	1.0896	—	—
240	0.68	22.75	1.0924	10	0.05
240	6.80	28.25	1.1178	414	0.24
240	68.00	30.95	1.1307	660	4.85
240	680.00	34.85	1.1498	1084	52.30
240	6300.00	37.25	1.1618	1394	530.0

Source: Reproduced from Ref. 16 by the permission of Academic Press.

[a]Total number of latex beads per sample was approximately 1.4×10^{10}.

[b]Rebanded coated latex established reference conditions; approximately—1: 535 Ag molecules carried per bead, 2: 504 Ag molecules carried per bead, 3: 545 Ag molecules carried per bead.

Table 5.4 Density of Complexes Formed by Interaction Between Bovine IgG Coated Latex Beads and Anti(Sheep IgG) at 23°C, 1.0% w/v NaCl, pH 8.2 (Each Sample Contained 10 μg of Coated Latex in a Total Incubation Volume of 0.2 ml)

Concentration of Ab Incubated (μg/ml)	Wt % Sucrose at Band Center	Density of Complex (g/ml)
0	22.0	$\rho_c = 1.0886$
10	22.8	$\rho_x = 1.0922$
100	22.7	1.0918
1000	23.0	1.0931
		Ave. $\rho_x = 1.0924$
		S.D. $\pm$ 0.0007

Source: Reproduced from Ref. 16 by the permission of Academic Press.

The adsorption coefficient with intermolecular interaction, K', can be related to the adsorption coefficient without interaction, K, in such a way that

$$K' = K \exp\left(\frac{-\Gamma\xi}{RT}\right) \qquad (5.3.10)$$

Equation 5.3.10 was obtained from (5.3.6) and (5.3.9). Thus, using (5.3.10) to modify the familiar double-reciprocal form of Langmuir's adsorption isotherm, the following form can be obtained:

$$\frac{1}{\overline{C}_s} = \frac{1}{K'(\overline{C}_s)_{max}} \frac{1}{\overline{C}_L} + \frac{1}{(\overline{C}_s)_{max}}$$

$$\approx \frac{1 + \Gamma\xi/RT}{K(\overline{C}_s)_{max}} \frac{1}{\overline{C}_L} + \frac{1}{(\overline{C}_s)_{max}} \qquad (5.3.11)$$

The second part of (5.3.11) was obtained by a series expansion of exp ($\Gamma\xi/RT$), omitting terms higher than the first order.

A breakpoint in each isotherm of Figs. 5.6a and 5.6b indicates that a two-mode mechanism occurred in the adsorption. For the solid support concentration used, liquid phase concentrations below $(\overline{C}_L)_{transition}$ produced Langmuir-type adsorption, and at $(\overline{C}_L)_{transition}$ or greater, the adsorption mechanism included the effect of molecular interaction. Since the adsorption isotherms in each region were defined by relatively straight lines, their characteristic parameters were determined independently using (5.3.11).

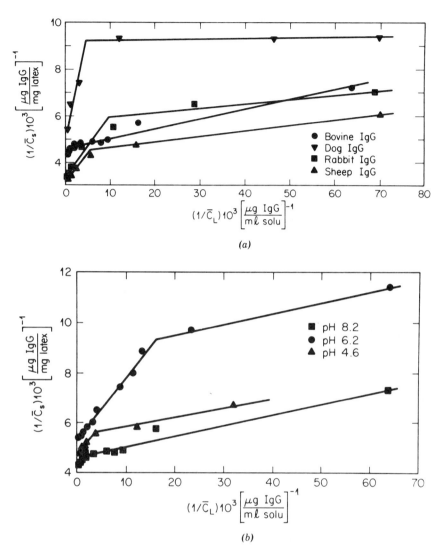

Fig. 5.6 Adsorption isotherms of immunoglobulins on latex beads in double reciprocal forms. (*a*) At 24°C, pH 8.2, and 1.0% w/v NaCl. (*b*) With three different pH values at 24°C and 1.0% w/v NaCl of bovine IgG. *Source:* Reproduced from Ref. 15 by the permission of John Wiley.

The parameters K, $(\overline{C}_L)_{\text{transition}}$, $(\overline{C}_s)_{\text{transition}}$, ΔG_I, and $(\overline{C}_s)_{\text{max}}$ for the four immunoglobulins and for bovine immunoglobulin in three different pH buffer solutions are presented in Table 5.5a and 5.5b respectively.

Interaction Isotherms

For a long time it has been considered that antibody-antigen reactions involve the combination of specific sites by which very large aggregates may be attained. To allow the existence of aggregates of this kind, one must necessarily assume that the antibody and antigen molecules are multivalent with respect to each other. If one is to require further that the antibody-antigen molecular ratio of these very large aggregates be variable and no less than unity, then one must consider antibody molecules to be bivalent and antigen molecules to be bivalent or greater. It should be noted that the existence of univalent antibody molecules in the system is still permitted. Heidelberger and Pedersen [69] demonstrated that a specific precipitate formed between an antigen and its antibody can be dissolved by adding to it a sufficient excess of the antigen in solution. Inhibition to precipitation in regions of antigen excess and also antibody excess are of fundamental importance in understanding antibody-antigen reactions. Our experimental model is the case of antibody excess, since the reactant antigen was already immobilized on polystyrene latex beads. A method to evaluate the equilibrium between free antibody and the complex formed by reacting with the immobilized antigen was used. In dealing with multivalent antigen which may combine simultaneously with a number of molecules of antibody, the mathematical formulation of the reaction is formidable. In simplifying the model, the assumption was made that the free energy of combination of an antibody molecule with a combining site of the antigen is the same for all such sites and is not affected by the number of antibody molecules that have already combined with the antigen. This can hardly be strictly true, but it is adequate as a first-step approximation. If A represents an individual antibody site, G an antigen site, and the complex formed by the interaction is AG, the reaction may be written as

$$A + G \underset{k_2}{\overset{k_1}{\rightleftharpoons}} AG \tag{5.3.12}$$

in which k_1 and k_2 are the forward and reverse reaction rate constants, respectively. If we let the association constant for the formation of a single antibody-antigen bond be K_{12} and the number of combining sites on the antigen molecule be m, the ratio r of antibody molecules combined with an antigen molecule can be written as the Langmuir adsorption isotherm [68]

$$r = \frac{mK_{12}[A]}{1 + K_{12}[A]} \tag{5.3.13}$$

Table 5.5a Adsorption Parameters for Four Immunoglobulins on Latex Beads (at 24°C, pH 8.2, 1.0% w/v NaCl)

Immunoglobulins	k	$(\bar{C}_L)_{transition}$ (µg IgG/ml Soln.)	$(\bar{C}_s)_{transition}$ (µg IgG/mg Latex)	k'	$\Delta G_I \times 10^3$ (cal/µg)	$(\bar{C}_s)_{max}$ (µg IgG/mg Latex)
Dog IgG	3.220	227	108	0.005	2.266	196
Rabbit IgG	0.296	106	167	0.012	0.089	294
Sheep IgG	0.181	182	200	0.011	0.056	323
Bovine IgG	0.110	500	213	0.017	0.022	238

Source: Reproduced from Ref. 15 by the permission of John Wiley & Sons, Inc.

Table 5.5b   Effect of pH on Adsorption Parameters of Bovine Immunoglobulin on Latex Beads (at 24°C, 1.0% w/v NaCl)

pH	k	$(\overline{C}_L)_{transition}$ (μg IgG/ ml *Solun.*)	$(\overline{C}_s)_{transition}$ (μg IgG/ mg *Latex*)	k'	$\Delta G_1 \times 10^5$ (cal/μg)	$(\overline{C}_s)_{max}$ (μg IgG/ mg *Latex*
4.6	0.151	270	178	0.019	-2.58	213
6.2	0.204	63	167	0.023	-2.95	185
8.2	0.110	500	213	0.017	-2.24	238

Source: Reproduced from Ref. 15 by the permission of John Wiley & Sons, Inc.

and the association coefficient K_{12} is defined as

$$K_{12} = \frac{k_1}{k_2} = \frac{[AG]}{[A][G]} \tag{5.3.14}$$

where the brackets denote concentration, $[A]$ is the concentration of free and unbound antibody, $[G]$ is the concentration of free and unbound antigenic determinants or haptenic group or ligands, and $[AG]$ is the concentration of joined units. In an intact antibody and an intact antigen system, the concentrations are usually expressed either in terms of bound antibody sites or bound antigenic determinants. If one denotes (1) n and s as the valence of antibody and of antigen, (2) A and G as the total number of antibody and antigen molecules in the system, (3) a_b and g_b as the fraction of bound antibody and antigen molecules, and (4) i and k as the fraction of sites occupied in the bound antibody molecules and antigen molecules, the concentrations in (5.3.14) can be expressed as

$$[A] = nA - ia_b nA \tag{5.3.15a}$$

$$[G] = sG - ia_b nA \tag{5.3.15b}$$

$$[AG] = ia_b nA \tag{5.3.15c}$$

Substitution of (5.3.14) into (5.3.13) gives the association constant K_{12}:

$$K_{12} = \frac{ia_b}{(1 - ia_b)(sG - ia_b nA)} \tag{5.3.16}$$

Furthermore, if we define the quantity f as a ratio between the total number of antibody molecules bound and the total number of antigen molecules in the system and also define d as the number of free antibody molecules in the system, (5.3.16) can be written as

$$\frac{f}{d} = sK_{12} - nfK_{12} \qquad (5.3.17)$$

in which

$$f = ia_b \frac{A}{G} \qquad (5.3.18a)$$

$$d = (1 - ia_b)A \qquad (5.3.18b)$$

The experimental data presented in Table 5.3 are shown in the Scatchard form in Fig. 5.7. In using (5.3.17), the antigenic particle was defined as an immunoglobulin (IgG) immobilized as antigen on latex beads. Thus, the valence of the antigenic particle depends on the number of IgG molecules attached to

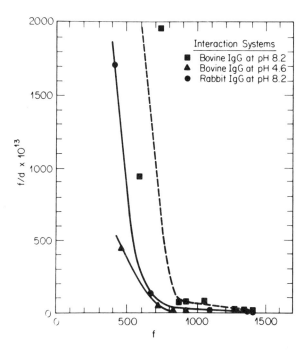

Fig. 5.7 Scatchard plot of interaction between immunoglobulin (Ag) and its specific antigenic immunoglobulin (Ab) immobilized on latex beads. *Source:* Reproduced from Ref. 16 by permission of Academic Press.

the bead and the number of determinants available for interaction on each antigen molecule. Since antibodies are known to be bivalent ($n = 2$), the valences on antigen s may be evaluated from the experimental data plot together with (5.3.17) by extrapolating (f/d) = 0. Thus, one obtains $s = 2f$. However, f was defined in terms of i as

$$f = i \frac{\text{number of antibody molecules bound}}{\text{number of antigen molecules in system}}$$

where i has been defined as the fraction of the binding sites of bound antibody molecules that have been filled. With the antibody molecules behaving bivalently (both sites filled on each molecule) in the intersection, i was equal to 1.0 and s was equal to twice the graphic abcissa determined. With the antibody molecules behaving monovalently (one site on each molecule filled) in the interaction, i was equal to 0.5 and s was equal to the graphic abcissa determined. As the interaction progressed, i varied and (5.3.17) was nonlinear. For estimating the antigenic valence and the equilibrium constant for the interaction process as the antigenic determinants became saturated, only the linear region of the curve corresponding to monovalent antibody behavior, $i = 0.5$, was used.

The values of the equilibrium constant were determined from the slope of the curves in Fig. 5.7 using only the region of the curves representing monovalent antibody binding $f > 800$. The free-energy changes associated with these interactions were calculated from the relation $\Delta G = RT \ln K$. These results are summarized in Table 5.6. All interactions were observed to proceed spontaneously, due to the relatively large negative free-energy changes for each system, which also reflect the biospecificity of the interaction between the antigenic determinants and the antibody combining sites.

In summary, the density-gradient isopycnic banding centrifugation combined with a carrier technique demonstrated here can be useful for obtaining quantitative information on the immobilization of enzymes, microorganisms, active biological fractions, or whole cells and on antigen–antibody interactions for microactive mass measurements.

4   CENTRIFUGAL FAST ANALYZERS

The combination of high-resolution separation methods which produce large numbers of fractions for analysis, the requirement for precise kinetic assays, the lability of many biological materials, the requirement in many situations for minute quantities of sample per assay, the use of more complex analytical procedures, and the trend toward the demand for more reliable and

Table 5.6 Thermochemical Properties of Immunoglobulin Reaction Systems

Antigen	Antibody	pH	$K \times 10^{-6}$	ΔG (kcal/mole)	Antigenic Valence
Bovine IgG	Anti-bovine IgG (rabbit) IgG	8.2	4.1	−9.1	1400
Bovine IgG	Anti-bovine IgG (rabbit) IgG	4.6	5.9	−9.3	920
Rabbit IgG	Anti-rabbit IgG (goat) IgG	8.2	1.8	−8.6	1400

Source: Reproduced from Ref. 16 by the permission of Academic Press.

definitive results through the use of statistical data reduction routines point toward the need for new fast computer-interfaced microanalytical systems that provide more experimentalist interaction with the experiment in progress and minimize the tedium involved in the actual analysis. Based on this concept, the Centrifugal Fast Analyzer (CFA) or the GeMSAEC Fast Analyzer (an acronym derived from two of the major sources of support, the National Institute of·General Medical Sciences and the United States Atomic Energy Commission Fast Analyzer) was developed by N. G. Anderson [1-14] while he was at the Oak Ridge National Laboratory developing the zonal centrifuges and ancillary systems for tissue fractionation and analysis [70]. This analyzer was successively developed by C. D. Scott and his associates also at Oak Ridge National Laboratory.

The centrifugal fast analyzer is made possible by the invention of methods for measuring, transferring, mixing, and measuring the absorbances of a large number of samples over very short intervals [1]. Three key problems are simultaneously solved by incorporating the analytical systems into centrifuge rotors: (1) accurate measurement of volume using centrifugal force to fill, debubble, and accurately level menisci in small vessels; (2) quantitative transfer of fluids using centrifugal force; and (3) rapid measurement of the absorbance of a large number of samples by rotating them rapidly past a beam of light. Basically, the analyzer arose from the idea of performing analyses in parallel. For example, a large number of individual reactions are simultaneously initiated, thereby establishing identical starting times for all of these reactions. The reactions then proceed under similar conditions of time, temperature, and reaction composition. The number of parallel analyses can be large enough to make possible the inclusion of standards in each set, thereby allowing measurements to be made before reactions have run to completion. To be generally useful, fast analyzers must be adaptable to a large number of different procedures. This concept differentiates the centrifugal fast analyzer from the continuous-flow and discrete-sample analyzers, which are based on analytical reactions that are sequentially initiated, processed, and monitored.

The centrifugal fast analyzer is a unique photometric system that utilizes the centrifugal field generated by a spinning, multicuvette rotor to transfer and mix a parallel series of samples and reagents into their respective cuvettes. The rotary motion of the rotor is utilized to move the cuvettes through the optical path of either a spectrophotometric or fluorometric optical system. By operating in this manner, the analyzer combines the continuous referencing advantages of a double-beam instrument with the operational simplicity of a single-beam system. The resulting signals obtained from the photodetector of the analyzer's optical system can be displayed on an oscilloscope set for spectrophotometric measurements. An electrical signal proportional to the light transmittance of the solution contained in a cuvette is displayed on the oscilloscope. The number of signals displayed is equal to the number of cuvettes contained in each rotor. The relative transmittance values may then be converted to absorbance values, which, after statistical evaluation, may be used for quantitative calculations. Because of the rotary motion of the cuvette assembly, these optical measurements can be made rapidly and repetitively. Thus, the data obtained from successive revolutions of the rotor may be accumulated and statistically averaged.

This technique of signal averaging further improves the signal-to-noise performance of the analyzer. The overall effect of combining continuous referencing with signal averaging is that the analyzer is a multicuvette, high-performance spectrophotometer or photometer. Thus, such analyzers are unaffected by minor signal drift in their systems and are capable of accurate, precise, and sensitive measurements of absorbance changes occurring in several simultaneous reactions. If fluorometric measurements are made, the signals for the case are proportional to the fluorometric intensities of the solutions contained in the various cuvettes instead of transmittance or absorbance of the solutions.

These fast analyzers are analytical instruments which provide output data in a form and at a rate suitable for direct input into a computer. They have been successfully interfaced with small computers, and the computer software necessary for routine and research operation has also been developed. In addition, over 20 clinical chemical procedures have been developed or adapted for use with the fast analyzers.

Recent developments include the design and fabrication of a miniature fast analyzer which provides the optical performance and parallel analysis advantages of the larger analyzer with the additional advantages of portability, space and weight economy, and a further decrease in sample and reagent column requirements. To capitalize on the portability advantages of the miniature analyzer, a portable miniature data system has also been developed and evaluated [71]. Concurrent with the development of the miniature system has been the design and fabrication of disposable plastic rotors that contain prepackaged and preloaded reagents and which are capable of accepting, automatically processing, and analyzing whole blood samples.

In addition to the development of the portable miniature system, a fluorescence detector is being developed to broaden the range of applications of fast analyzers. Concurrent with this is the development and adaptation of various fluorometric analyses. Computer programs based on the Michaelis–Menten rate expression have also been developed.

In summary, major efforts in the development of the centrifugal fast analyzer have been in automating the manual operations for routine use of the system, the sample and reagent loading, and the analysis start cycle; in developing new instrumental concepts that will allow greater sample throughput and overall reduction in equipment size; in improving temperature control and monitoring and developing a more compact analyzer system using smaller samples; in developing new chemical procedures to incorporate photometric techniques with the analyzer to increase the scope of analyses; in developing more methods in biochemical, enzymatic, and immunological assays; and in optimizing computer software.

Instrumental Systems

The centrifugal fast analyzer concept has resulted in the development of analytical systems by the Oak Ridge National Laboratory researchers and others. Three commercial models are now available; they are shown in Fig. 5.8. Their instrument names and manufactures are listed in Table 5.7

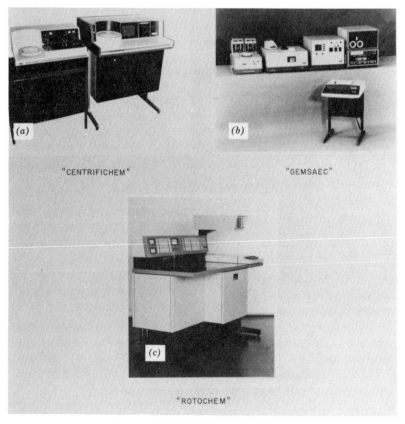

Fig. 5.8 Three commercial centrifugal fast analyzers: (a) CentriChem. (b) GeMSAEC. (c) RotoChem. Illustration courtesy of Oak Ridge National Laboratory.

Table 5.7 List of Commercial Centrifugal Fast Analyzers

Instrument Name	*Manufacturer*	*Address*
Centrifi Chem	Union Carbide Corp.	Tarrytown, New York
GEMSAEC	Electro Nucleonics, Inc.	Fairfield, New Jersey
Rotochem	American Instrument Co.	Silver Springs, Maryland

together with their addresses. In general, all three types of machines are similar in that they are based on Anderson's original concept. Specially, they differ as to number of cuvettes per rotor, optical configuration, means of temperature control and monitoring, and type of data system. A centrifugal fast analyzer system can be thought of as consisting of the following: (1) an anaiytical module, (2) a data acquisition and processing system, and (3) a mechanized sample-reagent loader.

1. *Analytical Module.* The analytical rotor and removable transfer disk, a mechanical drive, an optical system (spectrophotometric, photometric, or fluorometric), a system to monitor and control the temperature of the rotor, and a means of synchronizing the photodetector output of the analyzer with data acquisition devices.

2. *Data Acquisition and Processing Module.* A centrifugal fast analyzer produces data at a very fast, generally milliseconds rate. The rate of data generation per minute is equal to the product of rotor speed in rpm and the number of cuvettes in a rotor. For instance, for a 15-cuvette rotor spinning at a speed of 900 rpm, the data generation rate is 13,500 data points/min, or 225 data points/sec. Thus, the centrifugal fast analyzer yields output data in a form and at a rate suitable for direct input into a small computer [3, 6] and has been successfully interfaced with small digital computers [72, 73]. The simplest and least expensive approach to processing the data from a centrifugal fast analyzer is to photograph the oscilloscope display of the cuvette transmission signals. The operator can then manually digitize the data by measuring each individual transmission signal using a calibrated concentrations of dye or similar device. Although the oscilloscope display is a useful component of the analyzer, only the use of digital data systems permits full use of the capabilities of a centrifugal fast analyzer.

3. *Sample-Reagent Loader.* One of the basic procedures in the operation of a centrifugal fast analyzer is the loading of discrete aliquots of samples and reagents into their respective cavities in the transfer disk using some type of accurate and precise pipetting device. Various mechanized devices have been utilized to automatically load the transfer disks of a centrifugal analyzer [13, 74]. The precision of this loading system depends on the volumes dispensed

and varies from 0.25 to 0.92% [74]. The commercially available analyzers also incorporate mechanical loading devices as integral components of their system.

Detailed descriptions of those components in each module have been made elsewhere [12]. The advantages gained using a centrifugal fast analyzer may be summarized as: (a) improved analytical precision due to elimination of tedious, repetitive, manual steps; (b) capability of analyzing several samples during a single run; (c) decreased sample and reagent volume required per assay; (d) performance of all the analyses under identical conditions; and (e) ease in optimizing the analytical parameters in a short period of time, since in a single run the reactant concentration, pH, buffer concentration, and so forth, can be varied. Principal parts of the centrifugal fast analyzer and a typical display of transmittance data on an oscilloscope screen are shown in Figs. 5.9 and 5.10, respectively.

Methods of Photometric Analysis

The use of centrifugal force to rapidly add and mix each sample and reagent simultaneously into the appropriate cuvettes establishes a unique initial reaction starting time, with all reactions proceeding essentially in parallel under similar conditions of time, temperature, and reaction composition. The centrifugal fast analyzer, as a multicuvette spectrophotometer, can perform most of the functions of a spectrophotometer with good precision and accuracy, at a much higher sample analysis rate and on a microchemical scale requiring only 10 to 50 μl sample per assay. Among the types of photometric analysis that have been adapted and developed for use on the centrifugal fast analyzer by the group at Oak Ridge National Laboratory are colorimetric endpoint [6] and spectrophotometric enzymatic endpoint analysis [9, 75, 76], kinetic spectrophometric enzyme assay [77, 78], and kinetic substrate analysis [75-78].

The methods of assay developed may be classified as spectrophotometric titrimetry and reaction-rate methods of analysis. Both methods involve the monitoring of changes in absorbance rather than the measuring of absolute values of absorbance or transmittance. This feature minimizes the seriousness of any scattering, reflection, or reflection of radiation from the sample and simplifies the instrumentation required for absorbance measurements. In addition, spectrophometric titrations and reaction-rate methods of analysis make use of chemical reactions to introduce more specificity into the absorption measurements. In general, interfering substances are those which interfere with the chemical reaction rather than those which interfere by absorbing radiation at the same wavelengths as species being determined.

The basic spectrophotometric principles involving the two methods are briefly outlined below.

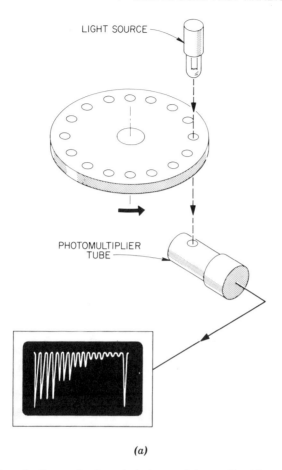

(a)

Fig. 5.9 (*a*) Schematic diagram showing principal parts of the centrifugal fast analyzer. (*b*) A 42-cuvette transfer disk: A, reagent cavity; B, sample cavity; C, mixing and transfer cavity; D, centering and indexing holes; and E, positioning tool used for carrying, placing, and removing transfer disk from rotor. Illustration courtesy of Oak Ridge National Laboratory.

Spectrophotometric Titrimetry [79]

Because all titrations involve the disappearance and formation of chemical species, the progress of a titration can often be monitored through measurement of the absorbance of the solution being titrated. This frequently permits more sensitive endpoint detection as well as the automation of titrations. Spectrophotometric titrimetry can be performed whenever (a) the substance being titrated, (b) the substance formed in the titration, or (c) the titrant itself exhibits distinctive absorption characteristics. In a centrifugal fast analyzer, the

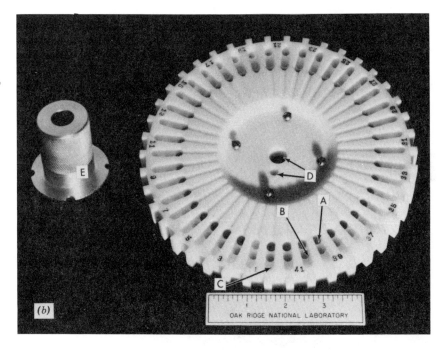

Fig. 5.9 Continued

volume of the titrant in each cuvette can be successively varied. In this case, a spectrophotometric titration curve (which is a plot of absorbance as a function of added titrant) consists essentially of two straight lines, their point of intersection being the equivalence point. In other situations in which none of the reactants or products absorbs radiation in the ultraviolet-visible region, it is sometimes feasible to use an acid–base, metallochromic, redox, or fluorescent indicator which does exhibit suitable absorption or luminescence properties. In these instances, the shape of the titration curve will depend on the nature of the indicator and on its interaction with reactant or product species.

When no indicator is employed, spectrophotometric titration curves having several different shapes are possible. Consider the general reaction

$$A + T \rightarrow P$$

in which A is the species being titrated, T is the titrant, and P is the product of the reaction. If just one of these three species—A, T, or P—absorbs radiation at the wavelength that is examined spectrophotometrically, the resulting titration curve will appear as shown in Figs. 5.11a, 5.11b, and 5.11c, respectively. However, if more than one species absorbs at a given wavelength, the shape of the titration curve will depend upon the relative molar absorptivities of the ab-

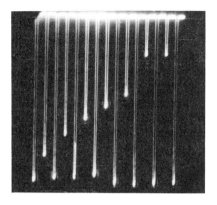

(b)   FLUORESCENCE

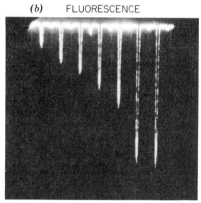

Fig. 5.10 Typical display of data on an oscilloscope screen as captured on photofilms. (*a*) Absorbance. (*b*) Fluorescence. Illustration courtesy of Oak Ridge National Laboratory.

sorbing components of the system, as illustrated in Figs. 5.11*d*, 5.11*e*, and 5.11*f*.

Note that, near the equivalence point in each figure, the experimental data do not fall perfectly on the two straight-line portions of the titration curve. This behavior is typical of spectrophotometric titration curves and is due to the fact that no reaction is ever 100% complete. The extent to which the experimental data deviate from the straight lines will increase as the equilibrium constant for the titration reaction decreases. In a manual spectrophotometric titration, one always takes experimental data on each side of the equivalence point, well away from the curved region, and then extrapolates the straight-line portions to their point of intersection to obtain the equivalence point. For greatest accuracy, it is necessary to correct the absorbance readings for the effect of dilution which occurs during the titration; otherwise, the titration curve

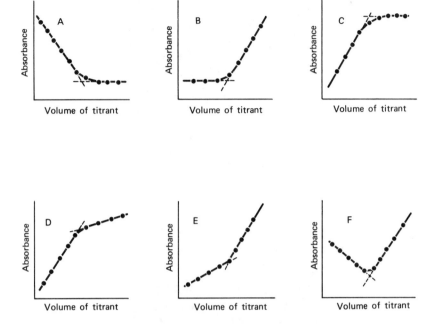

Fig. 5.11   Spectrophotometric titration curves for the hypothetical reaction A + T → P, in which A is the species being titrated, T is the titrant, and P is the product of the reaction. Curve A, only the species being titrated (A) absorbs. Curve B, only the titrant (T) absorbs. Curve C, only the product of the titration (P) absorbs. Curve D, A does not absorb, and the molar absorptivity of P is greater than that of T. Curve E, A does not absorb, and the molar absorptivity of T is greater than that of P. Curve F, P does not absorb and the molar absorptivity of T is greater than that of A. *Source:* Reproduced from Ref. 79, p. 659, Fig. 19-24, by permission of the authors and W. B. Saunders Co.

will show no straight-line portions, and extrapolation to the equivalence point will be uncertain. Of course, linearity in a titration curve presumes that Beer's law is obeyed.

Reaction-Rate Methods of Analysis [79]

Reaction-rate methods of analysis are finding increasing application in a number of areas, especially in the analysis of clinical samples by means of enzymatic reactions. These methods are based upon the principle that, if a species to be determined can react with another substance, the initial rate of reaction is approximately proportional to the initial concentration of the desired species. Therefore, measurement of the initial reaction rate permits an unambiguous determination of the initial reactant concentration. This type of analysis can be performed very rapidly, because it is unnecessary to wait until the reactants have come to equilibrium. This is especially important for slow reactions.

As an example, in the Michaelis-Menten equation for the generalized enzymatic reaction model,

$$S + E \underset{k_{-1}}{\overset{k_1}{\rightleftharpoons}} [SE] \underset{k_{-2}}{\overset{k_2}{\rightleftharpoons}} P + E$$

the rate equation for the disappearance of the substrate is given as [80]

$$\frac{-d[S]}{dt} = \frac{k_2[S][E_0]}{[S] + k_M} \tag{5.4.1}$$

where E is the enzyme, S is the substrate (reactant), and P is the product of the enzymatic reaction; k_M, the so-called Michaelis-Menten constant, is $(k_{-1} + k_{-2})/k_1$; $[E_0]$ is the initial concentration of the enzyme, and the bracket indicates the concentration of the substance inside the bracket.

DETERMINATION OF ENZYME

Two approaches can be employed to determine the initial concentration of enzyme $[E_0]$. One method is to saturate the enzyme with substrate; that is, the initial concentration of the substrate [S] is made very large compared with k_M. In this case, (5.4.1) takes the simple form

$$- \frac{d[S]}{dt} = k_2[E_0] \tag{5.4.2}$$

Therefore, the rate of the reaction remains constant for a considerable time until [S] falls to the level of k_M. To perform the actual determination, two reaction mixtures must be prepared; the first contains a known concentration of enzyme, the second contains the unknown concentration of enzyme, and each mixture has the same (large) substrate concentration. By comparing the reaction rates for the two systems, one can compute the unknown enzyme concentration.

However, it is not necessary that [S] be greater than k_M. If measurements are made of just the initial reaction rate, the concentration of substrate will not have time to change significantly. Therefore, all that is really required for the determination is for the two reaction mixtures to have the same initial substrate concentration (whatever its value), because the initial rate of reaction will be directly proportional to the enzyme concentration.

DETERMINATION OF SUBSTRATE

If, at the start of the reaction, [S] is chosen to be much less than k_M, (5.4.1) becomes

$$-\frac{d[S]}{dt} = \frac{k_2[S][E_0]}{K_M} = K[S][E_0] \qquad (5.4.3)$$

where $K = k_2/k_M$. Equation 5.4.3 indicates that the initial rate of reaction is proportional to both the initial enzyme and substrate concentrations. However, if the enzyme concentration is the same for all samples and standards to be analyzed, the value for the enzyme concentration can be incorporated into the constant K, so that (5.4.3) can be rewritten as

$$-\frac{d[S]}{dt} = K'[S] \qquad (5.4.4)$$

Thus, to determine the concentration of substrate in an unknown, two reaction mixtures are prepared, one containing the unknown and the other containing a known amount of substrate. Then, after addition of the same amount of enzyme, the initial reaction rate of each mixture is measured. The rate of reaction of the known mixture provides a value of the constant K', which can be used to find the concentration of the unknown sample from its initial reaction rate. This procedure succeeds because, when only the initial reaction rates are measured, the concentrations of two reactants remain virtually constant.

MEASUREMENT OF INITIAL REACTION RATES

For relatively slow reactions, manual mixing and stirring of the reactants are usually sufficient to initiate the reaction of interest. All the reagents and the sample can be added to a suitable sample cell, or cuvette, placed in the light path of a spectrophotometer. Then, an aliquot of the known reagent solution can be transferred into the cell, the solution stirred, and the initial reaction rate measured.

For faster reactions, special mixing techniques are often necessary to enable the reactants to be combined in a sufficiently short time. The centrifugal fast analyzer is most suitable for special quick mixing. The dynamic braking and acceleration feature of the centrifugal analyzer provides an excellent means of mixing.

For either slow or fast reactions, it is most convenient for interpretation of results to measure initial reaction rates in order to determine the concentration of the desired reactant. This minimizes the effects of temperature changes, side reactions, and other uncontrolled variables. When spectrophotometric monitoring of the reaction is employed and changes in the absorbance of the product are followed, it is possible to determine the initial reaction rate from the initial slope of a plot of absorbance versus time. For slow reactions, this plot can be simply displayed on a strip-chart recorder. For fast reactions, a high-speed recording device such as an oscilloscope or a computerized data acquisition system must be used.

For convenience, ease of automation, and reliability, many reaction-rate procedures employ sophisticated systems which provide a direct readout of the initial reaction rate or, if calibrated properly, of the reactant concentration itself.

Analysis by Luminescence Spectrometry

Although luminescence spectrometry is not used as much as absorption spectrophotometry, it provides the basis for some of the most sensitive molecular analytical techniques. *Luminescence* is the emission of radiation from a species after that species has absorbed radiation. The radiative loss of energy from a singlet to the ground state is termed fluorescence. If a molecule undergoes a transition from a triplet state to the ground state, it is termed phosphoresence. Depending on the specific molecule under investigation and its environment, the rates of the various processes competing for further energy loss are on the order of 10^7 to 10^9 sec^{-1} for fluorescence and 10^{-1} to 10^6 sec^{-1} for phosphorescence. Because of its spin-forbidden character, phosphorescence has a much longer life time than fluorescence.

There are two classes of luminescence spectrometers: those which employ filters and those which employ monochromators for spectral isolation. Filter instruments are called *fluorimeters* or *phosphorimeters,* depending on which luminescence process is being studied. Similarly, the monochromator instruments are termed *spectrofluorimeters* and *spectrophosphorimeters,* respectively. Obviously, instruments that utilize monochromators are more versatile and provide greater selectivity; because a narrower wavelength band is isolated for both excitation and observation of the luminescence spectrum, there is less chance of interference by other sample components.

Both fluorimetry and phosphorimetry techniques have several advantages over absorption spectrophotometric methods. First, fluorimetry and phosphorimetry offer greater selectivity and freedom from spectral interferences. This is because there are far more chemical species which absorb ultraviolet or visible radiation than those which fluoresce or phosphoresce. Moreover, in fluorimetry and phosphorimetry, one can vary not only the absorption wavelength but also the emission wavelength, so that it is possible to further reduce spectral interferences by judicious choice of both excitation (absorption) and luminescence wavelengths.

Second, fluorimetry and phosphorimetry are generally more sensitive than absorption methods. This is because in the luminescence techniques a direct measurement of the power of the emitted radiation can be made. By contrast, it is necessary in absorption methods to determine the difference between two large radiation levels—the incident power P_0 and the transmitted power P.

Because it is always easier to measure a small signal against no background than it is to measure the difference between two large signals, fluorimetry and phosphorimetry provide greater sensitivity than absorption spectrophotometry. This added sensitivity gives the luminescence methods still another advantage compared with absorption spectrophotometry. Whereas Beer's law plots are often linear over a ten- to hundred-fold range of concentration, it is not uncommon to find the relationship between luminescence power and concentration being linear over three or four orders of magnitude in concentration. Although this extended linear range is not necessary for quantitative analysis, it is often quite useful practically and requires fewer points on a calibration curve.

One additional advantage often enjoyed in phosphorimetry is that of *time resolution*. Because phosphorescence has a relatively long lifetime, it is possible to discriminate among different molecular species on the basis of their luminescence-time behavior. This provides an added dimension useful in both qualitative identification and quantitative determination of particular species by minimizing interferences from fluorescence and phosphorescence of other sample constituents as well as fluorescence from the analyte itself. In addition, scattering from the sample is eliminated through the use of time resolution.

Despite the obvious advantages of time resolution, phosphorimetry is not as widely utilized as fluorimetry. This fact is due to the added complexity of phosphorimetric instrumentation, the smaller number of species that phosphoresce, and the inconvenience of having to cool the samples to obtain adequate phosphorescence quantum efficiencies [81].

Most luminescence determinations are performed on samples of clinical, biological, or forensic interest, although inorganic luminescence analysis is not uncommon. Many drugs possess rather high quantum efficiencies for luminescence (the fraction of excited molecules that fluoresce), so that the determination by means of fluorimetry or phosphorimetry is both sensitive and practicable. Quinine, for example, can be detected at levels below one part per billion by weight and is often used as a calibration standard for fluorescence analysis. Many inorganic ions can be determined by means of fluorimetry or phosphorimetry if complexed with a luminescing organic ligand. Such ligands can be quite selective in their affinities for certain metals or nonmetals, so that luminescence methods for the determination of inorganic substances can be both specific and extremely sensitive. The procedure to adapt phosphorimetry for use with the centrifugal fast analyzer is not available yet at this time. The preparatory work involved in the adaption of general colorimetric or fluorometric procedures to use by a centrifugal fast analyzer is analogous to the preparations required to adapt such procedures to any other spectrophotometer or fluorometer.

Applications of Centrifugal Fast Analyzer Technology

A number of diverse applications of centrifugal fast analyzer technology have been developed by the researchers in Oak Ridge National Laboratory [6-14, 71, 74, 75, 78, 82, 83]. The development of new applications of centrifugal fast analyzer technology has been in the general areas of clinical enzymology, immunology, hematology, chemistry, and environmental analysis. In addition, certain enzymes are being investigated for their potential as biochemical markers of cancer using the centrifugal fast analyzers.

Enzymatic Analyses

Enzymes have been increasingly used as analytical reagents because of their biological sensitivity and specificity. Thus, a number of enzyme-based assays have been developed for the analyses of single substrates in biological fluids and tissues. Many of these assays are based on the enzymatic reduction or oxidation of nicotinamide adenine nucleotide, as an excess substrate in either the primary enzymatic reaction or a coupled enzymatic system, to quantitate the substrate [10, 84-86]. Both the equilibrium or endpoint and the reaction-rate methods have been used for this purpose.

In the equilibrium method, the reactants are mixed and the ensuing reaction is allowed to go to completion. The initial absorbance or fluorescence intensity and the final absorbance or fluorescence intensity of a reactant, or product, are measured, and the concentration of the unknown substrate is related directly to the change in absorbance (ΔA) or fluorescence intensity (ΔI) either by using a standard curve or by multiplying the signal change by a precalculated substrate factor (SF). In a spectrophotometric assay, the SF is calculated as [12]

$$SF = (MW)\frac{(V_T)(100)}{(V_s)(a)(b)} \qquad (5.4.5a)$$

in which MW is the molecular weight of substrate, V_T is total reaction volume, V_s is sample volume, a is molar absorptivity of the monitored component, and b is the cuvette path length (in cm). For a fluorometric assay, the substrate factor becomes

$$SF = \frac{(MW)(V_T)(RC)}{(MW_r)(RA)(V_s)10^4} \qquad (5.4.5b)$$

where RC is the concentration of the reference compound, MW_r is the molecular weight of reference compound, and RA is the micromolecular intensity ratio of the monitored compound to the fluorometric reference compound.

In the reaction-rate method, the initial rate of the reaction can be related to the concentration of the reactants. If the initial concentrations of all the reactants other than the species to be determined are kept in sufficient excess to ensure that their initial concentrations do not change appreciably during the measurement of initial rate, the rate equation for the nth-order reaction can be rewritten in analogy with (5.3.4) as

$$-\frac{d[S]}{dt} = K'(T)[S]_0^n \qquad (5.4.6)$$

The advantage of using a centrifugal fast analyzer as a parallel analyzer is evident from (5.4.6). When several reactions having common characteristics are proceeding simultaneously in one rotor under similar reaction conditions, including temperature, the analytical variability in the relationship between $-d[S]/dt$ and $[S]_0$ due to $K'(T)$ is eliminated.

When the reaction is first order, $n = 1$, a linear relationship exists between reaction rate and concentration. However, it has been shown [87] that, in order to maintain a linear relationship between the initial reaction rate and substrate concentration, it is necessary to measure the reaction rate during the initial 1 to 2% of the reaction. In practice, this is often a difficult measurement, especially if the reaction is relatively fast. In such a case the centrifugal fast analyzer offers a particular advantage in that it can be used to obtain a series of very early absorbance measurements during the first few seconds of a reaction, and, through the use of its data processing unit, reaction rates around the initiation period can be extrapolated and calculated with accuracy.

For reactions of higher order ($n > 2$), the analytical implications of (5.4.6) are more complex. However, a higher order reaction-rate method still has analytical application in that the initial reaction rate is a known function of the initial reactant concentration.

For a first-order or pseudofirst-order reaction, (5.4.6) can be integrated and rearranged for the t_1 and t_2 interval, which gives

$$[S]_0 = \frac{[S]_2 - [S]_1}{e^{-K't_2} - e^{-K't_1}} \qquad (5.4.7)$$

When t_1 and t_2 are precisely and accurately defined and all other reaction variables are held constant, the initial substrate concentration $[S]_0$ is linearly related to the change in substrate concentration during the time period $\Delta t = t_2 - t_1$. Since the change in absorbance ΔA or in fluorescence intensity ΔI is generally proportional to the change in substrate concentration, either of these changes can be used to calculate $[S]_0$. Thus, this reaction-rate method is a "fixed-time" approach to enzyme-catalyzed reaction-rate analysis, which is ideally suited for use with a centrifugal fast analyzer, since t_1 and t_2 can be precisely and accurately defined for all the reactions contained in one rotor. By including a calibration standard in each analytical run, a factor can be obtained and used to convert either ΔA or ΔI in the time interval Δt to the initial concentration of the analyte.

A number of biochemical and enzyme assays that have been adapted or developed for use with centrifugal fast analyzers are listed in Table 5.8.

Table 5.8 Assays Developed and Adapted to the Centrifugal Fast Analyzer

a: Biochemical Assays

Adenosine triphosphate	Glycerol
Albumin	Hemoglobin
Blood urea nitrogen	Inorganic phosphate
Bilirubin	Magnesium
Calcium	Morphine
Cholesterol	Serum ammonia
Cortisol	Total protein
Creatinine	Triglycerides
Glucose	Uric acid

b: Enzyme Assays

Oxidoreductases	*Transferases*
Alcohol dehydrogenase	Adenyl kinase
Cytochrome C reductase	Alanine aminotransferase
Glycerol dehydrogenase	Aspartate aminotransferase
Lactate dehydrogenase	Creatine phosphokinase
Isocitrate dehydrogenase	Galactose 1-puridyl transferase
Galactose-6-phosphate dehydrogenase	Glucokinase
Glucose-6-phosphate dehydrogenase	Glucuronyl transferase
β-Glucuronidase	Glutamic-oxaloacetic transaminase
Glutamate dehydrogenase	Glutamic-pyruvic transaminase
Glutathione peroxidase	Glutamyl transpeptidase
Glutathione reductase	Glycerol kinase
Glyceraldehyde phosphate dehydrogenase	Hexokinase
6-phosphate gluconate hydrogenase	Phosphoglucerate kinase
Hydroxybutyric dehydrogenase	Pyruvate kinase
Indophenol oxidase	Serine–pyruvate transaminase
Malic dehydrogenase	Trans–sulfurase
Malic enzyme	
Sorbitol dehydrogenase	

Table 5.8   (Continued)

Lyases	Others
Aldolase-fructose 1-phosphate	Adenosine dimerase
Aldolase-fructose 1,6,-diphosphate	Diphosphoglyceromutase
Catalase	Fatty acid synthetase
Enolase	Glucose phosphate isomerase
Glyoxalase I	Triosephosphate isomerase
Glyoxalase II	Uridinediphosphate glucose epimerase

Hydrolases
Acetylcholine esterase
Acid phosphatase
Adenosine triphosphatase
Alkaline phosphatase
Cholinesterase
Fructese diphosphatase
β-Galactosidase
Lysozyme
Trypsin

As an example, an enzymatic assay using the ATP-hexokinase reaction for the determination of glucose with the minature centrifugal fast analyzer by the Oak Ridge Researchers [83] is presented in the following: Sample (2 μl) and reagent (20 μl) were loaded into a conventional sequential transfer rotor (total volume 122 μl). Data acquisition parameters were: delay time, 0.2 sec (this does not include a time interval of approximately 0.28 sec during which the rotor and its contents were accelerated to 4000 rpm, and dynamic braking was applied with subsequent reacceleration of the rotor); run speed, 3500 rpm (17 msec/revolution); observation interval, 0.3 sec; and an average of five readings per data point. Data were acquired and all computations performed by the microprocessor unit of the portable analyzer. The first 30 measurements obtained by the analyzer (corresponding to approximately 0.5 to 9.5 sec after the rotor first began to accelerate) are illustrated in Fig. 5.12; data are shown for a reagent blank and a 200-mg glucose/100 ml aqueous sample—both versus a water reference. A final measurement was obtained on the samples after equilibrium had been attained (approximately 30 min at 25°C).

The data of Fig. 5.12 are summarized in Table 5.9. In the figure, the first data points appear to be high; this may be due to a transient power drain on

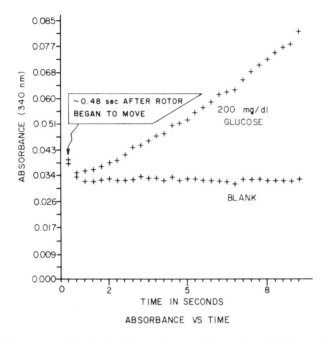

Fig. 5.12 Progress curve for the hexokinase-glucose reaction monitored by the portable centrifugal fast analyzer (total time elapsed ≈ 9.2 sec. Illustration courtesy of Oak Ridge National Laboratory.

the system from the mixing sequence and can thus be corrected. The first few data points were omitted in the calculation of the data of Table 5.9. The optical precision of the system is shown in the measurement of the reagent blank. The measurement precision within cuvette was <1%, and the average blank ($n = 5, \overline{A} = 0.0323$ optical density units (ODU)) was precise to within ±1.4 to 1.8%. At higher absorbances (e.g., the final absorbance of the 400 mg/100 ml sample), the overall precision was approximately 0.2% ($n = 5, \overline{A} = 1.2132$ ODU). The ratio of the change in absorbance of the 400 mg/100 ml sample to that of the 200 mg/100 ml sample was 1.995 (theoretical $= 2.0$). The linear least-squares analysis subroutine of the microprocessor unit was used to estimate the initial absorbance (A_0) of the reactions, and this was used, together with the measured final absorbance (A_f), to estimate the change in absorbance. The results of this analysis agreed very well with the average change in absorbance measured at equilibrium (see Table 5.9).

In summary, the optical performance of the analyzer was excellent. The microprocessor unit performed faultlessly for the data acquisition and manipulation, and the feasibility of performing moderately fast kinetic

Table 5.9 Enzymatic Determination of Glucose on Portable Centrifugal Fast Analyzer

Cuvette Number	Glucose Concentration (mg/100 ml)	A_0^a Extrapolation	A_f^b	A_f^c Average	$A_f - A_0$	Average	$\bar{A}_f - \bar{A}_f$ Blank
2		0.0325	0.0323	0.0323	−0.0002	0.0001	0.0000
3		0.0328	0.0328	±0.0004	0.0000		
4	0	0.0313	0.0317	(1.36%)	0.0004		
5		0.0323	0.0327		0.0004		
6		0.0325	0.0322		−0.0003		
7		0.0257	0.6203	0.6243	0.5946		
8		0.0267	0.6280	±0.0045	0.6013	0.5960	0.5920
9	200	0.0279	0.6203	(0.72%)	0.5924	±0.0034	
10		0.0283	0.6230		0.5947	(0.57%)	
11		0.0331	0.6301		0.5970		
12		0.0240	1.2177		1.1937		
13		0.0247	1.2100	1.2132	1.1853	1.1893	1.1809
14	400	0.0206	1.2127	±0.0028	1.1921	±0.0035	
15		0.0245	1.2122	(0.23%)	1.1877	(0.29%)	
16		0.0257	1.2133		1.1876		

Source: Reproduced from Ref. 83, p. 33, ORNL/TM-5533.

[a] y-Intercept for linear least-squares analysis of absorbance (y) versus time (x) for the time interval 1.9 to 9.2 sec.

[b] Data taken after 30-min reaction time at 25°C.

[c] Data reported as mean ± standard deviation (coefficient of variation given in parentheses).

analyses with great sensitivity was established (note in Fig. 5.12 that 30 data points representing an absorbance change of 47 millilabsorbance units were obtained in 9.2 sec, with a standard deviation of the slope of only 1.7%).

Immunological Analyses

Centrifugal fast analyzers have also been used for immunological analyses. Among many potential applications, three have been developed, which include measurement of antigen–antibody aggregation using a light-scattering monitor, qualitative monitoring of red blood cell agglutination, and fluoroimmunoassays using fluorescent-tagged bioreagents [12].

MEASUREMENT OF AGGREGATION

The interaction of antigen with certain specific antibodies results in the formation of an aggregation complex that scatters light at detectable levels above solvent–antibody background scatter. The relationship between the light scatter from these aggregating protein systems and the mechanism of antigen–antibody interaction is complex; consequently, this technique has been reduced in importance as a means of determining the mechanisms of antigen–antibody aggregation [88]. However, there is a direct relationship between the change in light-scatter intensity (ΔI) and the antigen concentration, which, when proper calibration standards are employed, can be used for the rapid quantitation of specific protein in biological samples without prior fractionation [89–93]. The detectable concentration of immunoglobulins is of the order of 1×10^{-9} mole/liter with current state-of-the-art fluorometric instruments that do not employ laser optics.

The application of light-scatter measurement analytical procedures to the centrifugal fast analyzer has followed the development of right-angle fluorescence optics for the miniature system. The analytical parameters that need to be considered are antibody titer, dynamic range of antigen concentration with reference to antigen excess, reaction volume, and time of analysis.

MEASUREMENT OF AGGLUTINATION

The dynamic braking and acceleration features of the miniature centrifugal fast analyzer can be utilized to develop assays involving agglutination. For example, the rotor of the miniature analyzer can be accelerated to 4000 to 5000 rpm and maintained at this centrifugal speed for several minutes. The rotor can then be braked from 4000 to 0 rpm in less than 1 sec, providing an excellent means of mixing or resuspending packed particles. The combination of centrifugal addition of antisera (reagent) to washed cells (sample) and the above features of the miniature centrifugal fast analyzer provides the basis of easy, reliable agglutination measurements.

The principle is simple. Properly titered antiserum is incubated for approximately 1 min in the slowly spinning rotor, and then the mixture is centrifuged for 2 min at 4000 rpm. Subsequently, the packed material in the rotor is then subjected to the braking of the rotor from 4000 to 0 rpm in less than 1 sec of elapsed time. If the reaction between the antiserum and washed cells is negative, little or no agglutination occurs; consequently, the rapid braking operation results in the resuspension of the nonagglutinated or partially agglutinated particles. However, if the reaction is positive, the agglutinated particles do not resuspend upon rapid braking of the rotor. To determine whether the reaction is negative or positive, the rotor is rotated at 500 rpm after the rapid braking operation, and an optical measurement is made. If the reaction is negative, the resuspended particles result in a solution of high absorbance. If it is positive, the agglutinated particles are centrifuged from the optical path of the rotor, and the resulting solution has a low absorbance. Therefore the presence or absence of absorbance provides a simple means of optically distinguishing agglutinated, partially agglutinated, and nonagglutinated particles such as red blood cells.

FLUOROIMMUNOASSAY

The area of analytical immunology is in a period of rapid development, and considerable interest has been devoted in developing sensitive immunological assays that are not dependent on the use of radioisotopes as tracers. Consequently, a great deal of effort has been undertaken to utilize fluorescent-labeled antigen or antibody in the development of fluoroimmunoassays. The advantages of such an analytical approach are the use of stable fluorescence tracers and the direct measurement of bound antigen–antibody without the prior need for a separation step as is required with radioimmunoassay systems. Two approaches are being investigated, one of which involves fluorescence polarization. Dandliker et al. [94, 95] and Tengerdy [96] have suggested the potential use of this approach for antibody quanitation, and recently Spencer et al. [97] have demonstrated the use of an automated flowing-stream fluorescence polarization apparatus for the determination of insulin antibodies. A second approach is to use fluorescent-labeled antigens in competitive binding assays, and Aalberse [98] has recently demonstrated the successful quantitation of human IgG using fluorescent-labeled antibodies to IgG. The sensitivity of this approach, 5 to 10 ng/ml, is two or three orders of magnitude more sensitive than light-scatter procedures and potentially useful for the quantitation of human IgE.

The centrifugal fast analyzer equipped with a fluorometric monitor offers the potential of coupling dynamic multicuvette fluorescence measurements with a centrifugal field, thereby providing some interesting opportunities and possibilities relative to the use of fluorescent-labeled antigens and solid-phase coupled antibodies for fluoroimmunoassays. Investigation of the sensitivity requirement of fluorescence instrumentation for use in fluoroimmunoassays has been reported by Tiffany et al. [82] in the concentration range of haptens and antigens as low as 10^{-9} to 10^{-11} M. The capability of measuring monolabeled fluorescein isothiocarbamylinsulin in the concentration range of 3×10^{-10} to 2×10^{-9} M with the fluorometric centrifugal fast analyzer has been demonstrated [82]. This is two orders of magnitude above the concentration range needed to quantitate insulin in human samples by fluoroimmunoassays but should be useful for antibody quantification. The fluorometric centrifugal fast analyzer can also be used to make multicuvette fluorescence polarization measurements, which adds an additional mode of operation in the adaptation of immunological analytical procedures to the centrifugal fast analyzers.

Enzymes as Biochemical Markers of Malignancies [83]

A number of enzymes and isoenzymes have been shown to be altered in tumor tissue and in the physiologic fluids of subjects with malignancies. In general, such changes, whether originated by the tumor or induced by organ damage due to the malignant lesion, have not been diagnostically useful for neoplastic disease because of their nonspecificity. However, once the presence of a malignancy has been established, serum enzyme measurement can be extremely useful in monitoring the patient's progress [99, 100]. Caution must be exercised, however, in the interpretation of these changes in activity levels when using serum enzymes in this manner. Enzymes are generally selected on the basis of their sensitivity, and they may well be sensitive to other acute or chronic illnesses which may develop as a result of either the malignancy or its treatment while the malignancy is under surveillance.

Recently, there has been a great deal of interest in the search for biological markers that may be used as sensitive and noninvasive measures of malignant cell population. After surgical or chemotherapeutic treatment of a malignancy, such biochemical markers could be used to determine whether the treatment has successfully arrested tumor growth, as shown by decreasing or stable concentrations of the biological markers. Several of the serum enzymes may be useful in this context. No biochemical marker has been discovered that is specific for malignant disease. The primary value of markers has thus far been to monitor the progress of diagnosed tumors. Again, no one marker is sufficient for monitoring malignant cell populations in all cases; therefore, a battery of enzyme assays has been recommended [99, 100].

For example, serum levels of exoenzymes such as gamma-glutamyl transpeptidase (GGTP), leucine aminopeptidase (LAP), and 5'-nucleotidase (5'-Nase) are elevated above the upper limit of the normal range in carcinoma involving the hepatobiliary duct system [101]. A battery of enzymes may include the exoenzymes mentioned above plus the less specific enzymes glutamate–oxaloacetate transaminase (GOT, or 2-oxoglutarate aminotransferase), alkaline phosphatase (AP), and lactate dehydrogenase (LDH), all of which are of some utility in the diagnosis of cancer [102]. The latter three enzymes have already been adapted to the centrifugal fast analyzer in the enzymatic assays.

GAMMA-GLUTAMYL TRANSFERASE

Gamma-glutamyl transferase catalyzes the transfer of the gamma-glutamyl group from gamma-glutamyl peptides to other peptides or amino acids. Although found in nearly all tissues, the enzyme has its highest activity in the kidney, followed by the pancreas, liver, and spleen [101]. Clinically, the enzyme has been found to be a very sensitive indicator for diseases of the hepatobiliary tract and pancreas.

LEUCINE AMINOPEPTIDASE

A distinction must be drawn between "classical" leucine aminopeptidase (LAP), utilizing leucinamide as a specific substrate, and the group of amino acid arylamidase enzymes active toward synthetic substrates such as L-leucine-p-nitroanilide [101, 103]. It is the latter group of enzymes that is of diagnostic value in clinical chemistry, not "true" LAP.

So-called LAP can be detected with varying activity in nearly all human tissues, occurring principally in the mucosa of the small intestine and in the pancreas [101, 104]. Leucine aminopeptidase is elevated in disseminated malignant disease, including carcinoma of the head and neck, the gastrointestinal tract, bronchus, prostate, geritourinary tract other than prostate, hematologic malignancies, and malignant melanoma [105]. It may also be elevated in patients with nonmalignant disease, especially of the liver [105]. Leucine aminopeptidase activity also rises progressively as pregnancy advances, reaching several times the upper limit of the normal range at the end of pregnancy [106]. This elevation is due primarily to a heat-labile placental LAP, which may be differentiated from normal serum LAP (heat-stable) by incubation at 60°C for 30 min [106]. Thus, by itself, elevated LAP activity in serum lacks specificity as a diagnostic test for the presence of tumor. In this respect, it has been indicated that elevated LAP in 24-hr urine specimens, as opposed to sera, may be of greater specificity in estimating the extent of the tumor and the success of treatment [105].

5'-NUCLEOTIDASE

5'-Nucleotidase (5'-Nase) has a wide distribution in tissue but is highest in liver, lung, brain, and kidney [101]. 5'-Nase has been frequently used as a biochemical marker for plasma membrane, since one form of the enzyme is largely restricted to the plasma membrane of liver cells and bile canaliculi [107].

The clinical significance of elevated 5'-Nase activity is due to its high specificity for hepatobiliary disease [101]. 5'-Nase and LAP appear to be equally sensitive for the detection of hepatic metastases and exceed the sensitivity of nonspecific alkaline phosphatase (AP) [101]. In contrast to LAP,

5′-Nase is unaffected in pregnancy [101]. Both AP and 5′-Nase are elevated in liver disease, while 5′-Nase is rarely elevated in bone disease [107]. Thus, an elevated serum AP level in the presence of a normal 5′-Nase level is usually an indication of bone disease, and where both are raised, hepatobiliary disease is virtually certain [107]. Serum 5′-Nase rather than AP should be estimated in children or pregnant women suspected of having liver disease [108].

The assay of serum leucine aminopeptidase (LAP) in connection with a battery of other exoenzyme assays greatly enhances its utility as a marker for cancer. The properties of those enzymes are shown in Table 5.10. Details of the assay procedures are available elsewhere [83, 101–108].

ALKALINE PHOSPHATASE

Alkaline phosphatase (AP) exists in normal serum as a number of isoenzymes. The primary isoenzymes, named for the tissue with the greatest specific activity, are bone, liver, and intestinal AP. Increased total AP activity may reflect bone proliferation, including primary tumors such as osteogenic sarcoma and parathyroid adenoma or carcinoma, and neoplasms such as carcinoma of breast or prostate that metastasize to bone [109]. Activity may also be elevated in hepatic disease (especially biliary obstruction), pregnancy (placental AP during the third trimester), or renal infarction [110].

It has been previously reported that an assay of 5′-Nase may be used in conjunction with nonspecific AP to determine the origin (liver or bone) of elevated AP levels [107]. It has been suggested that either LAP or GGTP may be used in a similar manner [102]. In addition, the kinetic properties of the various isoenzymes differ enough to enable estimation of the individual activities without using laborious electrophoretic techniques [110]. Much work has been reported on the so-called Regan isoenzyme as a potential aid in monitoring therapy and disease progress in cancer [100, 102]. This isoenzyme appears to be identical with the heat-stable, L-phenylalanine-sensitive AP found in the human placenta [102].

Environmental Applications

There has been in recent years considerable public concern about environmental quality, particularly with respect to the presence of toxic and carcinogenic contaminants in air and water. In response, the U.S. Congress has enacted several pieces of legislation aimed at solving current problems and preventing future abuses. The chief enforcement arm, the Environmental Protection Agency, has the responsibility of defining hazardous materials, setting maximum tolerance limits, issuing standards, and requiring that appropriate tests be made. Individual states, moreover, may require compliance with additional regulations. It is evident that to meet these diverse requirements there is a serious need for rapid, reliable, and inexpensive analytical methodology.

Table 5.10 Properties of Three Biliary Tract Enzymes

	LAP	5'-Nase	GGTP
Technical facility of assay	Simple (400 nm)	Suitable UV	Simple
Age dependence of serum activity	Significant until near puberty	Increased, but less significant until near puberty	Adult activity reached by 4 months
Pregnancy effect on serum activity	Significant	Not affected	Not affected
Specificity for hepatobiliary disease	High, but not exclusive	Nearly exclusive	Least specific of the group, yet high
Sensitivity for hepatic neoplasms	Greater than AP, especially in the anicteric patient	Same	Same
Existence of multiple molecular forms (molecular heterogenicity)	Yes	Yes	Yes

Source: Reproduced from Ref. 83, p. 137, ORNL/TM-5865/v2.

The centrifugal fast analyzer is ideal for this purpose in evaluating environmental quality. Several analytical methodologies have been developed for the analysis of pollutants in air and water [83, 111]. In many cases, standard methods could be adapted to the centrifugal fast analyzer, in others new methods were developed. In general, analyses performed with the centrifugal fast analyzer were faster, more precise, and more accurate than with conventional instrumentation. Analytical methods developed for use with the centrifugal fast analyzer are for the following elements and compounds: SO_2, O_3, Ca, Cr, Cu, Fe, Mg, Se(IV), Zn, Cl^-, I^-, NO_2^-, PO_4^{3-}, S^{2-}, and SO_4^{2-}, using either the spectrophotometric method or the turbidimetric, kinetic (including initial rate, fixed time, and variable time), and chemiluminescence methods. The determination of inorganic nutrients commonly found in the waste water and of adenosine phosphate and the field measurement of pH have also been investigated with the centrifugal fast analyzer [83, 111].

A listing of the specific analyses developed for those elements and compounds together with their concentration ranges is shown in Table 5.11. Methods developed for the others are as follows.

SILICA AND PHOSPHATE

These two major inorganic nutrients in natural waters must frequently be determined in the presence of each other. The colorimetric determination of these species is usually achieved through the formation of heteropolymolybdates. Often, to enhance the sensitivity of the assays, the heteropolymolybdates (yellow) are subsequently chemically reduced to yield a "heteropoly blue" or "molgbdenum blue" complex [112]. Under the usual conditions for the determination of phosphate as the molybdenum blue complex, soluble silica is generally not a serious interferent [113]. On the contrary, phosphate may be a serious interferent in the determination of silica; however, the addition of tartaric acid almost completely prevents formation of the phosphate complexes while leaving the silicon complex largely unaffected [114]. Thus, kinetic differentiation between silica and phosphate is also possible [87, 115]. Ingle and Crouch [87] have determined silica by the rate of formation of the unreduced heteropoly acid and have subsequently determined phosphate in the same sample by measuring the rate of reduction to the molybdenum blue complex.

The Ingle and Crouch procedure has been modified and adapted for use with the centrifugal fast analyzer by Bostick et al. [116]. The modification includes a modification of the reagent to permit a more rapid analysis and a change in monitoring absorbance wavelength from 400 to 340 nm. The authors report that the sensitivity of the procedure is approximately four times as great when monitored at 340 nm, and, with the modified reagent, reactive silica could be determined either by an initial-rate kinetic approach

Table 5.11 Analyses of Pollutants Adapted to Centrifugal Fast Analyzer

Substance	Method of Analysis	Concentration Range (ppm)	Rel. Std. Dev. (%)
SO_2	West-Gaeke Method, fixed-time kinetic	$<1^a$	1
O_3	Reaction with eugenol, fixed-time kinetic	$<1^a$	2
Ca(II)	As calcium oxalate, turbidimetric	10-100	10
Cr(III), Cr(VI)	Catalytic luminol method, chemiluminescence	0.05-0.6	1
Cu(II)	Catalytic method, variable-time kinetic	0.1-1	5
Fe(III)	Reaction with thiocyanate, spectrophotometric	0.1-3	3
Mg(II)	As $Mg(NH_4)PO_4$, turbidimetric	5-70	5
Se(IV)	Catalytic method, variable-time kinetic	0.1-1	5
Zn(II)	Reaction with PAR, spectrophotometric	0.2-1	3
Cl^-	Reaction with $Hg(SCN)_2$, spectrophotometric	5-75	2
I^-	Catalytic method, initial-rate kinetic	0.02-0.1	10
NO_2^-	Diazotization and coupling, spectrophotometric	0.1-1	1
PO_3^{3-}	Reduced phosphomolybdate method, fixed-time kinetic	1-14	5
S^{2-}	Synthesis of ethylene blue, spectrophotometric	0.1-1	1
SO_2^{2-}	Reaction with Ba(II) or CAD, turbidimetric	5-50	2

Source: Reproduced from Ref. 111, p. 18, ORNL-NSF-EATC-15.
[a] Dependent on sample collection time.

or by an equilibrium absorbance measurement. Each approach yielded a calibration curve that was linear for a broad range of silica concentrations and had a near-zero intercept. When the effect of phosphate interference was investigated, the superiority of the initial-rate procedure was definitely established. A typical result of the experimental run is presented in Fig. 5.13. The procedure is based on measurement of the rate of the formation of the heteropolymolybdates and extrapolation of the data to obtain a measure of the initial absorbance. In the case of the phosphate reaction, the extrapolated "initial" absorbance represents the endpoint absorbance for a very fast reaction. This extrapolation procedure is analogous to that which has been previously employed with centrifugal fast analyzers for the two-point fixed-time kinetic analysis of glucose [78, 117]. The substrate is enzymatically linked with the production of NADPH, and the reaction is monitored at 340 nm. The first few seconds of the reaction are monitored to obtain (by extrapolation) an estimate of the initial absorbance (A_0); then a final reading is made to obtain the final absorbance $(A_f - A_0)$, which is directly related to substrate concentration.

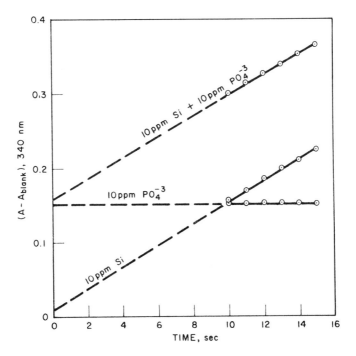

Fig. 5.13 Simultaneous determination of silica and phosphate. The slope of the absorbance-vs.-time data is a measure of the silica content of the sample, and the absorbance intercept is a measure of the phosphate content. Illustration courtesy of Oak Ridge National Laboratory.

Several of the inorganic nutrients may be similarly enzymatically linked to the oxidation–reduction of NADH or NADPH.

SULFATE

The determination of trace quantities of sulfate ion (SO_4^{2-}) in water has also been developed for use with the miniature centrifugal fast analyzer by a novel direct, kinetic spectrophotometric method [83]. The method is based on the zirconium–methylthymol blue reaction [118].

One difficulty in the evaluation of the kinetic data is that the time interval for which the reaction rate is apparently first order for the sulfate ion varies with the age of the zirconium reagent (occurring at longer intervals as the zirconium ages and becoming more highly polymerized). In addition, within a run, the linear segment of the reaction progress curve occurs at shorter time intervals as the concentration of sulfate ion is increased. In order to automate the selection of the time interval used for computation of reaction rate, a linear search program is used. The minimum acceptable correlation coefficient selected (0.9993) permits discrimination against the initial rapid increase in absorbance due to the presence of nonpolymerized zirconium in the reagent. Following this initial rapid increase, the reaction is governed by the rate of depolymerization catalyzed by the sulfate ion [83]. This permits a single set of observation conditions to be used for a variety of sample concentrations and reagent ages.

The precision in the determination of reaction rate using the linear search subroutine is 1 to 2%, or 0.2 to 0.3 ppm SO_4^{2-}. The detection limit (two times the standard deviation of the rate measurement of a 2 ppm SO_4^{2-} standard divided by the slope of the calibration curve) is ~ 0.3 ppm.

ENZYMATIC DETERMINATION OF AMMONIA

An enzymatic procedure to determine microquantities of ammonia in water using the centrifugal fast analyzer has also been developed [83]. L-glutamate dehydrogenase, ADP, α-ketoglutarate, and NADH were combined as a single reagent in 0.30 M Tris buffer, pH 8.6. The combined reagent was found to be stable for at least three days when stored at 2 to 4°C. Reagent, 30 μl, plus 25 μl diluent and 45 μl sample plus 30 μl diluent are loaded into the appropriate chambers of the rotor. The rotor is brought to 30°C and placed on the analyzer; the reaction is initiated by acceleration of the rotor to 4000 rpm, followed by dynamic breaking and adjustment to a run speed of 1000 rpm. The reaction is monitored by the decrease in absorbance at 340 nm due to consumption of NADH as a function of time. Since the reaction is highly selective (no competing reactions) and since the time required to approach equilibrium is relatively short, the equilibrium endpoint assay has been used as an absolute method.

ENZYMATIC DETERMINATION OF NITRATE

Biological reduction of nitrate is quite common [119, 120], and several attempts have been made to utilize microorganisms or enzyme fractions as a reagent for the determination of nitrate. Garner et al. [121] used a microorganism isolated from the rumen of a sheep to reduce nitrate to nitrite, then analyzed the nitrite formed via the diazotization-coupling reaction with sulfanilamide and N-(1-naphthyl)ethyldiamine. In a similar manner, Hill et al. [122] utilized the bacterium *Pseudomonas oleovorans.*

The majority of recent applications of enzyme fractions in the reduction of nitrate to nitrite have involved the nitrite reductase from *Escherichia coli* [123-126]. When this organism is grown under anaerobic conditions, nitrate is primarily used as an electron acceptor (in the place of oxygen), and the enzyme isolated under these conditions is referred to as "respiratory" nitrate reductase (EC 1.9.6.1), in contradistinction to the "assimilatory" nitrate reductase (EC 1.6.6.1) predominant in other organisms in which nitrate is used in the synthesis of nitrogen-containing cell constituents [120]. The commercially available respiratory enzyme [123] uses dithionite as an electron donor and the radical cation of methyl viologen as a cofactor. Both of these reagents are unstable and oxygen sensitive.

One method for enzymatic measurement of nitrate is based on monitoring NADPH consumption at 340 nm. Nitrate reductase (reduced NADP:nitrate oxidoreductase), EC 1.6.6.1, catalyzes the reduction of nitrate to nitrite:

$$\text{NADPH} + \text{NO}_3^- \underset{\text{reductase}}{\overset{\text{nitrate}}{\longleftrightarrow}} \text{NADP} + \text{NO}_2^- + \text{H}_2\text{O} \qquad (5.4.8)$$

The enzyme isolated from *Neuropora crassa* is a molybdoenzyme and requires FAD as a cofactor [127-129]. The determination of nitrate is by monitoring the decrease in absorbance at 340 nm due to the oxidation of NADPH. Thus, the assay would be amenable to the concept of a "nutrient rotor," whereby a number of related inorganic species, including soluble silica and orthophosphate [116], and ammonia [83] could be analyzed simultaneously from a single water sample within one multicuvette rotor. All assays could be conducted under a single set of observation conditions including monitoring the wavelength (340 nm).

DETERMINATION OF ADENOSINE PHOSPHATES

Measurements of adenosine triphosphate (ATP) and other adenosine phosphate compounds, such as adenosine monophosphate (AMP), are useful indices of microbial biomass and microbiological activity. Various photometric procedures for the determination of these compounds are available [130]. Currently, the most frequently utilized method for the estimation of ATP content involves the luciferin (LH_2)-luciferase (E)-ATP bioluminescent reaction [130]:

$$E + LH_2 + ATP \rightarrow E \cdot LH_2 - AMP + \text{pyrophosphate} \qquad (5.4.9)$$

$$E \cdot LH_2 - AMP + O_2 \rightarrow \text{oxyluciferin} + CO_2 + AMP + \text{light} \qquad (5.4.10)$$

This reaction is very sensitive—less than 1 pg ATP can be detected with sophisticated instrumentation [130]. The method, however, is subject to interference by a number of common ionic species (Ca^{2+}, K^+, Na^+, Rb^+, Li^+) creating many methodological difficulties [131, 132]. Bostick and Ausmus [133] have modified the well-known hexokinase-glucose-6-phosphate dehydrogenase procedure for ATP for adaptation to the centrifugal fast analyzer which permits a very sensitive kinetic measurement of ATP + ADP (~ 3.1 pmole). Longer reaction times and fluorimetric measurement of the NADPH produced further enhance the sensitivity of the procedure. Their procedures [133] are presented below.

Near Ultraviolet Photometric Methods

A convenient spectrophotometric procedure for the determination of ATP is based on the reaction with hexokinase (HK, EC 2.7.1.1) and glucose-6-phosphate dehydrogenase (G-6-PDH, EC 1.1.1.49):

$$\text{glucose} + ATP \xrightleftharpoons{HK, Mg^{2+}} \text{glucose-6-phosphate} + ADP \qquad (5.4.11)$$

$$\text{glucose-6-phosphate} + NADP^+ \xrightleftharpoons{G-6-PDH} \text{6-phospho-}\delta\text{-lactone} + NADPH + H^+ \qquad (5.4.12)$$

where $NADP^+$ is the oxidized form of β-nicotinamide adenine dinucleotide phosphate. In this case, an increase in absorbance or fluorescence is monitored. A distinct advantage of the spectrophotometric procedures is that they are absolute methods, with the substrate concentrations estimated directly from the molar absorptivity of the monitored species (NADPH, $\epsilon_{340\,nm} = 6.22 \times 10^3$ liters/(mole $\cdot$ cm). Luminescent procedures (and kinetic spectrophotometric procedures) require that sufficient standards be run to establish an empirical calibration curve. Reaction (5.4.11) is highly specific for ATP, which cannot be replaced by other nucleotides [134]. The specificity of hexokinase for glucose is low. Other hexoses could be phosphonylated and would not be detected by the specific indicator enzyme, glucose-6-phosphate dehydrogenase. The great excess of glucose in the reagent, however, effectively obviates interference by nonglucose hexoses.

The other adenosine phosphates may also be determined spectrophoto-metrically. The pyruvate produced in the reaction

$$ADP + PEP \rightleftharpoons ATP + \text{pyruvate} \qquad (5.4.13)$$

or by the coupling of reaction (5.4.13) with

$$AMP + ATP \rightleftharpoons 2ADP \qquad (5.4.14)$$

may be monitored by the enzyme lactate dehydrogenase (LDH, EC1.1.1.27) as [84]

$$\text{pyruvate} + NADH + H^+ \overset{LDH}{\rightleftharpoons} NAD^+ + \text{lactate} \qquad (5.4.15)$$

Again, a decrease in absorbance must be measured, and other nucleoside diphosphates interfere, since they may be used by the enzyme pyruvate kinase to produce pyruvate.

Far-Ultraviolet Photometric Methods

The procedure has been proposed as a measure of adenosine monophos-phates [84]:

$$AMP \overset{AP}{\rightleftharpoons} \text{adenosine} + \text{phosphate} \qquad (5.4.16)$$

$$\text{adenosine} + H_2O \overset{ADA}{\rightleftharpoons} \text{inosine} + NH_3 \qquad (5.4.17)$$

where AP is alkaline phosphate (EC 3.1.3.1) and ADA is adenosine deami-nase (EC 3.5.4.4). The indicator reaction (5.4.17) may be monitored at 260 to 265 nm due to the difference in molar absorptivity of adenosine and inosine [136, 137]. Alternately, the ammonia liberated in (5.4.17) may be determined with the enzyme L-glutamate dehydrogenase (GLDH, EC 1.4.1.3):

$$NH_3 + \alpha\text{-ketoglutarate} + NADH \overset{GLDH}{\rightleftharpoons} \\ \text{L-glutamate} + NAD \qquad (5.4.18)$$

Again, the consumption of NADH can be monitored by spectrophotometric [135, 136] or fluorimetric [137] procedures. The ammonia produced may also be determined by the colorimetric Berthelot reaction [138]. Adenosine monophosphate may be determined specifically, in a manner similar to (4.5.17) by the use of adenylic acid deaminase (Schmidts deaminase, EC 3.5.4.6):

$$5'\text{-AMP} + H_2O \rightleftharpoons 5'\text{-IMP} + NH_3 \qquad (5.4.19)$$

where 5'-IMP = 5'-inosine monophosphate. Coupling (5.4.19) with myokinase and hexokinase procedures allows estimation of ATP and ADP [139].

MEASUREMENT OF pH WITH THE CENTRIFUGAL FAST ANALYZER

The deleterious effect of sulfate ion in natural water systems is, in part, a function of the pH of the solution. The measurement of pH is thus an important adjuvant to the determination of sulfate ion.

With the use of suitable indicator dyes, one may accurately determine pH by optical measurements in the visible spectral region. Table 5.12 lists several suitable indicator dyes and the corresponding pH range for the conversion of the acidic form to the basic form of the indicator.

SYMBOLS

$[A]$	Concentration of free and unbound antibody
a_i	Activity coefficient of component i
Ab	Antibody
Ag	Antigen
$[AG]$	Concentration of antigen–antibody joined unit
A_i	Quantity as defined in (5.2.23)
B^+	Univalent cation
c_i	Concentration of component i in molarity
c_b	Concentration at cell bottom
c_m	Concentration at cell meniscus
c_p	Concentration at plateau region
c_0	Initial concentration ($t = 0$)
d	Number of free antibody molecules
D	Diffusivity
D_{ij}	Multicomponent diffusivity
D_p	Diameter of particle
E	Enzyme
f	Total number of antibody molecules bound per total number of antigen molecules
F	Reaction sites of antibody
F_z	Repulsive force
G	Gibbs free energy function
G	Arbitary function
$[G]$	Concentration of free and unbound antigenic determinant
H	Reaction sites of antigen
i	Fraction of sites occupied by antibody
IgG	Immunoglobulin

Table 5.12 Possible Indicators for Colorimetric Measurement of pH

Approximate pH Range	Indicator	Maximum λ, nm		Color Change	
		Acidic	Basic	Acidic	Basic
2.5–4.0	2,4-Dinitrophenol	—	—	Colorless	Yellow
4.0–5.4	Bromocresol green	444	617	Yellow	Blue
5.2–6.8	Chlorophenol red	—	—	Yellow	Red
5.2–6.8	Bromocresol purple	433	591	Yellow	Purple
6.2–7.6	Bromothymol blue	433	617	Yellow	Blue
6.8–8.4	Phenol red	433	558	Yellow	Red
7.2–8.4	Cresol red	434	577	Yellow	Red
7.8–9.6	Thymol blue	430	596	Yellow	Blue

Source: Reproduced from Ref. 83, p. 118, ORNL/TM-6012/vl. (Data source Ref. 140.)

j	Fraction of filled binding antibody
k	Fraction of sites occupied by antigen
k	Proportionality constant
k_i	Reaction rate coefficient of the ith reaction
K_I	Adsorption coefficient with repulsive interaction between immobilized molecules
K_M	Michaelis-Menten constant
K	Adsorption coefficient as defined in (5.3.5)
K_{12}	Association coefficient as defined in (5.3.14)
m_i	Molality of component i
m_L	Mass per bead
M	Molecular weight
M_i	Molecular weight of component i
n	Refractive index
$\underline{n}$	Valences of antibody
n_L	Number of molecules per bead
N	Number of molecules
P	Product
PX_z	Polyelectrolyte
r	Radial variable in a cylindrical coordinate
r	Langmuir adsorption isotherm as defined in (5.3.13)
r^*	Radial location of a step function
r_b	Radial distance of cell bottom
r_m	Radial distance of meniscus
r_p	Radial position of plateau
R	Gas constant
s	Valences of antigen
S	Svedberg's sedimentation coefficient as defined in (4.2.2)
S	Substrate
$[S]$	Concentration of substrate
SE	Substrate-enzyme complex
t	Time
T	Temperature
T	Titrant
u	A function
v	A function

$\bar{v}_i$	Partial specific volume of component i
v_c	Volume of carrier
v_L	Volume of load
$\mathbf{v}_r$	Radial velocity
X^-	Univalent anion
y	Variable
z	Surface charge

Greek Symbols

$\gamma_i^{(c)}, \gamma_i^{(m)}$	Activity coefficient of component i in molarity (c) and molality (m) concentration
Γ	Binding coefficient as defined in (5.2.31a)
Γ	Energy required to overcome repulsive force as defined in (5.3.8a)
$|\Gamma_k|$	Quantity as defined in (5.2.60b)
μ	Dynamic viscosity
μ_i	Chemical potential of component i
$\mu_{i\kappa}$	Quantity as defined in (5.2.60a)
ξ	Reduced solid-phase concentration as defined in (5.3.8b)
σ	Quantity as defined in (5.2.73)
ρ	Density
ρ_H	Density of hybrid particle
ρ_L	Density of load
ω	Angular velocity

Subscripts

a	Arbitrary reference point
b	Bottom
L	Load, liquid phase
m	Meniscus
p	Plateau
p	Pressure
s	Solid phase
w	Weight average
z	z-Average
0	Quantity evaluated at reference condition or $t = 0$

max   Maximum

min   Minimum

Superscripts

0   Quantity evaluated at reference condition

app   Apparent quantity

Other Symbols

< >   Average quantity

[ ]   Concentration of the quantity inside the bracket

References

1. N. G. Anderson, *Anal. Biochem.*, **28**, 545 (1969).
2. N. G. Anderson, *Anal. Biochem.*, **32**, 59 (1969).
3. N. G. Anderson, *Science*, **166**, 317 (1969).
4. N. G. Anderson, *Clin. Chim Acta*, **25**, 321 (1969).
5. N. G. Anderson, *Anal. Biochem.*, **31**, 272 (1969).
6. D. W. Hatcher and N. G. Anderson, *Am. J. Clin. Pathol.*, **52**, 645 (1969).
7. N. G. Anderson, *Am. J. Clin. Pathol.*, **53**, 778 (1970).
8. D. N. Mashburn, R. H. Stevens, D. D. Willis, L. H. Elrod, and N. G. Anderson, *Anal. Biochem.*, **35**, 98 (1970).
9. C. A. Burtis, W. F. Johnson, J. E. Attrill, C. D. Scott, N. Cho, and N. G. Anderson, *Clin. Chem.*, **17**, 686 (1971).
10. T. O. Tiffany, C. A. Burtis, and N. G. Anderson, "Fast Analyzers For Biochemical Analysis," in *Methods in Enzymology*, Vol. 31, S. Fleischer and L. Packer, Eds., Academic, New York, 1974.
11. T. O. Tiffany, "Centrifugal Fast Analyzers in Clinical Laboratory Analysis," in *CRC Critical Reviews in Clinical Laboratory Sciences*, J. W. King and W. R. Faulkner, Eds., The Chemical Rubber Co., Cleveland, Ohio, 1974.
12. C. A. Burtis, T. O. Tiffany, and C. D. Scott, "The Use of a Centrifugal Fast Analyzer for Biochemical and Immunological Analyses," in *Methods of Biochemical Analysis*, D. Glick, Ed., Vol. 23, Wiley, New York, 1976.
13. N. G. Anderson, C. A. Burtis, J. C. Mailen, C. D. Scott, and D. D. Willis, *Anal. Lett.*, **5**, 153 (1972).
14. C. A. Burtis, J. C. Mailen, W. F. Johnson, C. D. Scott, T. O. Tiffany, and N. G. Anderson, *Clin. Chem.*, **18**, 753 (1972).
15. R. K. Genung and H. W. Hsu, *Biotech. Bioeng.*, **20**, 1129 (1978).
16. R. K. Genung and H. W. Hsu, *Anal. Biochem.*, **91**, 651 (1978).
17. Spinco Division of Beckman Instruments, Inc. Publication, *SPINCO SB-189E*, Palo Alto, Calif., 1975.

18. C. H. Chervenka, *A Manual of Methods for Analytical Ultracentrifuge,* Spinco Division, Beckman Instruments, Inc., Palo Alto, Calif., 1969.
19. J. W. Williams, K. E. Van Holde, R. L. Baldwin, and H. Fujita, *Chem. Rev.,* **58**, 715 (1958).
20. H. Fujita, *Mathematical Theory of Sedimentation Analysis,* Academic, New York, 1962.
21. H. Fujita, *Foundation of Ultracentrifugal Analysis,* Wiley, New York, 1975.
22. J. W. Williams, *Ultracentrifugation of Macromolecules: Modern Topics,* Academic, New York, 1972.
23. C. Tanford, *Physical Chemistry of Macromolecules,* Wiley, New York, 1967.
24. W. J. Archibald, *J. Phys. Colloid Chem.,* **51**, 1204 (1947).
25. R. L. Baldwin and K. E. Van Holde, *Fortschr. Hochpolym.-Forsch.,* Bd. 1.S., **451**, 511 (1960).
26. K. E. Van Holde and R. L. Baldwin, *J. Phys. Chem.,* **62**, 734 (1958).
27. W. D. Lansing and E. O. Kraemer, *J. Am. Chem. Soc.,* **57**, 1369 (1935).
28. W. D. Lansing and E. O. Kraemer, *J. Am. Chem. Soc.,* **58**, 1471 (1936).
29. J. W. McBain, *J. Am. Chem. Soc.,* **58**, 315 (1936).
30. R. J. Goldberg, *J. Phys. Chem.,* **57**, 194 (1953).
31. J. L. Sarquis and E. T. Adams, Jr., *Arch. Biochem. Biophys.,* **163**, 442 (1974).
32. H. Morawetz, *Macromolecules in Solution,* Interscience, New York, 1965.
33. M. Wales, *J. Phys. Colloid Chem.,* **52**, 235 (1948).
34. M. Wales, J. W. Williams, J. O. Thompson, and R. H. Edward, *J. Phys. Colloid Chem.,* **52**, 983 (1948).
35. M. Wales, F. T. Adler, and K. E. Van Holde, *J. Phys. Colloid Chem.,* **55**, 145 (1951).
36. M. Wales, *J. Phys. Colloid Chem.,* **55**, 282 (1951).
37. A. Tiselius, *Z. Phys. Chem.,* **124**, 449 (1926).
38. K. O. Pedersen, *Z. Phys. Chem.* (Leipzig), **A170**, 41 (1934).
39. O. Lamm, *Ark. Kemi Mineral, Geol.,* **17A**, (25), (1944).
40. J. S. Johnson, K. A. Kraus, and G. Scatchard, *J. Phys. Chem.,* **58**, 1034 (1954).
41. J. S. Johnson, G. Scatchard, and K. A. Kraus, *J. Phys. Chem.,* **63**, 873 (1959).
42. T. Svedberg, *Z. Phys. Chem.,* **127**, 51 (1927).
43. T. Svedberg and K. Pederson, *The Ultracentrifuge,* Clarendon, Oxford, 1940.
44. W. J. Archibald, *Ann. N.Y. Acad. Sci.,* **43**, 211 (1942).
45. F. H. Miller, *Partial Differential Equations,* Wiley, New York, 1947.
46. R. Trautman and V. N. Schumaker, *J. Chem. Phys.,* **22**, 551 (1954).
47. H. K. Schachman, *Ultracentrifugation in Biochemistry,* Academic, New York, 1959.
48. E. T. Adams, Jr., *Biochemistry,* **4**, 1646 (1965).
49. E. T. Adams, *Fractions,* No. 3, (1967).
50. P. W. Chun, S. J. Kim, J. D. Williams, W. T. Cope, L. H. Tang, and E. T. Adams, Jr., *Biopolymers,* **11**, 197 (1972).
51. J. L. Sarquis and E. T. Adams, Jr., *Arch. Biochem. Biophys.,* **163,** 442 (1974).
52. L. H. Tang and E. T. Adams, Jr., *Arch. Biochem. Biophys.,* **157**, 520 (1973).
53. C. A. Weirich, E. T. Adams, Jr., and G. H. Barlow, *Biophys. Chem.,* **1**, 35 (1973).

54. M. Meselson, F. W. Stahl, and J. Vinograd, *Proc. Natl. Acad. Sci.,* 43, 581 (1957).
55. J. Vinograd and J. E. Hearst, *Fortschr. Chem. Organ. Naturst.,* 20, 372 (1962).
56. J. M. Singer and C. M. Plotz., *Am. J. Med.,* 21, 882 (1956).
57. N. G. Anderson and J. P. Breillatt, Jr., *Nature,* 233, 112 (1971).
58. N. Bloomfield, M. A. Gordon, and F. Dumont, *Proc. Soc. Exp. Biol. Med.,* 114, 64 (1963).
59. G. Dezelic, N. Dezelic, and N. Telisman, *Eur. J. Biochem.,* 20, 553 (1969).
60. G. Dezelic, N. Dezelic, and N. Telisman, *Eur. J. Biochem.,* 23, 575 (1971).
61. B. O. Duboczy and F. C. White, *Am. Rev. Resp. Dis.,* 100, 364 (1969).
62. I. Oreskes and J. M. Singer, *J. Immunol.,* 86, 338 (1961).
63. J. Singer, C. M. Plotz, E. Prader, and S. K. Elster, *Am. J. Clin. Path.,* 28, 611 (1957).
64. J. M. Singer, G. Altmann, A. Goldenberg, and C. F. Plotz, *Arthritis Rheumat.,* 3, 515 (1960).
65. J. M. Singer, *Am. J. Med. Rev.,* 31, 766 (1961).
66. J. M. Singer, G. Altmann, I. Oreskes, and C. M. Plotz., *Am. J. Med.,* 30, 772 (1961).
67. R. K. Genung, Ph.D. Dissertation, University of Tennessee, Knoxville, Tenn., 1976.
68. I. Langmuir, *J. Am. Chem. Soc.,* 40, 1361 (1918).
69. M. Heidelberg and K. O. Pendersen, *J. Exp. Biol. Med.,* 65, 393 (1937).
70. N. G. Anderson, in *The Development of Zonal Centrifuges,* N. G. Anderson, Ed., *Natl. Cancer Inst. Monogr. 21,* U.S. Government Printing Office, Washington, D. C., 1966.
71. J. E. Mrochek, C. A. Burtis, W. F. Johnson, M. L. Bauer, D. G. Lakomy, R. K. Genung, and C. D. Scott, *Clin. Chem.,* 23, 1416 (1977).
72. J. M. Jansen, Jr., *Clin. Chem.,* 16, 515 (1970).
73. M. T. Kelley and J. M. Jansen, Jr., *Clin. Chem.,* 17, 701 (1971).
74. C. A. Burtis, W. F. Johnson, J. C. Mailen, and J. E. Attrill, *Clin. Chem.,* 18, 433 (1972).
75. T. O. Tiffany, W. F. Johnson, and M. E. Chilcote, *Clin. Chem.,* 17, 715 (1971).
76. D. L. Fabiny-Byrd and G. Ertingshausen, *Clin. Chem.,* 18, 841 (1972).
77. D. L. Fabiny-Byrd and G. Ertingshausen, *Clin. Chem.,* 17, 696 (1971).
78. T. O. Tiffany, J. M. Jansen, Jr., C. A. Burtis, J. B. Overton, and C. D. Scott, *Clin. Chem.,* 18, 829 (1972).
79. D. G. Peters, J. M. Hayes, and G. M. Hieftje, *Chemical Separations and Measurements,* Saunders, Philadelphia, 1974.
80. L. Michaelis and M. L. Menten, *Biochem. Z.,* 49, 333 (1913).
81. J. D. Winefordner and P. A. St. John, *Anal. Chem.,* 35, 2212 (1963).
82. T. O. Tiffany, M. B. Watsky, C. A. Burtis, and L. H. Thacker, *Clin. Chem.,* 19, 87 (1973).
83. Oak Ridge National Laboratory, *Technical Memorandum 5533,* Feb. 1976; *5865,* Aug. 1976; *6012,* March 1977.
84. H. U. Bergmeyer, Ed., *Methods of Enzymatic Analysis,* Academic, New York, 1970.

85. G. C. Guilbault, *Enzymatic Methods of Analysis*, Pergamon, Oxford, 1970.
86. S. Udenfriend, *Fluorescence Assays in Biology and Medicine*, Vol. II, Academic, New York, 1969.
87. J. D. Ingle and S. R. Crouch, *Anal. Chem.*, **43**, 697 (1971).
88. W. B. Dandliker and S. A. Levison, *Immunochemistry*, **5**, 171 (1967).
89. L. M. Killingsworth and J. Savory, *Clin. Chem.*, **17**, 936 (1971).
90. L. M. Killingsworth, J. Savory, and P. O. Teague, *Clin. Chem.*, **17**, 374 (1971).
91. L. M. Killingsworth and J. Savory, *Clin. Chem.*, **18**, 335 (1972).
92. L. M. Killingsworth and J. Savory, *Clin. Chem.*, **19**, 403 (1973).
93. I. Eckman, J. B. Robbins, C. J. A. van der Hamer, and J. H. Scheinberg, *Clin. Chem.*, **16**, 58 (1970).
94. W. B. Dandliker, S. P. Halbert, M. C. Florin, R. Alonso, and H. C. Shapiro, *J. Exp. Med.*, **122**, 1029 (1965).
95. W. B. Dandliker and V. A. deSaussure, *Immunochemistry*, **7**, 799 (1970).
96. R. P. Tengerdy, *J. Lab. Clin. Med.*, **70**, 707 (1967).
97. R. D. Spencer, F. B. Toledo, B. T. Williams, and N. L. Yoss, *Clin. Chem.*, **19**, 923 (1973).
98. R. C. Aalberse, *Clin. Chim. Acta*, **48**, 109 (1973).
99. E. H. Cooper, R. Turner, L. Stede, A. M. Neville, and A. M. Mackay, *Br. J. Cancer*, **311**, 111 (1975).
100. A. M. Neville and E. H. Cooper, *Ann. Clin. Biochem.*, **13**, 283 (1976).
101. J. G. Batsakis, "Serum Enzymes and Cancer: Their Use in Hepatic Metastases," in *Proc. 1st Invitational Symposium on the Serodiagnosis of Cancer*, Armed Forces Radiobiology Research Institute, Bethesda, Md., 1974, pp. 1–18.
102. M. K. Schwartz, *Clin. Chem.*, **19**, 10 (1973).
103. W. Nagel, F. Willig, and F. H. Schmidt, *Klin. Wochenschr.*, **42**, 447 (1964).
104. T. Samorajski, C. Rolsten, and J. M. Ordy, *Am. J. Clin. Pathol.*, **38**, 645 (1972).
105. R. W. Phillips and E. R. Manildi, *Cancer*, **26**, 1006 (1970).
106. S. Mizutani, M. Yoshino, and M. Oya, *Clin. Biochem.*, **9**, 16 (1976).
107. D. M. Goldberg, *Digestion*, **8**, 87 (1973).
108. W. Van der Slik, J. P. Persign, E. Engelsman, and A. Riethorst, *Clin. Biochem.*, **3**, 59 (1970).
109. M. Schwartz, M. Fleisher, and O. Bodansky, *Ann. N.Y. Acad. Sci.*, **166**, 775 (1969).
110. Calbiochem. Co., *Calbiochem® Alkaline Phosphatase Reagents*, Doc. No. LO 3002, La Jolla, Calif., 1975.
111. Oak Ridge National Laboratory, *Report ORNL-NSF-EATC-15*, Oak Ridge, Tenn., Dec. 1975.
112. D. F. Boltz, *Colorimetric Determination of Nonmetals*, Interscience, New York, 1958, pp. 29 and 47.
113. I. M. Kolthoff and P. J. Elving, *Treatise on Analytical Chemistry*, Part II, Vol. 2, I. M. Kolthoff and P. J. Elving, Eds., Interscience, New York, 1962, p. 107.
114. APHA, AWWA, and WPCF, *Standard Methods for Examination of Water and Wastewater*, 13th ed., APHA, New York, 1971.
115. L. G. Hargis, *Anal. Chim. Acta*, **52**, 1 (1970).
116. W. D. Bostick, C. A. Burtis, and C. D. Scott, *Anal. Lett.*, **9**(1), 65 (1975).

117. W. Hasson, J. R. Penton, and G. M. Widdowson, *Clin. Chem.*, **20**, 15 (1974).
118. R. V. Hems, G. F. Kirkbright, and T. S. West, *Talanta*, **16**, 789 (1969).
119. B. Schuknecht, *Environ. Lett.*, **9**, 91 (1975).
120. P. D. Boyer, H. Lardy, and K. Myrback, *The Enzymes*, Vol. VII, Academic, New York, 1963.
121. G. B. Carner, J. S. Baumstark, M. E. Muhrer, and W. H. Pfander, *Anal. Chem.*, **28**, 1589 (1956).
122. R. M. Hill, H. Pivnick, W. E. Engelhard, and M. Bogard, *J. Agr. Food Chem.*, **7**, 261 (1959).
123. Worthington Biochemical Corporation, *Nitrate Reductase* (E. coli), *Cytochrome: Nitrate Oxidoreductase*, Freehold, N.J., 1975.
124. A. L. McNamara, G. B. Meeker, P. D. Shaw, and R. H. Hageman, *J. Agr. Food Chem.*, **19**, 229 (1971).
125. D. R. Senn, Ph.D. Dissertation, University of Georgia, Athens, Ga., 1975.
126. W. R. Hussein and G. G. Guilbault, *Anal. Chim. Acta*, **76**, 183 (1975).
127. A. Nason and H. J. Evans, *J. Biol. Chem.*, **202**, 655 (1953).
128. R. H. Garrett and A. Nason, *Proc. Natl. Acad. Sci. USA*, **58**, 1603 (1967).
129. B. Venneland and C. Jetschmann, *Biochim. Biophys. Acta*, **229**, 554 (1970).
130. M. J. Cormier, D. M. Hercules, and J. Lee, *Chemiluminescence and Bioluminescence*, Plenum, New York, 1973, p. 461.
131. A. Lundin and A. Thore, *Anal. Biochem.*, **66**, 47 (1975).
132. J. W. Patterson, P. L. Brezonik, and H. D. Putnam, *Environ. Sci. Technol.*, **4**, 569 (1970).
133. W. D. Bostick and B. S. Ausmus, *Anal. Biochem.*, **88**, 78 (1978).
134. R. K. Crane, *Enzymes*, **6**, 47 (1962).
135. W. D. Bostick, R. C. Lovelace, and C. A. Burtis, *Water Res.*, **10**, 887 (1976).
136. J. Bottsma, P. A. Roseland, and D. M. Goldberg, *Clin. Chem. Acta*, **41**, 219 (1972).
137. R. J. Spooner, P. A. Roseland, and D. M. Goldberg, *Clin. Chem. Acta*, **65**, 47 (1975).
138. A. Belfield, G. Ellis, and D. M. Goldberg, *Clin. Chem.*, **16**, 396 (1970).
139. H. G. Albaum and R. Lipshitz, *Arch. Biochem. Biophys.*, **27**, 102 (1950).
140. L. Meites, Ed. *Handbook of Analytical Chemistry*, McGraw-Hill, New York, pp. 3-35.

Chapter **VI**

LIQUID CENTRIFUGATION
IN PRACTICE

The liquid centrifuge has played an essential role in almost every advance in molecular and cell biology, and it is now one of the most basic and valuable pieces of equipment available in biology laboratories. Over the years, the laboratory centrifuge, particularly the high-speed version, the ultracentrifuge, has undergone extensive development. However, despite daily use in many laboratories, the full potential of liquid centrifuges is rarely realized. The purpose of this chapter is to present a discussion of all aspects of the methodologies for the separation and fractionation of biological particles by centrifugation.

For convenience, as discussed previously, centrifuges may be categorized as preparative or analytical, although the techniques for which they are employed do not fall necessarily into these divisions. That is, analytical investigations may be performed by using a preparative instrument, and it is possible, using separation cells in the analytical ultracentrifuge, to prepare quantities of material at the nanogram level.

Particles in a solution under the normal force of gravity sediment at rates that depend mainly on their mass differences. The centrifuge hastens this process by applying a very much higher force field than is provided by the earth. It is customary to indicate the intensity of centrifugation by the "relative centrifugal field" (RCF), the "*g* value" or the "centrifugal effect." This simply represents the accelerating field as a multiple of the earth's gravitational field. A simple bench-model centrifuge capable of 3000 rpm at a distance of 90 mm from the axis of rotation gives an accelerating field approximately 900 times greater than that of the earth, 900 *g*.

The centrifugal force field, or accelerating field (F_c), at 90 mm from the axis of rotation without the specific charge of ionic species and neglecting the Coriolis force is given by (2.7.6):

$$F_c = \omega^2 r \qquad (2.7.6)$$

where r is the distance from the axis (i.e., 90 mm) and ω is the angular velocity (sec^{-1}), which is calculated as follows:

$$\text{rate of revolution} = 3000 \, \text{rpm} = \frac{3000}{60} \, \text{rps}$$

Since one revolution is equivalent to 2π radians, the angular velocity is

$$\omega = \frac{(2\pi)\,(3000)}{(60)} = 314 \, \text{sec}^{-1}$$

thus, if r is represented in meters, the units of acceleration are meters per second per second $= \text{m sec}^{-2}$; thus, the accelerating field F_c is

$$F_c = (314)^2 (0.09) \, \text{m sec}^{-2} = 8.883 \times 10^3 \, \text{m sec}^{-2}$$

The earth's gravitational field is $9.81 \, \text{m sec}^{-2}$, thus, the relative centrifugal field (RCF), or the relative centrifugal acceleration to the earth's gravitational acceleration, is

$$F_c = \frac{8.883 \times 10^3}{9.81} g = 905g$$

A word of caution must be given, however, about reading g values from the literature. There are at least three positions from where values can be measured, namely, the top, the middle, and the bottom of the centrifuge tube. This can make quite a difference to the separation achieved. If one uses an angle-head rotor, it should be noted that the type of centrifuge head employed will affect the RCF values obtained. It should also be noted that the duration of centrifugation, as well as the RCF, is important in determining effective separations. In many cases, RCF time is constant, or nearly so. Thus, a given separation may be achieved at a lower RCF simply by increasing the centrifugation time. This cannot be taken to extremes, however, for there is usually a critical RCF below which a given particle will not sediment, since opposing diffusion forces are then greater and the particle diffuses as quickly as it is sedimented by the centrifugal field.

The particles separated in preparative centrifuges may have sedimentation coefficients, defined by (4.2.4), which range over six orders of magnitude [1]. While zonal centrifuges now allow high-resolution separations to be made over a very wide range of particle sizes, no *single* zonal centrifuge experiment can achieve separations over more than a fraction of this range. Angle-head or swinging-bucket centrifugation is therefore often a necessary prelude to high-resolution zonal separations, especially where a minor component is to be concentrated and resolved.

1 SEPARATIONS BY SWINGING BUCKET AND ANGLE-HEAD ROTORS

There are two main types of rotor, in addition to those used for zonal centrifugations: swinging bucket and fixed angle-head rotors. Typical examples of these are summarized in Tables 1.1 and 1.2, respectively. There is also some variation in the angle at which tubes are held in the angle-head rotor. Recently, Beckman Instruments, Inc., has introduced vertical tube rotors. In the vertical tube rotor, the axes of the tubes are parallel to the rotor's axis of rotation. At speed, particles sediment only a very short distance—across the diameter of the tube. Since the tubes in a vertical tube rotor are located near the edge of the rotor where the centrifugal force is high, it is possible for particles to sediment across the tube in as little as one fifth the time it would take to sediment from the top to the bottom of a swinging bucket tube.

Selection of Rotors

The selection of a rotor for a specific purpose is usually made on the basis of the amount to be centrifuged and the techniques to be used. It is often possible to centrifuge a larger volume for a longer time at a lower rotor speed, that is, $\omega^2 t$ is a constant within limits. The length of the liquid column also affects the time of the run. If a very quick separation of a given volume is required, the largest possible rotor and tubes should be used. This results in a shorter column length of liquid, and hence shorter distances for the particles to move, though this does result in some sacrifice in purity. The angle of the head also affects the rate of solution clearance; the nearer the tube is to being at right angles to the direction of centrifugation, the quicker the separation. This is because a steeper angle results in a shorter path length.

For general-purpose centrifugation, the angle of the tubes is usually 35° to the vertical. This has been found to be the most suitable angle for large, slow-diffusing molecules. However, for small fast-diffusing molecules, the greatest efficiency is achieved with tubes orientated at 15 to 20° to the vertical. Since the advent of the use of the angle heads for isopycnic cesium chloride centrifugation, it has been found that the angle used also affects the steepness of the gradient formed; that is, the smaller the angle, the shallower is the gradient obtained, thus increasing resolution and the usefulness of the technique. A summary of comparison of rotor types for different centrifugation techniques [2] is presented in Table 6.1.

The tubes used in these rotors are cylindrical, usually with rounded bottoms. They are most frequently made of polypropylene, polycarbonate, or stainless steel, for use in high-speed centrifugation. One of the disadvantages of cylindrical tubes is the wall effect; that is, the particles sediment radially,

Table 6.1 Comparison of Rotor Types for Different Centrifugation Techniques

Type of Centrifugation	Type of Rotor		
	Fixed Angle Rotors (FA)	Swinging Bucket Rotors (SW)	Vertical Tube Rotors (VT)
Differential centrifugation (pelleting): Separates particles from a mixture by sedimenting those of larger mass to the bottom of the tube, leaving a portion of the smaller particles in the supernatant.	A "button" or "pellet" accumulates in the most centrifugal portion of the tube. Easy to recover pellet and supernatant. Holds more tubes and larger volumes than SW or VT rotors.	Seldom used for this technique because of long distance to sediment and smaller volumes compared to FA and VT rotors.	Faster than FA or SW rotors for the same volume and rotor speed. Material is sedimented against the entire outer tube wall, and may be difficult to collect in a small volume. The sediment may also redistribute during reorientation.
Rate zonal separations: Separates particles based on differences in sedimentation rate (essentially particle size) in a density gradient (usually sucrose).	Seldom used. Good resolution but particles may be lost by wall effects as zones impact the sloping walls of the tube. Best results are obtained with thick-wall polycarbonate or polyallomer tubes used without tube caps.	Best resolution, particularly for multi-component samples and small S-value particles; least wall effects. Longest run time.	Faster than FA or SW rotors. Holds more tubes then SW rotors. Less resolving power than SW rotors: zones increase in volume (less concentrated) as sedimentation occurs, due to geometry of container and diffusion (with small particles). Wall effects may result in sample loss as zones pass through midpoint of tube.
Isopycnic separations: Separates particles based on differences in particle density (essentially particle composition) in a density gradient (usually cesium chloride solutions).	Large volume between band centers. Band volumes are larger than SW rotors. Holds more tubes and larger volumes than SW or VT rotors. Longer run times than VT rotors.	Band volumes are small (more concentrated). Least volume between band centers. Longer run times than FA or VT rotors.	Faster than FA or SW rotors. Largest volume between band centers provided bands are near center of tube. Holds more tubes than SW rotors. Band volumes are larger (less concentrated).

Source: DS-494A, Beckman Instruments, Inc. Spinco Division, with permission.

318

whereas the walls are parallel. This means that, due to the bouncing of some particles off the walls, any boundary formed will not be sharp. Sector-shaped tubes have been tried but are not yet a commercial item.

The tubes, like the rotors, should be equilibrated before the run. They usually have some type of closure, which serves to minimize evaporation, to maintain sterile conditions and to prevent tube collapse. The tubes plus closures are normally balanced to within 0.1 g of each other. Although this is not strictly necessary in every case, it does help to prolong the life of the spindle. If the tube is less than half full, it is advisable to fill it with a light inert liquid, usually liquid paraffin, to minimize the risk of tube collapse, especially on long runs.

Density-Gradient Solutions

As discussed previously, density gradients are employed in centrifugal separations in a number of different ways. The shape of a gradient normally refers to its concentration profile, that is, the variation of concentration along the tube. One has to note that concentration does not always have a linear relationship with either density or viscosity. A density-gradient shape is usually characterized by a density profile and a viscosity profile as demonstrated in (4.2.10a) and (4.2.10b). Sedimentation rate varies inversely with viscosity, and so particles are slowed down by the increased viscous drag as they move through the gradient. In practice, this is more important than density in determining centrifugation time as demonstrated in Chapter IV, Section 2. Density is the factor which determines the stability of the gradient; the density gradient must be steep enough to prevent convection and to maintain the stability of zones throughout centrifugation, which is also discussed in detail in Chapter IV, Section 2.

For example, if a mixture contained three components with densities of 1.12, 1.14, and 1.35 g/cm^3, the ideal gradient would be S-shaped. The gradient would be quite shallow through the region of the first two components in order to separate them farther in a given time. The gradient would become quite steep to encompass the gap between these two components and the third component, with a continuous gradient within a short distance down the tube. The gradient would then become shallower through the region of the third component.

Preparation of Density Gradients

In forming a density-gradient solution, there are two techniques available, namely, self-forming (diffusion) techniques and preformed (mechanical) techniques.

SELF-FORMING GRADIENTS

The quickest way to make a density gradient is to pipette a dense and a light solution into a centrifuge tube and gently stir the interface with a sawtooth wire. This is not reproducible but may be adequate for some preliminary experiments and may be satisfactory for stabilizing sample layers for separations employing differential pelleting. A more satisfactory method is to introduce several layers of solution of differing densities into a tube as discussed for reorienting isodensity layers in Chapter IV and allow them to diffuse into a continuous gradient, usually over 18 to 24 hr. It is important to leave enough time for this diffusion, as steep regions in the gradient can produce spurious peaks in the distribution of sedimenting particles. The shape of the gradient can be checked by measuring the refractive index of each fraction. A variant of this method is to use only two layers.

PERFORMED GRADIENTS

More recently, a number of mechanical devices have been developed to produce gradients that are smooth, reproducible, and of accurately known characteristics.

There are three basic types of gradients that represent simple mathematical functions and are easy to prepare with simple laboratory equipment. These are (a) linear, (b) convex-exponential, and (c) concave-exponential gradients. All three types can be produced by continuously adding a dense gradient solution of concentration C_D from a reservoir to a mixing chamber containing a light-density gradient solution of C_L, while at the same time displacing the mixture into the gradient tube. The arrangement is shown schematically in Fig. 6.1. The gradient produced by the device is linear-with-volume if the chambers have equal diameters, as shown in Fig. 6.1(a). With chambers of different diameters, convex or concave gradients are produced. These are shown in Figs. 6.1b and 6.1c.

A concentration profile of density gradients produced by this device was mathematically obtained by Noll [3]. The initial concentration in the mixing chamber $(C_L = C_i)$ changes as a function of the volume added and gradually approaches that in the reservoir $(C_L \rightarrow C_D)$. The concentration gradient in the mixing chamber changes with the volume (dV) added from the reservoir if the volume of the mixture V_m contracts by the amount dV_m. Thus, the concentration of gradient solution in the mixing chamber becomes

$$\frac{dC_L}{dV} = \frac{C_D - C_L}{V_m} \tag{6.1.1}$$

Integrating and taking into account that $dV = -\beta\, dV_m$ together with the

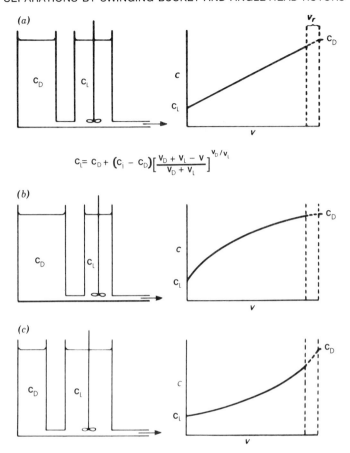

$$c_i = c_D + (c_i - c_D)\left[\frac{v_D + v_L - v}{v_D + v_L}\right]^{v_D/v_L}$$

Fig. 6.1 Schematic diagram of gradient forming devices. (a) Linear. (b) Convex. (c) Concave.

initial conditions $V_m = V_i$ and $C_L = C_i$, one obtains the concentration in the gradient as a function of the volume remaining in the mixing chamber:

$$C_L = C_D + (C_i - C_D)\left(\frac{V_m}{V_i}\right)^\beta \qquad (6.1.2)$$

where

$$\beta = \frac{V_D}{V_L} \qquad (6.1.3)$$

The quantities V_D and V_L are initial volume of dense and light-density gradient solutions. It is more convenient to express the concentration of the density gradient as a function of the volume (V) accumulating in the centrifuge tube. Thus, since $V_i = V_D + V_L$ and $V_m = V_D + V_L - V$, one rewrites (6.1.2) as follows:

$$C_L = C_D + (C_i - C_D) \left[\frac{V_D + V_L - V}{V_D + V_L} \right]^{(V_D/V_L)} \qquad (6.1.4)$$

Tubes can be filled by gravity or by using a peristaltic pump to control the flow. The following cautions are to be taken: (1) There will always be a residual volume in the apparatus, so the density of the gradient will not reach the density of the denser solution. (2) Very vigorous stirring increases the effective head of pressure in the mixing vessels, thus distorting the gradient. (3) The connecting channel needs to be quite wide to allow rapid equilibration of the liquid levels. (4) For the liquid levels to be at equilibrium initially, the heights (the volumes, if the vessels are identical) must be in inverse proportion to their densities. With viscous solutions, it is important to use a pump to ensure consistency in flow rate as the viscosity changes.

Simple apparatuses of this type are marketed by several companies, including the major centrifuge manufacturers; large-scale versions for use with zonal rotors are also available.

Anderson [4] has described a machine in which the pistons are pushed by rotating arms of variable lengths. This system produces a variety of S-shaped gradients. Anderson [5] has also introduced a dense layer of 80% sucrose into the bottom of the gradient columns to serve as a cushion to prevent large particles from reaching the bottom of the tube.

An excellent way to ascertain whether the gradient has the proper characteristics is the method developed by Holter and co-workers [6] and Weber [7]. They placed the gradient-filled tubes in the optical system of an electrophoresis apparatus and observed the schlieren pattern.

There are the more sophisticated, programmable gradient generators which rely on changing the relative pumping rates from two reservoirs. These are, in general, large-volume devices for zonal rotors, but the Dialagrad (ISCO), for example, is also available in a small-volume version. The Ultrograd (LKB) is a gradient programmer that can be used for large or small volumes, depending on the pump. For zonal rotors, such a gradient generator can also be used to load the sample and overlay and can be used as a pump for unloading the rotor. The program is determined by a cut-out template (Ultrograd, Beckman Model 141 and MSE gradient pump) or by setting a number of dials (Dialagrad).

Loading and Running

The sample zone must be layered on top of the gradient gently enough to prevent mixing. Frequently, samples can be loaded on to gradients by using a pasteur pipette, although it is more difficult to obtain a stable sample zone if it is required to achieve maximum resolution. Qualitatively, there are three aspects to this stability. First, the concentration of the sample particles must not cause the density of the sample suspension to exceed the density at the top of the gradient. If it does, drops will stream down the gradient until they reach equilibrium. Secondly, the sample zone should not be bounded by sharp changes in osmotic pressure. Such steps encourage diffusion of the solvent and can broaden the sample zone. Thirdly, the relative rates of diffusion of the sample particles, the solute, and the solvent should not be such that the zone becomes unstable. For example, a narrow zone of 1% bovine serum albumin (BSA) layered over 2% sucrose will be stable initially; but since sucrose diffuses much faster than BSA, it will have diffused into the BSA zone before the BSA has diffused appreciably. The result, therefore, is a layer containing 2% sucrose and 1% BSA on top of 2% sucrose only, which is obviously unstable. This phenomenon is called, variously, streaming, droplet formation, and turnover effect, and so on, as discussed under loading capacity in Chapter IV.

Temperature also plays a part in influencing the shape of the density gradient as well as the rate of sedimentation, and relatively low temperatures are often required for labile materials; the rate of sedimentation changes by approximately 3% between 4 and 20°C. Maintenance of a constant temperature is also essential to minimize convection. In general, to minimize convection, the gradients should be at the temperature at which the sample is to be centrifuged. Tube gradients to be run in the cold are best made in a cold-room or are stored for 2 to 3 hr in a refrigerator at the required temperature. For a uniform temperature run, the controls should be set to the temperature at which the gradient is believed to be at the start of the run.

Fractionation and Sample Recovery

When the sample has been centrifuged by one of the techniques discussed in Chapter IV, Section 2, a method must be found to remove the contents of the tube or rotor without disrupting the separation achieved. Unfortunately, techniques for the removal of samples are far from perfect, although with care, disruption may be kept to a minimum.

After centrifugation, the first step in analysis is to determine the positions of the particles in the tube. Many substances, especially viruses, can be seen with the naked eye. For materials that cannot be seen directly, Anderson [8] has devised a simple apparatus for observing the changes in refraction index produced by the material. A card ruled with parallel lines set at an angle of 45° is viewed through the centrifuge tubes, which are immersed in water in a parrellel-sided vessel. Two-component systems are easily distinguished, but more complex systems are difficult to resolve. For materials which absorb in the ultraviolet region, quartz tubes can be used to hold the sample during centrifugation, and the tubes can then be scanned in a spectrophotometer. For materials that cannot be seen, it will be necessary to separate the contents of one of the tubes into small fractions, and then to analyze each fraction chemically in order to locate the particles. Thus, the postcentrifugation step may be conveniently divided into three steps: (1) unloading of contents from the tube, (2) collection of fractions or sample recovery, and (3) analysis. Each step is discussed in somewhat more detail below.

Unloading from the Tube

Unloading of the contents from a tube is probably the most critical step. From the time the centrifuge stops, the resolution achieved can only be maintained by great care on the part of the operator. All mechanical handling procedures should be kept to a minimum. It is also important to keep the tubes at the temperature of the run to avoid blurring of the resolution achieved by convection currents. The contents of the tube should be removed as soon as possible after the run. The technique used depends somewhat on the type of centrifugation performed, as well as the apparatus available. The unloading techniques are outlined below.

DECANTATION

This procedure is most generally used after normal differential centrifugation. The contents are simply separated into two fractions—the pellet and the supernatant. For a tube that has been spun in a swinging bucket rotor, the supernatant is simply poured out. When an angle-head rotor is used, the pellet formed is at an angle to the bottom of the tube and not at right angles to the vertical walls, as is the case when a swinging bucket rotor is used. In this case, the tube is turned 180° so that when pouring is carried out, the fluid is carried across the pellet and not away from it. If this is not done, disruption of the pellet is likely to occur.

The advantage of this method is that it is quick and simple. The main disadvantage is that some of the pellet tends to resuspend and flow out with the supernatant. Also, the layer of liquid immediately above the pellet is best discarded since it contains the components of both the supernatant and the pellet.

ASPIRATION

This method is also mainly used after differential centrifugation. Here, the supernatant is removed manually by means of a Pasteur pipette, although an automatic version is available commercially. The tip of the pipette is kept in the meniscus so that a mixture of air and solution is withdrawn. A Pasteur pipette made with a hole at the tip at right angles to the place of the normal hole (which is sealed) is especially useful for removing the liquid just above the pellet without disturbing it. Alternatively, if the tube is wide enough, a bent tip may be used.

This method is less likely to cause cross contamination of fractions, and the liquid just above the pellet may be readily discarded. However, some materials are very prone to denaturation due to surface phenomena, and if any are present in the supernatant, this technique should not be used.

TUBE SLICING

This procedure may be used after most of the centrifugal techniques. The tube is sliced horizontally, a device for which is commercially available, so that the fractions are physically separated. It is most generally used where a large number of tubes is involved and only a few fractions are required.

TUBE PIERCING

Tube piercing for the removal of the contents of the tube after density-gradient centrifugation is now a common procedure. In the simplest case, the tube is punctured with a hypodermic needle and the contents are removed with a syringe. If the position of the required band is known unequivocally, then the tube can be pierced in the appropriate position and the band removed. However, one of the commercial tube piercers now more often used enables the whole of the contents of the tube to be analyzed after fractionation. Figure 6.2 shows an ISCO Model 184 tube-holding and tube-piercing mechanism. A centrifuge tube held in this unloading device is first pierced from beneath by turning the knurled screw in the bottom holder; this action causes the insertion of a tapered syringe needle through which a dense gradient solution is pumped. The contents of the tube are lifted through the conical top and into an exit line, thus allowing the contents to be collected in volume increments; but care must be taken to avoid letting air bubbles into the system.

WITHOUT PIERCING THE TUBE

For certain applications, it may be desirable to remove the contents without piercing the bottom of the tube. This can be accomplished through the use of a special fitting in the universal flow cell. A dense "chase" solution is introduced into the centrifuge tube from the top through a length of hypodermic tubing, as shown in Fig. 6.3.

Once the positions of the particles have been determined, separate zones can be drawn off for further analysis. All methods of fraction removal cause some zone intermixing. The method of fraction removal chosen, therefore, will depend in part on the level of the particles in the tube. The particles should usually be moved as small a distance as possible from their original position to the point of exit. In some systems, however, a contaminating material may lie between the zone of interest and the nearest point of exit; the longer path would then be preferred. But no matter what method of fraction removal is used, or where the tube is punctured, the material should be removed slowly and carefully.

Fig. 6.2   ISCO Model 184 tube piercer, shown with a centrifuge tube and flow cell in place. Courtesy of ISCO.

Fraction Collecting

Fractions may be collected by various unloading methods from tubes. The type of fraction collecting required depends very much on individual needs. The fractions to be collected will vary from one or two drops to hundreds of cubic centimeters. Quite often, the fractions are collected manually as in the case of zonal centrifugation since the volumes are so large and the pump rate relatively fast. Automatic fraction collectors are now commercially available either through the use of a timed mode or a volumetric mode during the changeover of tubes on fraction collectors. Several devices have now been designed to shut off the flow during this period. In the middle range (0.5 to 50 cm^3), most commercial collectors are found to be adequate. In the case of very small fractions, these can easily be collected manually. Fraction collector preparations should be made prior to the start of the scanning process or spectrophotometric analysis of the sample. However, they need not be instituted until liquid first appears from the discharge tube.

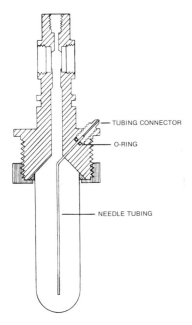

TUBING CONNECTOR

O-RING

NEEDLE TUBING

Fig. 6.3 Device for unloading without piercing the tube. Courtesy of ISCO.

Analysis

In density-gradient separation, the analysis of the contents of the tube may be conveniently divided into the analysis of the sample and the analysis of the density-gradient solution or of the density-gradient profile.

Sample Analysis

The most common method of sample analysis or detection is that of spectrophotometry. The absorption of each fraction may be measured after collection, or automatically with a flow-through cell before collection. In the manual operation there are few problems, apart from the tedium and dilution factor. When using a flow-through cell, however, several factors should be taken into account. First, there are the problems of concentration and volume. These affect the path length of the flow-through cell to be used. If the sample is very dulute, then a cell with as long a path as is compatible with the volume is required; with concentrated samples, the opposite is true. Then there is the direction of flow. If the light end of the gradient is pushed out first, then the flow should be from the bottom to the top of the cell, and in the opposite direction if the more dense end is pushed out first. However, when the flow is from top to bottom, it is advisable to fill the flow-through system with a more dense solution first, otherwise the droplets will form at the top and fall through the cell instead of flowing, so that no optical change will be recorded apart from a little turbulence. The recorder used in conjunction with the monitor should have a reasonable selection of speeds to accommodate various flow rates and volumes.

Generally, monitoring is carried out at a single wavelength, usually that most appropriate for the sample being examined. A typical fractionation of density-gradient contrifuge tubes and light absorbance monitoring system is shown in Fig. 6.4. If the pump being used has a variable speed, the initial flow of the dense chase solution should be at a slow flow rate to avoid disturbing any pellet in the tube and to maintain a sharp interface between the tube contents and the chase solution. Once this interface is established, the final scanning speed should be set on the pump. Suggested scanning speeds are 3.0 to 6.0 ml/min for large tubes (e.g., $1\frac{1}{4} \times 3\frac{1}{2}$ in. tubes), 1.5 to 3.0 ml/min for medium-sized tubes (e.g., 1×2 in. tubes), and 0.2 to 1.0 ml/min for small tubes (e.g., $\frac{1}{2} \times 2$ in. tubes). Faster flow rates than these will lead to complications for the following reasons:

1. If a constant-speed recorder is used, it may give a compressed scanning curve for high syringe flow rates, resulting in poorer apparent resolution of peaks.

2. If a fast flow rate is used with very dense solutions, pressure may build up in the system and leaks may occur.

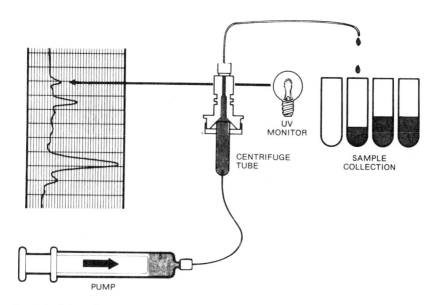

Fig. 6.4   Schematic diagram showing the unloading of density-gradient centrifuge tubes and light absorbance monitoring. Courtesy of ISCO.

3. Turbulence may occur at higher flow rates, causing serious problems.

4. The recorder needle moves slowly, and since the rate of change in optical density will increase with faster speeds, recording inaccuracies and poorer apparent resolution may result.

After the pump flow rate has been set as described above, the scanning process is started by turning the pump on. If the pump has a volumetric counter, it may be useful to either zero the counter prior to starting the scanning process (if it is resettable) or note its reading.

Another popular method of analysis is that used for radioactive samples. The scintillation medium of Bray [9] is especially suitable for direct use with solutions containing sucrose, although many people prefer to separate fractions and precipitate the sample before counting. Other methods include tests for enzymic activity and electron microscopy, depending on the sample.

Analysis of the Density Gradient

Unfortunately, at present, few commercially available instruments exist for the automatic on-line determination of the density of the gradient. There are a few prototypes, but these have yet to be proved. This means that determination of the true density gradient must wait until after the fractions have been collected, and hence only average values are obtained.

There are two main methods for the estimation of the density gradient. The first involves direct density measurements using a pycnometer, and the second is based on the change in refractive index. Since this quantity is proportional to the concentration, one of the appropriate empirical equations for the substance may be employed. The relationship between the density of aqueous sucrose solutions and refractive index may be found by experimentation. Table 6.2 presents the specific gravity and refractive index of aqueous sucrose solutions at 20°C.

The density of various gradient solutions may be estimated by using the following formula:

$$\rho^T = a n_D^T - b \tag{6.1.5}$$

where ρ^T is the density of a gradien solution at temperature T (°C), n_D is the refractive index using the sodium D line, and the coefficients a and b for various gradient solutes are given in Table 6.3.

Equation 6.1.5 applies to solutions of the density-gradient solutes in distilled water and does not take into consideration the other components, namely, buffers, salts, and so on, normally present in gradients. Although the latter are usually present in such low concentrations that they do not contribute significantly to the density, they make a significant contribution to the refrac-

Table 6.2 Specific Gravity (20°/4°C) and Refractive Index of Aqueous Sucrose Solutions at 20°C

Sucrose (%)	g/liter	Sp. gr.	n	Sucrose (%)	g/liter	Sp. gr.	n
0		0.9982	1.3330	43	512.6	1.1920	1.4056
1	10.02	1.0021	1.3344	44	526.8	1.1972	1.4076
2	20.12	1.0060	1.3359	45	541.1	1.2025	1.4096
3	30.30	1.0099	1.3374	46	555.6	1.2079	1.4117
4	40.56	1.0139	1.3388	47	570.2	1.2132	1.4137
5	50.89	1.0179	1.3403	48	584.9	1.2186	1.4158
6	61.31	1.0219	1.3418	49	599.8	1.2241	1.4179
7	71.81	1.0259	1.3433	50	614.8	1.2296	1.4200
8	82.40	1.0299	1.3448	51	629.9	1.2351	1.4221
9	93.06	1.0340	1.3464	52	645.1	1.2406	1.4242
10	103.8	1.0381	1.3479	53	660.5	1.2462	1.4264
11	114.7	1.0423	1.3494	54	676.0	1.2519	1.4285
12	125.6	1.0465	1.3510	55	691.6	1.2575	1.4307
13	136.6	1.0507	1.3526	56	707.4	1.2632	1.4329
14	147.7	1.0549	1.3541	57	723.3	1.2690	1.4351
15	158.9	1.0592	1.3557	58	739.4	1.2748	1.4373
16	170.2	1.0635	1.3573	59	755.6	1.2806	1.4396
17	181.5	1.0678	1.3590	60	771.9	1.2865	1.4418
18	193.0	1.0721	1.3606	61	788.3	1.2924	1.4441
19	204.5	1.0765	1.3622	62	804.9	1.2983	1.4464
20	216.2	1.0810	1.3639	63	821.7	1.3043	1.4486
21	227.9	1.0854	1.3655	64	838.6	1.3103	1.4509
22	239.8	1.0899	1.3672	65	855.6	1.3163	1.4532
23	251.7	1.0944	1.3689	66	872.8	1.3224	1.4558
24	263.8	1.0990	1.3706	67	890.1	1.3286	1.4581
25	275.9	1.1036	1.3723	68	907.6	1.3347	1.4605

26	288.1	1.1082	1.3740	69	925.2	1.3409	1.4628
27	300.5	1.1128	1.3758	70	943.0	1.3472	1.4651
28	312.9	1.1175	1.3775	71	961.0	1.3535	1.4676
29	325.4	1.1222	1.3793	72	979.0	1.3598	1.4700
30	338.1	1.1270	1.3811	73	997.3	1.3661	1.4725
31	350.8	1.1318	1.3829	74	1016.0	1.3725	1.4749
32	363.7	1.1366	1.3847	75	1034.0	1.3790	1.4774
33	376.7	1.1415	1.3865	76	1053.0	1.3854	1.4799
34	389.8	1.1463	1.3883	77	1072.0	1.3920	1.4825
35	402.9	1.1513	1.3902	78	1091.0	1.3985	1.4850
36	416.2	1.1562	1.3920	79	1110.0	1.4051	1.4876
37	429.7	1.1612	1.3939	80	1129.0	1.4117	1.4901
38	443.2	1.1663	1.3958	81	1149.0	1.4184	1.4927
39	456.8	1.1713	1.3978	82	1169.0	1.4251	1.4954
40	470.6	1.1764	1.3997	83	1188.0	1.4318	1.4980
41	484.5	1.1816	1.4016	84	1208.0	1.4386	1.5007
42	498.4	1.1868	1.4036	85	1229.0	1.4454	1.5033

Source: Iscotables.

Table 6.3 Coefficients for Calculation of Density from Refractive Indexes of Solutions by (6.1.5)

| Solutes | Coefficients | | Temperature T°C | Density Range |
	a	b		
CsCl	10.9276	13.593	20	1.22–1.90
	10.2402	12.648	25	1.00–1.38
	10.8601	13.497	25	1.25–1.90
Cs_2SO_4	12.1200	15.166	25	1.10–1.40
	13.6986	17.323	25	1.40–1.80
CsBr	9.9667	12.2876	25	1.25–1.35
Cs acetate	10.7527	13.4247	25	1.80–2.05
Cs formate	13.7363	17.4286	25	1.72–1.82
NaBr	5.8761	6.839	20	1.00–1.42
	5.8880	6.852	25	1.00–1.52
NaI	5.3330	6.118	20	1.10–1.80
	5.3283	6.103	25	1.00–1.29
K Br	6.5065	7.683	20	1.00–1.37
	6.4786	7.643	25	1.00–1.37
KI	5.7317	6.645	20	1.00–1.40
	5.6581	6.543	25	1.00–1.22
	5.8356	6.786	25	1.10–1.70
LiBr	4.9169	5.555	25	1.00–1.30
RbBr	9.1750	11.2410	25	1.15–1.65
Sucrose	2.7329	2.6425	20	1.00–1.20 (0°C)
Ficoll	2.381	2.175	20	1.0 –1.20
Iothalamate	3.904	4.201	25	1.0 –1.20
Renografin	3.5419	3.7198	25	1.0 –1.20 (4°C)
Metrizamide	3.350	3.462	20	1.0 –1.30
	3.453	3.601	20	1.0 –1.30 (5°C)

tive index, and a correction for this must be included. It is usually done in the following manner:

$$n_{corrected} = n_{observed} - (n_{buffer} - n_{water}) \qquad (6.1.6)$$

One of the disadvantages of using either method is that the sample also contributes to a certain extent. However, when its position in the gradient is known, allowance can be made.

2 FRACTIONATIONS BY LIQUID CENTRIFUGATIONS

While working with purified preparations of alfalfa mosaic virus in the analytical ultracentrifuge, Bancroft and Kaesbert [10] noted three peaks with sedimentation coefficients of 73, 89, and 99 S. They were unable to separate these three fractions by standard techniques such as ammonium sulfate fractionation, calcium phosphate adsorption, and pH adjustment. However, by using a sucrose gradient in a zonal experiment, they were able to separate the fractions and then to demonstrate that the 99-S component was the infective one.

In a study of bacteriophage ϕ X-174, Sinsheimer [11] noted a second and larger component on the ultracentrifuge pattern, in addition to the infective component. Although he was unable to separate these two components by ammonium sulfate fractionation or electrophoresis at various pH levels, he was able to resolve the two particles by density-gradient centrifugation. By using an isopycnic gradient of cesium chloride, he [12] showed that DNA form phage ϕ X-174 differs in density from native DNA. In a separate study with a cesium chloride gradient, Kozinski and Szybalski [13] detected the single DNA molecule in the parental ϕ X-174. To distinguish between the parental and progeny-phage DNA, they used 5-bromodeoxyuridine as a label. The labeled and unlabeled phages reached equilibrium at different density levels.

The above examples clearly demonstrate the specificity of density-gradient centrifugal separation. The power of density-gradient centrifugation in separating and revealing the properties of biological particles has led to its use in solving many research problems with a wide variety of materials.

Factors affecting idealized particle sedimentation in a density-gradient solution are studied in Chapter IV and are summarized there. The conclusions there may be applied to macromolecules, or even to particles as large as whole cells or nuclei. However, membrane-enveloped particles such as mitochondria and cell organelles are extremely sensitive to changes in osmotic strength of the medium and rapidly deteriorate in media of lower tonicity. Considerations in the choice of gradient solute used to make the density gradient are important.

Rate Centrifugation (Velocity Sedimentation)

Rate separations can be done in tubes in swinging-bucket rotors or in zonal rotors [14]. The work of Charlwood [15] and others suggested that fixed-angle rotors can also be used, and Vedel and D'Aoust [16] achieved good separations of RNA in a fixed-angle rotor. On the other hand, Castaneda, Sanchez, and Santiago [17] found serious wall effects when separating polyribosomes and recommended the use of swinging-bucket rotors.

For the best resolution (the completeness of separation of the zones from one another), the density-gradient profile was analyzed and was found to be given by (4.2.38a) and (4.2.38b). The gradient should have the minimum density and viscosity compatible with the separation required, so that the centrifugation is rapid and diffusion is kept to a minimum. Since the sample particles must be more dense than all parts of the gradient, their presence increases the density of this region of gradient. The narrower the band for a given load, the greater the increase and the steeper the gradient must be to support it; therefore, a compromise is usually necessary. Equations 4.2.38a and 4.2.38b represent this compromise. The resolution also depends on the width of the sample zone relative to the radial path length. This is particularly important if the particles to be separated have similar sedimentation coefficients. In this case, one should use a narrow sample zone and a long path length; gradients in long tubes have the additional advantage that they are more stable during handling. The relative sedimentation rates of particles with different densities are also affected by the density of the surrounding medium.

The length of centrifugation time in rate runs is also important. The optimum rate-centrifugation time in a dimensionless form is given in (4.2.41). In order to have the best resolution for ideal particles, (4.2.41) was obtained. In practice, to minimize both the inactivation of labile particles and band broadening due to diffusion, the procedure should be as short as possible. However, in the case of mammalian cells, rapid acceleration can damage the cells by enucleation. Moreover, the integrity of osmotically sensitive membranes of both cells and organelles can be damaged by the hydrostatic pressure generated by centrifugation at very high speeds.

Isopycnic Centrifugation (Equilibrium Sedimentation)

Isopycnic centrifugation, which is used to fractionate biological materials according to their buoyant densities, has proved to be an extremely versatile technique in both molecular and cell biology. It consists of centrifuging the mixture to be fractionated in a density gradient formed from a solute of suitable substance in solvent, almost invariably in water, until each species of particle in the mixture reaches its isopycnic point, the point in the gradient at which the density of the solution equals the buoyant density of that species of particle. Thus, it is an equilibrium method, which means that, once the particles have formed equilibrium bands at their isopycnic points, there is no change in the distribution of the particles in the gradient no matter for how much longer centrifugation is prolonged. The technique in preparative centrifugation is one of the simplest techniques and requires no sophisticated equipment besides the ultracentrifuge and rotor.

Since its introduction for the separation of DNA by Meselson, Stahl, and Vinograd [18], this technique has been widely used for the fractionation of a host of biological macromolecules and particles. However, its use has been sometimes limited by the types of density-gradient solute available. For example, most cellular and subcellular structures are sensitive to both the ionic content and osmotic strength of the surrounding medium. In addition, the solutions of carbohydrate gradient media are extremely viscous at high concentrations and have only a very limited density range. However, with the recent development of other types of density-gradient media, such as Ficoll,

Metrizamide, methylcellulose, and so on, some of the limitations have been overcome.

There are two methods of setting up an equilibrium density-gradient sedimentation separation. The first involves forming the density gradient in situ and consists of preparing a uniform mixture of the sample to be fractionated with a solution of the solute which is to form the density gradient in suitable buffer. The mixture is then centrifuged until (a) an equilibrium gradient is formed by redistribution of the solute in the centrifugal field, and (b) the components of the sample have reached their isopycnic positions. The second is similar to that used for rate separations; it consists of preparing a gradient from solutions of the solute in a suitable buffer, layering the sample to be fractionated on the top of the gradient, and centrifuging until the components of the sample have reached their respective isopycnic positions.

It has been shown that the resolving power of the isopycnic banding by centrifuging a solute solution at a given speed can be increased by using a fixed-angle head rotor instead of a swinging-bucket rotor [19]. This is because the difference in density between the top and bottom of a gradient formed by centrifugation depends on the horizontal distance between these points during centrifugation. When the rotor is brought to rest after centrifugation is·complete and the tube is reoriented to the vertical position, the distance between the top and bottom of a gradient formed in a swinging-bucket rotor is unchanged. In a gradient formed in a fixed-angle head rotor, the corresponding distance is increased; the gradient is stretched over a long distance so that $d\rho/dr$ is decreased and, consequently, resolution is increased. The extent to which the resolving power of an equilibrium gradient can be increased by this means increases as the angle of the tube to the direction of the centrifugal field increases from 0° (swinging-bucket rotors) to 90° (vertical-tube rotors).

It should be noted that the resolving power of isopycnic gradient separations in zonal rotors is the same as in swinging-bucket rotors and is thus less than in fixed-angle head rotors.

3 SELECTION OF DENSITY-GRADIENT SOLUTES

The nature of the gradient solute used to form the density gradient is of paramount importance in determining whether a particular fractional separation will be successful. The properties of the ideal gradient solute are summarized in Chapter IV. Clearly, none of the solutes used in density-gradient centrifugation satisfies all of these criteria. In particular, the high osmolality of the dense ionic solutions such as potassium citrate, potassium tartrate, cesium chloride, and so forth, causes difficulties with many cell organelles, whereas the high ionic strength of these solutions results in the disruption of most nucleoprotein complexes. It is the latter property of dense solutions of salts (ionic gradient media) which limits their suitability as buoyant density-gradient media for the isopycnic banding of nucleoproteins such as ribosomes and chromatin. Thus, one has to search for nonionic gradient solutes.

The thermodynamic activity of salts in aqueous solutions varies widely from one salt to another and is dependent on the salt concentration. It is always high, so that the water activity of these solutions is correspondingly low. In practical terms, this means that the hydration of particles and macromolecules in a salt gradient is very strongly dependent on the nature and concentration of the salt, and is usually low as compared to their degree of hydration in water alone. This effect is particularly notable with nucleic acids which consequently have high buoyant densities in solutions of salts. Thus, for many purposes, only salts that are very soluble in water are capable of forming solutions dense enough to permit isopycnic banding of biological particles, particularly nucleic acids and nucleoproteins.

If salts are used as buoyant density-gradient solutions, one has only limited control over the ionic conditions under which an isopycnic centrifugal separation is done and no control over the ionic strength or osmolarity of any part of the density gradient. In addition, some of the salts (notably iodides) absorb ultraviolet light strongly and therefore cause difficulty with detection of samples in these gradients, whereas others (for example, iodides, sulfates, etc.) interact with commonly used liquid scintillation fluids and therefore cause difficulties with measurements of radioactivity in gradient fractions.

Commonly used density-gradient solutes are discussed in the following paragraphs strictly from a practitioner's point of view.

CESIUM CHLORIDE

This is by far the most commonly used salt for buoyant density-gradient experiments. This is partly for historic reasons: aqueous CsCl was the medium selected for the first studies of the analytical use of the isopycnic banding by Meselson, Stahl, and Vinograd in 1957 [18], and its thermodynamic parameters have been throughly studied and are well known [20]. CsCl is very soluble in water, giving very dense solutions, up to 1.91 g/cm^3 at 20°C. The relative viscosity of these solutions is close to unity. Very pure (optical grade) CsCl can be purchased, and solutions of CsCl do not absorb ultraviolet light; however, this grade of salt is very costly, which limits its use to experiments in the analytical ultracentrifuge. Lower grades of CsCl are considerably less expensive, although still far from cheap; their solutions have a measurable absorption in the ultraviolet, but this is usually not so great as to preclude the use of these grades for experiments in preparative ultracentrifuges.

Dilute solutions of CsCl do not quench the usual liquid sintillation fluids. However, the thermodynamic properties of CsCl are not ideal in all situations. The water activity of CsCl solutions is very low with the result that the buoyant density of nucleic acids in CsCl is very high.

CESIUM SULFATE

This is the next most widely used salt. For the most part, its properties are similar to those of CsCl, but they do differ in a number of important respects. Although the maximum density obtainable with Cs_2SO_4 in water (2.01 g/cm^3 at 25°C) is only slightly greater than that with CsCl, the water activity of Cs_2SO_4 solutions is greater, and the buoyant densities of the nucleic acids are considerably lower in Cs_2SO_4 than in CsCl solutions. However, most species of RNA molecules either aggregate or precipitate in Cs_2SO_4, so reducing the usefulness of this solute for separations of RNA.

OTHER SODIUM AND POTASSIUM SALTS

These have not been used as density-gradient solutes to any significant extent. The densities obtainable are quite high, but many of these solutions are rather viscous. Moreover, the equilibrium density gradients formed by these salts tend to be too shallow to be of value for most fractionations. Despite these limitations, preformed potassium tartrate gradients have proved useful for the isopycnic banding of viruses, mainly because some viruses appear to maintain their viability better in solutions of this salt than in the salts more often used as buoyant density-gradient solutes.

SUCROSE AND POLYSACCHARIDE SOLUTIONS

Sucrose solutions are the most commonly used density-gradient solutions. The most commonly used density range is 1.00 to 1.20 g/cm^3, which is sufficient to band membranes, membrane-associated organelles, and viruses. This range includes the density of most cells, although the high osmolarity of sucrose solutions makes this medium unsuitable for separating cells when their viability after seperation is important. However, sucrose gradients are of limited use for separating mitochondria, lysosomes, and other ogranelles isopycnically, since in hypertonic solutions they band at essentially the same densities [21], unless one or the other is specifically modified.

Isopycnic separations in sucrose gradients have become a standard procedure for obtaining many species of purified virus from a variety of cell types. However, the suitability of sucrose gradients for virus purification is directly related to the type of host cell and, to a much greater extent, to the species of virus, since a certain amount of cellular material usually either co-bands or aggregates with the virus.

There are two types of interaction between monosaccharides and disaccharides with biological substances. First, monosaccharides and disaccharides inhibit the activity of many enzymes [22, 23]. However, this inhibition is linear with solute concentration and is readily reversible on dilution, although the mechanism remains unknown. Secondly, sucrose solutions have been reported to damage the respiratory control of mitochondria [24] and disrupt other vesicular structures [25], probably as a result of their high osmolarity.

One of the main disadvantages of using monosaccharides and disaccharides as density-gradient solutes is the high osmolarity of concentrated solutions. Therefore, in order to overcome this particular problem, polysaccharide solutes have been used for the rate-centrifugal and isopycnic-centrifugal separations of osmotically sensitive particles, such as mitochondria and cells. The most commonly used solute is a polymer of sucrose, Ficoll (Phamacia Fine Chemicals AB, Uppsala, Sweden).

Ficoll gradients have been used for both isopycnic-centrifugal and rate-centrifugal separations. Many types of cells, including blood cells [27], fibroblasts [28], tumor cells [29], rat liver cells [30], and mouse liver cells [31], have been fractionated by isopycnic centrifugation in Ficoll gradients. Many types of osmotically sensitive organelles from animal [32], plant [33], and fungal [34] cells have also been separated using polysaccharidic density-gradient media. Separations in these media reduce the possibility of damaging organelles, and, in addition, their buoyant density is lower in polysaccharidic media than in hypertonic sucrose solutions, partly because they are more highly hydrated. However, the degree of resolution of organelles depends on the composition of the medium [35]. Some cells may phagocytize the polysac-charide particles. Polysaccharide substances can also cause the aggregation of specific cell types [36].

Polysaccharides, like mono- and disaccharides, interfere with the chemical estimations of glycogen, nucleic acids, and protein. However, dextrans and Ficoll inhibit enzymes to a lesser extent than equivalent concentrations of sucrose [23]. An additional problem with polysaccharides is that they cannot be removed by dialysis or ultrafiltration, so particulate material is generally recovered by diluting and centrifuging the fractions.

Sorbitol solutions have been reported making density-gradient solutions with densities as high as 1.39 g/cm^3 [26], which are highly dehydrated in aqueous media. In such solutions, macromolecules, particularly nucleic acids, are partially dehydrated and band at densities higher than are obtainable with these compounds.

IODINATED DENSITY-GRADIENT MEDIA

These have the combined stability and inertness of the saccharidic media with the flexibility of the large density range of the alkali metal halide solutions. Their use for centrifugal separations is relatively recent. With the exception of Metrizamide, all of these media; Iothalamate, Renografin, Metrizoate, and so on, are to some extent ionic. The ionic strength of gradients prepared from such compounds is low. They are the most versatile of the nonionic media because they can be used over a wide density range and are relatively inert and nonviscous.

In Metrizamide, DNA bands at 1.12 g/cm^3, approaching the density of fully hydrated DNA [37]. In ionic iodinated media, DNA bands at a slightly greater density than in Metrizamide [38–40], probably as a result of the ionic nature of these compounds reducing the water activity of the gradients. It is possible to vary the ionic environment of Metrizamide at will; and thus it has been shown that the buoyant density of DNA in Metrizamide is dependent on the ions present and the pH of the solutions [41, 42]. Metrizamide interacts only weakly with protein ($K_{diss} > 10^{-2} M$) [43, 44].

Detailed discussion on various density-gradient materials and their usage in various biological separations has been presented by Birnie and Rickwood [45].

4 APPLICATIONS OF DENSITY-GRADIENT CENTRIFUGATIONS: EXAMPLES

The difference between *density-gradient centrifugation* and *differential centrifugation* is that in the former the particles sediment through a medium of increasing density, whereas in the latter the medium is of uniform density. Rate centrifugation is the method of separating particles into discrete zones on the basis of differences in sedimentation rate; isopycnic or equilibrium centrifugation is the method of separating particles on the basis of differences in buoyant or banding density. As such, these terms embrace all types of centrifugation, including those with swinging-bucket, fixed angle-head, and zonal rotors.

Comments about separating various biological materials using the gradient technique together with examples are presented in the following.

Viruses

Harvesting virus has two objectives, namely, concentration and purification. If the starting material is a complete culture fluid, often two centrifugation steps are required; the first is a concentration and partial purification, and the second is predominantly a purification. After removal of large contaminants such as cells, nuclei, and cellular debris by an initial low-speed centrifugation, sucrose is mixed with the virus-containing fluid. This suspension can then be introduced into the rotor as part of a discontinuous gradient, or used as one component in the generation of a linear gradient, or simply introduced between an overlay and a cushion.

The choice of density medium in each case is largely dependent upon the species of virus and the consequence of its interaction with medium. Virus size and morphology, in particular the degree of pleomorphism, are critical factors which affect viruses in different media. Conditions of high osmotic pressure or high ionic strength produced by concentrated solutions of sucrose or CsCl, for example, can adversely affect those viruses which are extremely pleomorphic on the one hand or those which are salt sensitive on the other, resulting in altered virus structure, causing them to become inactivated.

Some plant viruses (for example tobacco mosaic virus) are also stable in solutions of high ionic strength, and thus buoyant density-gradient centrifugation in CsCl and other salts has been used for their purification, fractionation, and characterization [46]. Poliovirus bands isopycnically in CsCl without significant loss of infectivity [47]. Relatively little is known about the suitability of iodinated media for separation of viruses. There are some reports of success [48, 49] and some evidence that some species of virus are particularly sensitive to iodinated media [50].

The work of Reimer et al. [51] is presented here as an example of mass purification density-gradient zonal centrifugation.

Influenza Virus

Influenza virus was produced by standard techniques (Division of Biologics Standards, National Institute of Health, USPHS, Minimum Requirements: Influenza Virus Vaccines, Types A and B; and Public Health Service Regulations, Title 42, Part 73) in the allantoic cavity of 11-day-old embryonated chicken eggs. The strains used were: A/RP-8,A_1/Ann Arbor/1-57, A_2 Taiwan/1-64, A_2/Japan-170/62, B/Maryland/1-59, and B/Massachusetts/3-66.

Barium Sulfate Absorption-Elution

Prior to zonal centrifugation, virus was concentrated 3.5 times and partially purified by the barium sulfate.

Virus was absorbed from allantoic fluid (pH 7.8 to 8.0) at 0 to 3°C by the addition of potassium oxalate and dry barium sulfate, as shown in Table 6.4. The suspension was stirred for 90 min and allowed to stand overnight. Harvesting of the barium sulfate with absorbed virus was done in a 4-liter bowl Sharples continuous-flow centrifuge at 7000 rpm with a flow rate of 50 to 80 liters/hr.

Virus was eluted by resuspending the barium sulfate into one seventh the original volume of fluid having the composition shown in Table 6.4. On the following day, the barium sulfate was removed from the resuspended virus at room temperature by centrifugation at 800 × g for 10 min in an International PR-2 centrifuge. The final eluate containing the partially purified virus was adjusted to pH 8.0 and diluted approximately twofold with pyrogen-free distilled water to a final specific gravity of 1.050 ± 0.005.

Zonal Centrifugation

Two-step purification (S-ρ) processes were used. The first was the concentration of virus by rate zonal centrifugation using B-IV and B-XV rotors. The second was purification and concentration by the continuous-flow-with-banding process using B-IX and K-II rotors. The reorienting gradient techni-

Table 6.4 Amounts of Dry Barium Sulfate Used to Absorb Six Strains of Influenza Virus and the Composition of the Eluting Media to Remove the Virus from the Barium Sulfate

	Adsorption				Elution			
Strain	$BaSO_2$ Added per liter of Allantoic Fluid (g)	Oxalate (M)	Bone gelatin (%)	Na_3 citrate (M)	pH	NaCl (M)	Tris (M)	Tween 80 (M)
PR-8	40	None	0.2	0.25	7.2	None	None	None
Taiwan	50	0.16	0.2	0.25	7.2	1.0	0.4	0.16
Ann Arbor	50	0.16	0.2	0.25	7.2	1.0	0.4	0.16
Jap 170	50	0.16	0.2	0.25	8.8	1.0	0.4	0.002
B Mass	50	0.16	0.2	0.25	7.2	1.0	0.4	0.16
B Md	60	0.16	0.2	0.25	7.2	1.0	0.4	None

Source: Reimer et al. [51].

que was used. The K-II rotor (3.6 liters internal volume) and associated tubing was filled with the lighter of the two gradient solutions which were used to make the self-forming gradient by diffusion. Sucrose solutions with 0.01 M phosphate buffer and 0.02% gelatin at pH 8.0 were used as a density-gradient solution. The sucrose concentration in the lighter of the two gradient solutions was adjusted so that this solution's specific gravity was 0.005 higher than the specific gravity of the virus eluate (ca. 1.050) from the $BaSO_4$.

After the rotor was completely filled with light solution, it was accelerated from the rest to 2000 rpm, and the direction of fluid flow through the rotor was reversed several times to remove entrapped air. The rotor was then brought to rest by means of the air brake. An appropriate amount (usually about 1.8 liter) of the heavier of the two gradient solutions [60% (w/w) sucrose, buffered as described for the light solution] was then forced into the bottom of the rotor, displacing an equal volume of the lighter solution from the top. (Note that the K-II rotor is unique in having liquid seals and flow lines attached to both ends of the rotor.) The lower rotor inlet line was then closed and the rotor was smoothly accelerated to 2000 rpm to reorient the liquid density gradient from a vertical to a centrifugal direction. Fluid flow inboard of the imprisoned gradient was then established from the top to the bottom of the rotor and maintained while the rotor was smoothly accelerated to 20,000 rpm (400 to 4000 rev/min^2 maximal acceleration rate, to 20,000 rpm). Virus sample flow was then started while the rotor was further accelerated at a rate manually governed to prevent high hydrostatic back pressure [not exceeding 1.034×10^5 N/m^2 (15 psi)] in the input line. This back pressure is due to the presence of two fluids which differ in density by approximately 0.005 g/cm^3 in the 700-ml core taper volume. Displacement of a dense fluid by a light one results in a back pressure in a continuous-flow rotor where both fluid lines are brought back to the axis of rotation before leaving the rotor. Once 700 ml has been displaced from the rotor by the lighter virus sample, the rotor was accelerated to operational speed without developing hydrostatic pressure in the input line.

Generally, the K-II centrifuge was operated at 27,000 rpm with 2.124 m^3 (75 ft^3) of filtered dry air per minute at 1.724 N/m^2 at 25 psi. Acceptable clean-out of influenza virus occurred from the 1.050 specific gravity citrate solutions when the sample flow was adjusted to between 4 and 5 liters/hr.

Upon termination of flow, sample fluid remaining in the core taper volume (700 ml) was displaced by a sucrose gradient solution 0.005 g/cm^3 lighter than the sample. Rotation was then continued at full speed to allow virus to band for an additional 30 min.

The rotor was allowed to coast to rest before it was unloaded. The mass of the rotor is sufficiently large to ensure an even and gradual reorientation during deceleration. The gradient was collected as a discrete series of fractions from the bottom of the rotor by use of air pressure through the upper fluid line.

Gradient Shape

Gradient shape was determined by measuring the sucrose concentration with a Bausch and Lomb Abbé refractometer.

Absorbance

Absorbance between 250 and 350 nm was determined for all fractions in a Cary 15 recording spectrophotometer.

Other pathological properties, such as chicken cell agglutination (CCA) assays, mouse protection tests, hemagglutinin assays, immunodiffusion, and protein determinations, were also done for each fraction collected.

Results and Discussion

Figure 6.5 shows the hemagglutinin and optical density profiles for six strains of influenza virus with the use of the combined process and the K-II centrifuge. These are representative profiles from commercial production-

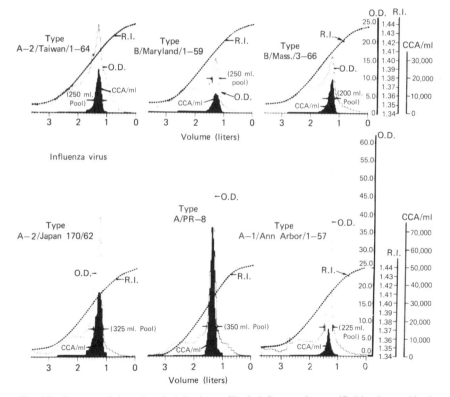

Fig. 6.5 Hemagglutinin and optical density profiles in influenza virus purified by the combined process of BaSO$_4$ absorption–elution, followed by isopycnic banding in the K-II zonal centrifuge. *Source:* Reproduced from Ref. 51 with permission of American Society of Microbiology.

sized lots, which utilized an average of 15,000 eggs each. These virus pools were diluted and further processed through formalin inactivation and bacteriological filtration to make monovalent and polyvalent vaccines. The larger K-II centrifuge (volume 3.6 liters), when used sequentially with other purification procedures, has routinely processed more than 150-liter volumes of allantoic fluid per day. Kilogram quantities of impurity have been rapidly separated from gram quantities of purified virus, which has been conveniently concentrated several hundredfold by the purification process.

A 105-liter volume of allantoic fluid containing Maryland B strain of influenza virus was processed through the B-IX rotor. Figure 6.6 shows the optical density and hemagglutinin profiles of fractions harvested from this continuous-flow isopycnic banding device. This crude egg harvest has been first cleared of gross debris in the Sharples Supercentrifuge with essentially all of the virus remaining in the flowing stream. This virus was then quantitatively captured in the B-IX rotor spinning at 40,000 rpm with a flow rate of 8 liters/hr. The bulk of soluble impurities remained in the stream flowing through the B-IX rotor and was discarded. As calculated from either the hemagglutinin test or the chick cell agglutination (CCA) assay, two thirds of the total virus originally present in the 105-liter pool of crude allantoic fluid was recovered in the sucrose gradient as a band having a volume of 200 ml. The density of the peak fraction was 1.187 g/cm³. The peak hemagglutinin titer was 320,000, as compared with 512 for the original allantoic fluid.

An 80-ml amount of this B-IX concentrate was further purified by a rate zonal process in the B-IV rotor after first removing the sucrose in a G-25

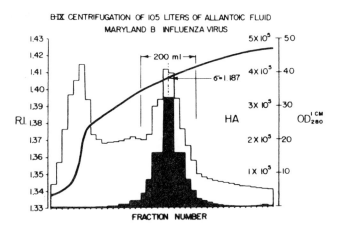

Fig. 6.6 Distribution of influenza virus (HA shaded area) and optical density in a sucrose gradient (refractive index) from a B-IX rotor. *Source:* Reproduced from Ref. 51 with permission of American Society of Microbiology.

Sephadex column. The rate zonal sedimentation profile of the B-IX concentrate was compared with a similar profile for a typical commercial 10X vaccine concentrate as shown in Fig. 6.7. Soluble impurities are found in the peak on the left; large particulate impurities are found in the peak on the right. The central peak, rising out of a background of small particles, contains the virus. The relation between the size of the virus peak and the impurity peaks demonstrates graphically the increase purity of the virus from the B-IX process.

Figure 6.8 shows an electron micrograph of a vaccine made with the combined process by K-II rotors.

Fractional Cleanout

Fractional cleanout of virus particles from the sample flow stream can be estimated by (4.2.62). Since the sedimentation coefficients of large bioparticles are not as commonly known as their diameter and isopycnic density, the sedimentation coefficient of the particles, S, may be expressed by integration of (4.2.3) with (4.2.25), which gives

$$S = \frac{D_p^2}{18} \left(\frac{\rho_p - \rho(r)}{\mu(r)} \right) \times 10^{13} \qquad (6.4.1)$$

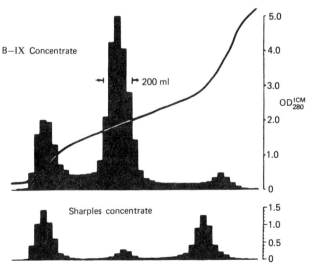

B–IV Rate–zonal purification of Maryland–B influenza virus

B–IX Concentrate

⊢ 200 ml

Sharples concentrate

OD$_{280}^{ICM}$

Fig. 6.7 Rate-zonal sedimentation profiles for the B-IX concentrate of Fig. 6.6, and a 10X Sharples commercial virus concentrate. The central peak contains the virus. *Source:* Reproduced from Ref. 51 with permission of American Society of Microbiology.

Fig. 6.8 Electron micrograph of a monovalent vaccine (Md.B) purified by the combined process × 37,000. *Source:* Reproduced from Ref. 51 with permission of American Society of Microbiology.

where D_p is the diameter of the particle (in cm), ρ_p is the density of the solvated particle, $\rho(r)$ and $\mu(r)$ (in poises) are respectively the density and viscosity of the density gradient at the banding position at operating temperature. Combining (4.2.62) and (6.4.1), the fractional cleanout can be expressed by

$$f = \frac{4\pi^3 r_1^2 L D_p^2 \left[\rho_p - \rho(r)\right] (\text{rpm})^2}{9\,\mu(r)\,Q\,10^3} \tag{6.4.2}$$

For the case, $D_p = 5.4 \times 10^{-6}$ cm (estimated from Fig. 6.8), $r_1 = 4.61$ cm (tapered K-II), $L = 75.08$ cm, $\rho_p = 1.187$ g/cm^3, $\rho(r) = 1.050$ g/cm^3, $\mu(r) = 0.0223$ poise, rpm = 2700 rpm, and $Q = 4$ to 5 liters/hr.

Then, the fractional cleanout of the particles is 52 and 42% for the sample flow rates of 4 and 5 liters/hr respectively. Predicted 100% cleanout flow rate will be 0.48 liters/hr theoretically.

Membranes

Centrifugation in zonal rotors has been widely used to isolate and fractionate membranes from a variety of tissues and cells. The fractionation methods used are almost as varied as the sources of the membranes, because each tissue presents its own problem, many of which have been solved in different ways.

Cell membrane-rich fractions from rat liver and other sources have been isolated by a combination of centrifugal methods which depend on sedimentation rate and on isopycnic separations, as presented in Chapter IV. Anderson [52] has reported that as the pH of the homogenizing medium is raised from 5.8 to 7.8, the efficiency of homogenization under controlled conditions increases and the amount of cytoplasm attached to cell membranes decreases. In rat liver homogenates prepared in slightly alkaline hypotonic solutions, a large number of cell membrane fragments may be produced by gentle homogenization which have sufficient size to sediment more rapidly than mitochondria. The initial centrifugal procedures that have been used to separate membranes and nuclei from the majority of lysosomes and mitochondria on a rate basis first used swinging bucket or angle-head rotors. This step may be repeated several times. To separate the membranes from nuclei, the preparation is then banded isopycnically in a sucrose gradient.

The work of Anderson et al. [53] using a B-XV titanium zonal rotor to isolate rat liver cell membranes by combined rate and isopycnic zonal centrifugation technique is presented as an example below.

Basis of Separations

Both mitochondria and cell membrane fractions are heterogeneous with respect to their sedimentation properties. However, the mean sedimentation coefficients appear to differ widely. The degree of overlap is unknown, however and may depend on the method of homogenate preparation, that is, on whether the membranes in the initial homogenate are broken down into large or small fragments. If a clean separation of cell membranes from mitochondria, lysosomes, microbodies, and endoplasmic reticulum fragments can be made in a zonal centrifuge rotor, then the centripetal portion of the gradient may be removed and replaced with a particle-free solution. The rotor may then be spun for a prolonged period of time to band the membranes sharply at their isopycnic points. Nuclei and nuclear fragments would be centrifuged to the rotor wall since their banding density exceeds the density of the centrifugal end of the sucrose gradient.

Rate Zonal Separation of Mitochondria and Membranes

PREPARATION OF HOMOGENATES

Fresh, unperfused rat liver from male Sprague–Dawley rats was diced with a scissors in cold 0.001 M NaHCO$_3$ adjusted to pH 7.6 with NaOH and homogenized in a loose-fitting Dounce-type homogenizer. Fifteen even strokes were used with each liver sample. The homogenate was diluted to a final volume of 30 ml per gram of liver and centrifuged for 15 min at 27,000 rpm at 0°C in an International Equipment Company PR-2 centrifuge using 250-ml Pyrex tubes in the No. 259 head. The supernatant was removed by gentle suction and discarded. The pellet was resuspended in 200 ml 0.001 M bicarbonate buffer and used as the sample suspension for zonal centrifugation.

ZONAL CENTRIFUGATION

A B-XV titanium zonal rotor was operated in a modified Spinco Model L centrifuge equipped with the temperature sensing and controlling device and $\int \omega^2 dt$ integrator. Gradients were formed in a Spinco Model 131 gradient pump. Rotor effluents were collected in either 40- or 10-ml fractions. Sucrose concentrations were determined refractometrically. Absorbance of the effluent stream was measured.

A membrane suspension (190 ml) was layered over a sucrose gradient (19 to 35% w/w sucrose, 500 ml), followed by an overlay (200 ml 0.001 M NaHCO$_3$), and centrifuged at 5000 rpm until $\omega^2 t = 2 \times 10^8$ sec^{-1}. The rotor was then unloaded and the relative amounts of mitochondria (visible granules) and cell membranes were then scored by phase contrast microscopy. The results are shown in Fig. 6.9. It was concluded that by unloading the first 750 ml (18^3/4 to 40-ml fractions) the bulk of the soluble microsomal and mitochondrial mass would be separated from the membranes. This procedure was found to be reproducible with liver preparations. An example is shown in Fig. 6.10.

After removal of the centripetal 750 ml the rotor was accelerated to 20,000 rpm and run until $\omega^2 t = 3.5 \times 10^9$ sec^{-1} at unloading. The membrane fragments were found to be sharply banded at an average of 41.04% (w/w) sucrose. (Experimental values were 40.2, 40.8, 40.8, 41.2, 40.8, 41.9, and 41.6.) Observations of the banded membranes in the phase contrast microscope revealed that very few visible granules were present. Electron microscopy confirmed the high purity of the preparation, as shown in Fig. 6.11.

Dynamic Similarity

If one has a B-XIV instead of a B-XV zonal rotor in his laboratory, the experimental conditions have to be modified according to the principle of

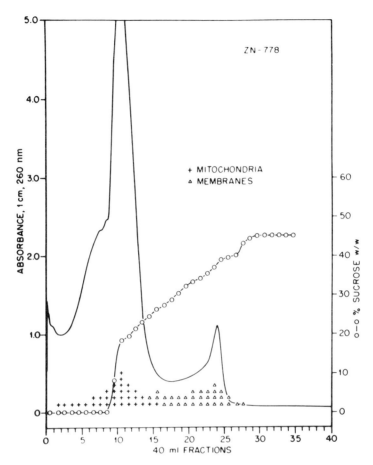

Fig. 6.9 Rate-zonal separation of rat liver cell membranes in a sucrose gradient in the B-XV titanium rotor. Relative quantities of mitochondria (+) and membrane (△) in recovered fractions as judged by phase contrast microscopy. *Source:* Reproduced from Ref. 53 with permission of the Wistar Institute of Anatomy and Biology and the author.

dynamic similarity, as discussed in Chapter IV. The characteristic differences between B-XV and B-XIV rotors are as follows:

Rotor	Radius (cm)	Volume (cm³)
B-XV	8.89	1665
B-XIV	6.601	630

The sample size, gradient volume, amount of overlay, the fractions to be re-

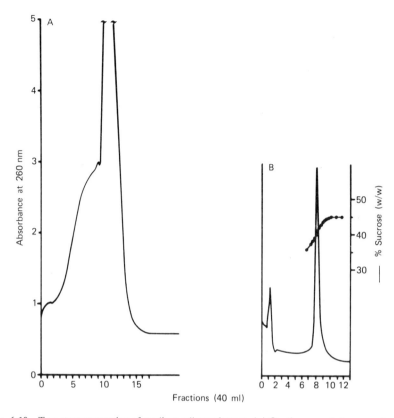

Fig. 6.10   Two-stage separation of rat liver cell membranes. (*a*) Overlay, sample layer, and part of gradient recovered after short rate-zonal centrifugation (*left*). (*b*) Banded cell membranes after centrifuging until $\omega^2 t = 6 \times 10^9$. Membranes banded sharply at 40.8% sucrose (w/w) (*right*). *Source:* Reproduced from Ref. 53 with permission of the Wistar Institute of Anatomy and Biology and the author.

moved, and so on, have to be scaled down by the ratio of volume of the rotors. Thus, one should modify various quantities to the following:

Expt. Conditions	B-XIV (ml)	B-XV (ml)
A membrane suspension sample	72	190
A sucrose gradient (19 to 35%)	190	500
An overlay (0.001 M NaHCO$_3$)	77	200
The first fraction to be unloaded	285	750

In order to have the same separation performed by a B-XV rotor for the same sample, the applied force field strength must be the same, that is, in

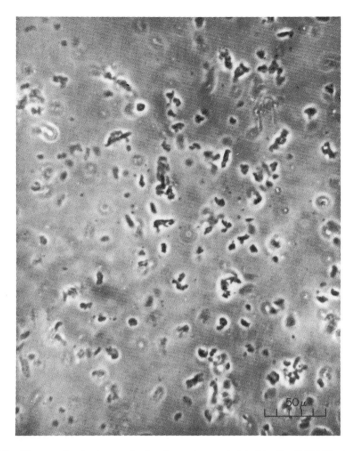

Fig. 6.11 Phase contrast micrograph of cell membranes isolated in the B-XV titanium rotor (ZU 768 Fraction 15). *Source:* Reproduced from Ref. 53 with permission of the Wistar Institute of Anatomy and Biology and the author.

each operation $\omega^2 t$ must have the same value. The force field of the first centrifugation at 5000 rpm $\omega^2 t = 2 \times 10^8$ and the second centrifugation, after removal of the centripetal 750 ml, at 20,000 rpm until $\omega^2 t = 3.5 \times 10^9$ in a B-XV or in a B-XIV rotor must be the same. This means that the reduced quantity $\mathbf{F} = [\omega^2 t \, \rho_o D_p{}^2 / \mu_o]$ must remain constant. However, in order to have the same sedimentation characteristic in both rotors, the reduced sedimentation coefficient (S) or the Taylor number $(\mathbf{Ta})$ has to be the same. For the separation of the same sample in two different size rotors, one obtains, from either of the reduced quantities, the following:

$$\omega_1 \, R_1 = \omega_2 \, R_2 \qquad (6.4.3)$$

Therefore, if one uses a B-XIV rotor, the speeds of the first and the second centrifugation have to be modified to 6740 and 27,000 rpm, respectively. Thus, the centrifugation time is reduced, but the quantity $\omega^2 t$ still remains the same.

DNA and RNA

The very large difference between the buoyant densities of RNA (> 1.9 in CsCl) and DNA (> 1.7 in CsCl) and between those of DNA and proteins (1.3 to 1.5 in CsCl) has led to the development of bouyant density-gradient methods for the large-scale isolation of DNA and RNA from tissues, cells, and cell organelles. There does not seem to be a universally applicable procedure whereby DNA from any organism, tissue, or organelle can be isolated and purified. There are, or course, many excellent methods [54], but no single one is ideally suited to all situations. The lack of general applicability is due to several factors, including (1) the different ways in which DNA is "packaged" within an organism; (2) the different types of proteins with which it may be associated or conjugated; (3) variation in the nature of this association or binding; (4) the proportion of DNA per unit dry weight of cellular material; (5) the relative amounts of the four main contaminants (protein, lipoprotein, polysaccharides, and RNA); and (6) variation in contaminating nucleases. Some or all of these factors may vary from organism to organism, and, indeed, large differences may be encountered between the tissues of an organism, which are further accentuated if organelles (nuclei, nucleoli, mitochondria, and chloroplasts) are studied.

Much less has been done to explore the fractionation of RNA by buoyant density-gradient centrifugation for the simple reason that it has been difficult to find a gradient solute suitable for this purpose. CsCl is not soluble enough to provide a solution denser than RNA except at 40 to 50°C, a temperature which is inconvenient from the point of view both of the RNA and the centrifuge. Concentrated solutions of cesium formate and cesium acetate are dense enough, but their use is hampered by their high viscosity. Cs_2SO_4 would be very suitable for the isopycnic banding of RNA were it not for the marked propensity of most RNA's to aggregate and precipitate in solutions of this salt.

The separational method consists of disrupting tissues, cells, or isolated nuclei in detergent and dense CsCl solution and centrifuging the mixture to equilibrium under conditions in which RNA pellets at the bottom of the tube, the DNA forms a band in the gradient, and the protein and carbohydrate form a pellicle at the top. The method is very tolerant of variations in procedure, provided that the basic requirements are satisfied.

Separation of native and denatured DNA, RNA, and DNA–RNA hybrid by Birnie [55] is presented here as an example. Cs_2SO_4 is the salt most frequently used for buoyant density analyses of DNA–RNA hybrids because DNA, RNA, and hybrid all band in Cs_2SO_4 gradients [56]. CsCl can also be used for this purpose, although RNA is too dense to band in CsCl gradients. However, neither of these salts is entirely satisfactory since they resolve native and denatured DNA poorly, an important consideration in studies of RNA-driven DNA-RNA hybriditation reactions using labeled DNA [57], in the analysis of which it is necessary to distinguish between RNA-DNA hybrids and reassociated DNA duplexes. Birnie [55] used NaI gradients for DNA to substitute for CsCl salts and reported that NaI gradients clearly distinguish between native DNA, denatured DNA, RNA, and DNA-RNA hybrid and that RNA is soluble in NaI solutions as long as care is taken to remove heavy metal ions from the NaI.

Materials and Methods [55]

[³H]DNA was isolated [58] from the nuclei of LS cells (a substrain of mouse NCTC strain L929, which grows in suspension) that had been grown for 18 hr in the presence of methyl-[³H]thymidine (22.4 mCi/μmole, 1 to 10 μCi/ml). It was denatured in 0.05 M NaCl by heating at 100° for 10 min. DNA-RNA hybrid was synthesized [59] by heat-denatured [³H]DNA (10 μg) and was incubated for 1 hr at 37°C in 50 mM Tris-HCl, pH 8, 2.5 mM MnCl$_2$, 1 mM dithiothreitol, and 0.1 mM EDTA with 0.8 mM ATP, GTP, CTP, and UPT. RNA polymerase (125 units) was prepared from E $scherichia$ $coli$ to fraction 6 of the Burgess [60] procedure, omitting the phosphocellulose step, in a total volume of 1.0 ml. The hybrid was isolated by chromatography on Sephadex G-50 in 0.05 M NaCl. [³H]RNA was isolated from LS cells grown for 18 hr in the presence of [5-³H]uridine (29.8 mCi/μmole, 1 μCi/ml). Nuclear and cytoplasmic fractions were prepared [61] and RNA extracted from both by shaking with phenol at 20°C in the presence of 6-aminosalicylate. The nuclear RNA preparation was digested exhaustively with DNAase I. Both RNA preparations were purified by chromatography on Sephadex G-100 in water; only the excluded peak was collected. $Neurospora$ $crassa$ nuclease was prepared as described by Linn and Lehman [62] omitting the phosphocellulose and hydroxyapatite chromatography steps.

NaI (May and Baker Ltd.) was dissolved in 0.1 M Tris-HCl, and 0.01 M EDTA, pH 8 (150 g in 90 ml), to give a solution of density 1.86 g/cm³. Na$_2$SO$_3$ (10 mg/100ml) was added to prevent oxidation of the NaI, and the solution was passed through a Millipore filter, then a bed of Chelex-100 resin (10 ml) at about 1 ml/min; Na$_2$SO$_3$ (10 mg/100 ml) was again added to the solution. CsCl (BDH Ltd.) was dissolved in the same buffer (150 g in 100 ml) to give a solution of density 1.8 g/cm³, filtered, and passed through Chelex-100 resin as for the NaI solution. These stock solutions were stored at room temperature. The labeled nucleic acids, together with 5 μg carrier mouse embryo DNA (native and denatured), were mixed with stock NaI or CsCl, and sufficient buffer was added to adjust the denisty of each mixture to that required as measured by its refractive index. The solution (4.6 ml) was placed in a 10-ml polypropylene tube, overlaid with paraffin, and centrifuged at 45,000 rpm and 20°C in the 10 × 10 titanium fixed-angle rotor of the MSE Superspeed 65 Mark II centrifuge for 63 to 66 hr. The rotor was allowed to come to rest without braking. The gradients were unloaded by upward displacement with saturated NaI or CsCl solution, respectively, and 0.2 ml fractions were collected. The refractive index of each fraction, from which its density was calculated, was measured; 0.1 ml of each fraction was counted after mixing it with 1 ml water and 10 ml Triton X-100 toluene-based scintillator. For NaI solutions, it was necessary to add one drop of mercaptoethanol to each vial to prevent formation of free iodine, a very effective quenching agent.

Results and Discussion

The density gradient generated in a NaI solution centrifuged in a fixed angle-head rotor is much shallower than that of CsCl or Cs_2SO_4 solutions under the same conditions. For example, the density gradient formed at 45,000 rpm in a NaI solution of initial density (ρ_i) 1.545 g/cm^3 was 1.514 to 1.584 g/cm^3, while that in a CsCl solution (ρ_i) 1.750 g/cm^3 was 1.702 to 1.820 g/cm^3; and that in a Cs_2SO_4 solution (ρ_i) 1.540 g/cm^3 was 1.430 to 1.760 g/cm^3.

The distributions of native and denatured DNA, RNA, and DNA–RNA hybrid after sedimentation to equilibrium in density gradients formed in situ from CsCl, Cs_2SO_4, and NaI are shown in Fig. 6.12.

Although NaI gradients are shallow, both denatured and native LS-cell DNA band sharply. The separation between native and denatured DNA is much greater, in terms of gradient volume, than in CsCl (or Cs_2SO_4) gradients, partly because of the shallow gradient, but mainly because there is a much greater difference between the buoyant densities of native and denatured DNA. This large difference is not due to iodination of the denatured DNA, since denatured DNA recovered from a NaI gradient (and centrifuged through Sephadex G-25 to remove NaI) banded in CsCl with the same buoyant density as denatured DNA which had not been dissolved in NaI.

Synthetic DNA–RAN hybrids band sharply in NaI gradients, in a position intermediate between those of native and denatured DNA. In contrast, in both CsCl and Cs_2SO_4 the buoyant density of DNA–RNA hybrid is intermidiate between those of denatured DNA and RNA.

There are two striking differences between density gradients of NaI and those of CsCl and Cs_2SO_4. These are, first, the much greater effect that denaturation has on the buoyant density of DNA and, second, the relatively small increase in buoyant density which occurs when one strand of the DNA duplex is replaced with an RNA strand. It appears that the buoyant density of both DNA and RNA is more dependent on secondary than on primary struc-

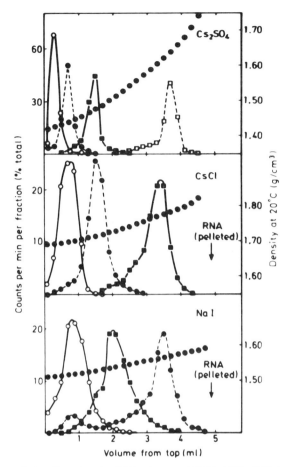

Fig. 6.12 Isopycnic banding of native DNA, denatured DNA, DNA–RNA hybrid, and RNA from mouse in density gradients formed from Cs_2SO_4, CsCl, and NaI. *Source:* Reproduced from Ref. 45 with permission of Butterworth Publishers, Inc., and the authors.

ture. The reason for this is unknown. In the case of DNA, it is not due to iodination of denatured strands nor to formation of heavy metal salts. A possible explanation is that one (or more) of the ring nitrogen atoms involved in base pairing in the duplex becomes available to form an iodide salt with a concomitant increase in density of the denatured strand. This could also occur, for example, in CsCl gradients, but would have a much less noticeable effect because of the low density of the chloride ion compared to that of the iodide ion.

Although NaI cannot replace CsCl or Cs_2SO_4 as a gradient material for all measurements of the buoyant density of nucleic acids, it has a number of practical advantages in some applications. The very large separation between native and denatured DNA in NaI gradients and in studies of DNA-RNA hybriditation NaI gradients are extremely useful examples.

Estimations of Molecular Weight

The apparent molecular weights of native DNA, denatured DNA, and DNA-RNA hybrid can be obtained from the isopycnic banding experiments using (5.2.73) together with the date presented in Fig. 6.12. The MSE 10 × 10 titanium fixed angle-head rotor has an angle of 35° from the axis of rotation and the distances from the rotational axis to the center of top and bottom of the centrifugal tube are 3.68 and 7.87 cm, respectively. Therefore, the abscissa in Fig. 6.12, volume from top (ml), may be converted into distance from the top at the inclined position, which gives 4.099, 4.518, 4.937, 5.356, and 5.775 cm for volumes 1, 2, 3, 4, and 5 ml, respectively.

Rearranging (5.2.73), one obtains the molecular weight of a macromolecule [$M^{(app)}$] in a density-gradient isopycnic banding:

$$M^{(app)} = \frac{RT}{\bar{\nu}_2 \, (d\rho/dr) \, \omega^2 r_o \sigma^2}$$

where R is the gas constant (8.314×10^7 ergs per degree K per mole), T is the absolute temperature, r_o is the center of mass of the macromolecular particle from the rotational axis, $\bar{\nu}_2$ is the partical specific volume of the macromolecules at the center of mass, $d\rho/dr$ is the slope of the density gradient solution, and σ^2 is the variance in a Gaussian distribution.

For the CsCl gradient, the quantities $\bar{\nu}_2$, $d\rho/dr$, r_o, σ^2 are evaluated from Fig. 6.12, which are listed in the following:

	r_o (cm)	$1/\rho_2 = \bar{\nu}_2$ (cm^3/g)	$d\rho/dr$ (g/cm^3.cm)	σ^2 (cm^2)
Native DNA	3.96	0.590	0.057	0.0289
Denatured DNA	4.31	0.583	0.057	0.0225
DNA-RNA hybrid	5.10	0.567	0.091	0.0484

The centrifugations were performed at 45,000 rpm and 20°C. Hence, one has $T = 293°$K and $\omega = 45,000 \times (2\pi/60) = 4712 \sec^{-1}$. Substituting the respective values in the $M^{(app)}$ formula, one obtains the molecular weight of various DNA as follows:

Native DNA	2.90×10^6
Denatured DNA	3.46×10^6
DNA-RNA hybrid	2.86×10^5

As mentioned previously, we emphasize that the density-gradient isopycnic banding method does not possess an advantage over ordinary sedimentation equilibrium in the determination of molecular weight, since heterogeneity in the sample causes the broadening of the distribution. In NaI gradients, this advantage is much more severe. The method presented above is an estimation of a crude molecular weight from a preparative centrifuge. For an exact estimation, one still has to rely on an ultraanalytical centrifuge with a purified sample.

Mitochondria, Lysosomes, and Peroxisomes

The separation of native lysosomes, mitochondria, and peroxisomes in linear sucrose gradients is exceedingly difficult because the isopycnic densities and sedimentation coefficients of those organelles overlap considerably. The use of tube gradients is thus very restricted and the greater resolving power of zonal rotors seems to offer greater scope for improving separations. Lysosomes and mitochondria are both membrane-enveloped structures, and in hypertonic sucrose solutions they both band at similar densities (1.19 to 1.22 g/cm^3). A recent study by Collot et al. [63] indicated that in Metrizamide, rat liver mitochondria and lysosomes also banded at the same density (1.140 to 1.145 g/cm^3), although peroxisomes were well separated from other organelles, banding at 1.230 g/cm^3.

The fractions obtained by differential centrifugation of liver homogenates in 0.25 M sucrose were summarized by Hogeboom et al. [64] as crude nuclear, mitochondrial (large-granule), microsomal (small-granule), and supernatant fractions. Study of the large-granule fraction by dark-background light microscopy confirmed that it was indeed rich in mitochondria. However, in the study of enzymic activity of the fraction, it was observed that the "free" activity of acid phosphatase in fresh tissue preparations under isotonic conditions was low, and that latent activity became manifest with severe treatment such as freezing and thawing under hypotonic conditions [65]. By 1955, as set down in a fascinating account by deDuve [66], this and other evidence had led the authors to propose that acid phosphatase is in a group of particles distinct from mitochondria, designated *lysosomes*. Thus, the lysosomes are believed to be intermediate in size between microsomes and mitochondria and to contain a variety of acid hydrolases capable of degrading most of the major constituents of the cells. The soluble phase obtained from homogenates of liver in 0.25 M sucrose is notably rich in catalase (E.C.* 1.11.1.6). This enzyme, together with certain others such as uricase and D-amino acid oxidase, is

---

*In 1961 the Commission on Enzymes of the International Union of Biochemistry introduced a classification system assigning each enzyme a code number, the first digit showing to which major class the enzyme belongs. E. C. stands for an Enzyme Commission number. Thus, each enzyme has a systematic name which begins with E. C. and a trivial name (normally its name before 1961).

associated not with lysosomes but with so-called *peroxisomes* (also termed uricosomes). These correspond to the microbodies long known to electron microscopists. They are similar in size to lysomoes and are hard to separate from them, but differ in having a dense core of osmophilic material. They contain oxidoreductases rather than hydrolases [67]. Catalase has definite latency like acid phosphatase, though differing in nature from the structure-linked latency of that enzyme [68]. The mitochondria fraction is rich in respiratory enzymes such as cytochrome oxidase (E.C. 1.9.3.1), whereas the microsonal fraction is rich in the enzyme glucose-6-phosphatase (E.C. 3.1.3.9). Each of the fraction is also quite rich in uricase (urateoxidase, E.C. 1.7.33).

Assay of fractions, and of the homogenate, for marker enzymes is a laborious but essential step in subcellular work. Suitable conditions for these enzymes are summarized by Reid [21], which are reproduced here in Table 6.5.

Studies on the distributions of acid phenyl phosphatase (lysosomes) and cytochrome oxidase (mitochondria) activities in rat liver brei fractionated in the zonal centrifuge by Schuel, Tipton, and Anderson [69, 70] are presented below as an example.

Materials and Methods

PREPARATION OF HOMOGENATES

Adult male Sprague–Dawley rats were used in this study. The animals were allowed to eat and drink freely until they were killed. They were decapitated, and the livers were perfused with cold Locke's solution prior to homogenization in a manually operated Potter and Elvehjem grinder with a Lucite pestle in cold 8.5% sucrose [52]. Large fragments of connective tissue were removed by filtration through several layers of cheese cloth.

FRACTIONATION IN THE ZONAL ULTRACENTRIFUGE

The liver brei was fractionated in a zonal ultracentrifuge rotor B-II at speeds of 10,000 to 30,000 rpm for 15 to 240 min [71]. A 1200-ml sucrose gradient, varying linearly with radius from 17 to 55% (w/w) with a "cushion" of 66% sucrose at the rotor edge, was employed. The tissue sample fractionated in the zonal ultracentrifuge consisted of 12.5 ml of a 25% (w/v) homogenate which contained approximately 3 to 4 g liver. All operations, including introduction of the sample layer in a short gradient (8.5 to 17% sucrose, w/w) and the recovery of the gradient with its concentric zones of separated particles, were accomplished while the rotor was rotating at 5000 rpm. The gradient was collected in 40-ml fractions for subsequent analysis.

VISUALIZATION OF MATERIAL IN THE GRADIENT

The presence of subcellular components was determined by continuously analyzing the gradient for ultraviolet absorbance as it emerged from the rotor. The recordings were made directly at 260 and 280 nm (without dilution) on the gradient as it passed through a quartz flow cell with a 0.2-cm light path. The data were converted in terms of the computed absorbance in a 1.0-cm light path.

Microscopic observations were made on the isolated fractions with a phase contrast microscope using a dark-medium objective.

ANALYSIS OF SUCROSE IN THE GRADIENT

The concentration of sucrose within the isolated fractions was analyzed refractometrically. These data were then used to calculate the densities of the sucrose solutions.

MEASUREMENTS OF ENZYME ACTIVITIES

Acid phosphatase activities were measured automatically with the Technicon AutoAnalyzer [72] by a modified King–Armstrong procedure [73, 74]. The module used was identical with the one recommended by Technicon Instruments Corp. for the clinical assay of acid phosphatase in human serum. The substrate was disodium phenyl phosphate (Technicon Instruments Corp., Chauncey, New York). In this method, the free phenol produced is condensed rapidly with 4-aminoantipyrine followed by oxidation with $K_3Fe(CN)_6$ under alkaline conditions. The reddish reaction product is measured at 505 nm. Sodium malonate (Eastman Organic Chemicals, Rochester, New York) was used to buffer over the range pH 2.6 to 6.6. However, the substrate disodium phenyl phosphate is itself a powerful buffer and contributes significantly to the buffer capacity of the system. The actual pH of the reaction mixture was determined after mixing the reagents in the same proportions employed in the automated assay system and incubation at 37°C. To release particle-bound enzymes, the nonionic detergent 0.2% Turgitol NPX[1] (Union Carbide Chemical Corp., New York, New York) was included in the substrate. Assays in conjunction with zonal ultracentrifuge runs were made in stream on the gradient as it was recovered from the rotor and also on each of the 40-ml fractions obtained.

Either cytochrome oxidase or succinic acid dehydrogenase should be a reliable chemical marker for the mitochondrial fraction of liver tissue, since it is generally considered that these enzymes are localized in the mitochondria [75, 76]. The cytochrome oxidase assay was chosen for automation using a Technicon Instrument module since it appeared to be sensitive enough and to utilize few components, involving basically only tissue sample, diluent, and substrate [77, 78]. The method uses a reduced cytochrome c as the substrate, and measures the decrease in intensity of its characteristic absorption band at 550 nm as the cytochrome is oxidezed by the enzyme.

Results

Assays of phosphatase activities were made in liver breis subjected to a wide range of centrifugal fields—10,000 to 30,000 rpm for 60 to 180 min. Under

Table 6.5 Suitable Conditions for Assay of Marker Enzymes

Enzyme	Succinate Dehydrogenase (mitochondria)	Acid Phosphatase[a] (lysosomes)	Uricase (peroxisomes)	Glucose-6-phosphatase (microsomes)	5'-Nucleoridase (plasma membrane)
Tissue equivalent (as fresh liver) per tube, for a lysosomal ("light-mitochondrial") fraction obtained by differential centrifugation	1 mg in 0.5 ml	(1) 10 mg[b] in 0.5 ml or (2) 2 mg in 0.5 ml	10 mg in 1 ml, in silica cuvette	10 mg in 0.5 ml	10 mg in 0.5 ml
Buffer[d]	0.25 ml of 0.5 M phosphate, pH 7.4, cont. 0.25 mg INT[e]	0.4 ml of 0.3 M 3,3-dimethyl-glutarate, pH 5.0	1 ml of 5 mM phosphate, pH 7.4, cont. 0.2 % (v/v) Triton X-100	0.4 ml of 0.3 M 3,3-dimethyl-glutarate, pH 6.4 cont. 25 mM EDIA	0.4 ml of 0.3 M tris-HCl, pH 7.8, cont. 12.5 mM MgCl$_2$
Substrate (pH-adjusted; sometimes permissible to add it with the buffer; omit in blanks)	0.25 ml of 0.3 M succinate (0.3 M malonate in blanks)	(1) 0.1 ml of 1 M β-glycero-phosphate or (2) 0.1 ml of 10 mM p-nitrophenyl-phosphate	1 ml of 10 mM phosphate, pH 7.4, cont. 86μg of Na$^+$ urate[f]	0.1 ml of 50 mM glucose-6-phosphate (Ba^{2+} salt)	0.1 ml of 50 mM 5'-NMP or 5'-UMP
Incubation conditions (preferably with shaking except for uricase)	20 min at 37°C	30 min at 37°C (10 min for "free" activity[b])	Follow F$_{292}$ for at least 5 min, preferably at 37°C	30 min at 37°C	30 min at 37°C

Final measurement on supernatant fluid after adding 1.5 ml of 6% trichloracene acid at 0°C	E_{490} on ethyl acetate extract (4 ml)	Estimate phosphate		Estimate phosphate	Estimate phosphate
Typical activity for whole homogenate, μmoles/g of liver/min	6	(1) 8 or (2) 20	4[c] at 20°C 9[c] at 37°C	15	10

Source: After E. Reid [21].

[a]Liberate latent activity by at least five freeze-thaws of the hypotonic tissue suspension.

[b]For "free" activity, use the equivalent of 50 mg of liver, with 0.25 M sucrose present.

[c]The amount of tissue needed is not easily predictable. P. Baudhuin, H. Beaufay, Y. Rahman-Li, O. Z. Sellinger, R. Wattiaux, P. Jacques, and C. deDuve, *Biochem. J.,* **92**, 179 (1964).

[d]Buffer pH values at temperature of incubation.

[e]2-(p-iodophenyl)-3-(p-nitrophenyl)-5-phenylterazolium chloride.

[f]Solution must be fresh; it becomes inhibitory when stored.

these conditions, separations of subcellular components are achieved by sedimentation equilibrium, the isopycnic banding, for the larger particles and by sedimentation rate still on the road to the equilbrium for smaller particles. The results of an experiment in which liver brei was subjected to 22,000 rpm for 120 min can be seen in Fig. 6.13. The phosphatase activities monitored at ph 4.8, 5.4, and 5.9 exhibited a similar distribution pattern. Large amounts of activity were associated with the soluble-phase mitochondria and membranous fragment fractions, while the microsomes and especially the nuclei appeared to contain low levels of activity. The activity measured at pH 4.1, however, exhibited a rather different pattern, with just trace amounts observed in the

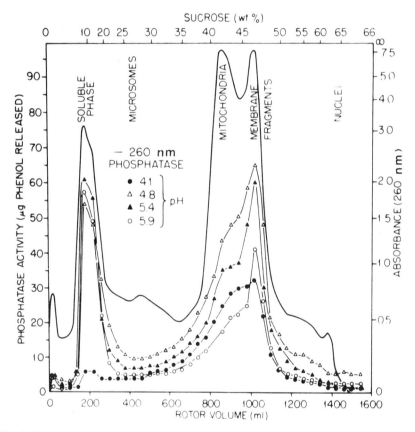

Fig. 6.13 Distribution of acid phenyl phosphatase activities at pH 4.1, 4.8, 5.4, and 5.9 in rat liver brei subjected to 22,000 rpm for 120 min in the zonal centrifuge. Assays made with sodium malonate buffers and in the presence of Turgitol NPX. *Source:* Reproduced from Ref. 69 with permission of the Rockefeller University Press and the author.

soluble phase, while most of the activity was associated with the particulate mitochondrial and memberanous fractions.

The results of an experiment in which rat liver brei was subjected to 10,000 rpm for 60 min are shown in Fig. 6.14. This treatment separated the components of the original brei into soluble-phase microsomal, mitochondrial, and membranous fragments and nuclear fractions. At this relatively low speed, the microsomes just barely separated from the soluble-phase and ribosomal material in the "sample zone" of the gradient, while the nuclei were thrown out into the cushion of 66% sucrose near the edge of the rotor. In most zonal centrifuge fractionations of rat liver brei, the nuclei were sedimented out to form a single peak in the "cushion" of 66% sucrose. However, occasionally, as was the case in this run, the nuclei were thrown out to form multiple peaks in the sucrose cushion. The mitochondria appeared to be distributed rather sharply about a peak at 42.7% sucrose, density 1.192 g/cm³. All of the cytochrome c oxidase activity was restricted to this region of the gradient, and no activity could be detected in the microsomal and nuclear fractions. In several experiments, cytochrome oxidase activity was assayed manually according to the procedure of Cooperstein and Lazarow [77]. The distribution of oxidase activity in the gradient measured in this way was the same as with the automated assay.

Under the conditions employed with the zonal ultracentrifuge rotor B-II, the larger subcellular components are essentially sedimented to their isopycnic position in the sucrose gradient. The data obtained in this study suggest that the isopycnic density of rat liver mitochondria is approximately 1.20g/cm³. Previous work using mitochondrial fractions obtained by differential centrifugation suggests that the isopycnic position in sucrose gradients is between 1.19 and 1.22 g/cm³ [79–81]. It must be recognized that the granules are subjected to an osmotic gradient during such experiments. These methods thus measure the density of particles at equilibrium with a solution of the same density and do not indicate what the density might be in solutions of lower osmotic pressure or in intact cells.

Almost without exception, other workers also observe that in sucrose gradients (approximately 15 to 55%) the lysosomes band centripetally to the mitochondria. The incorporation of a sharp density discontinuity within the gradient to arrest the movement of the lysosomes without appreciably affecting further sedimentation of the mitochondria has been widely adopted. Although the precise configuration of the gradient and the centrifugation conditions vary, the aim is to band the lysosomes around 30 to 32% sucrose with a steep gradient. Clearly, the difference in sedimentation rate between these subcellular organelles is so small that, even with the increased capacity and resolution of center-unloading zonal rotors, under normal conditions it is dif-

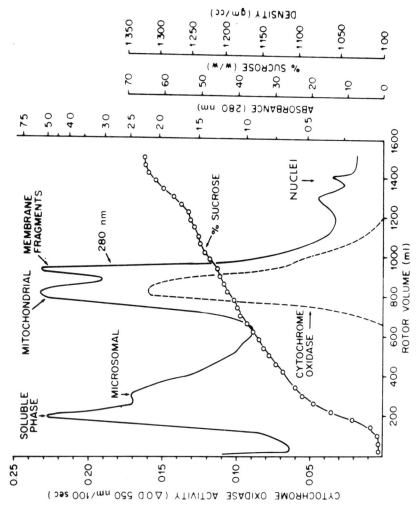

Fig. 6.14 Distribution of cytochrome oxidase activity in rat liver brei subjected to 10,000 rpm for 60 min in the zonal centrifugation. *Source:* Reproduced from Ref. 70 with permission of the Rockefeller University Press and the authors.

ficult to separate these components adequately by a single centrifugation without resorting to density perturbation.

If one does not have a B-II rotor in his laboratory, the procedures outlined for modification of experimental conditions for membrane separation by the principle of dynamic similarity may be followed.

Nuclei

The function of the nucleus is in the determination of heredity, and its role in protein synthesis has inspired intense interest in this organelle. Many biochemists in their studies of nuclei have used rat liver, which is one of the most commonly used materials for biochemical research. The liver is a complex organ. The main functional units are the parenchymal cells. The nonparenchymal cells are diploid. The heterogeneity in type of parenchymal cells and degree of ploidy is a major complication in the interpretation of biochemical experiments involving isolated nuclei.

An example of fractionation of purified rat liver nuclei by the low-speed zonal centrifugation to separate them into two main zones, one containing diploid and the other tetraploid nuclei [82], is presented below.

Material and Methods

ANIMALS

Norwegian Hooded rats fed ad libitum on Purina chow were used. The mice were albinos of the NIH strain.

ISOLATION OF NUCLEI

The nuclei were prepared from rat and mouse tissues by a modification of the method of Widnell and Tata [83]. Preliminary experiments indicated that it was essential to adjust the pH of all solutions containing sucrose to 7.4 if crenellation and clumping of the nuclei was to be avoided. Sufficient saturated $NaHCO_3$ solution was added to raise the pH to 7.4. The final concentration was approximately 1 mM. The medium for homogenization was 0.32 M sucrose, 3 mM with respect to $MgCl_2$, adjusted to pH 7.4. The nuclei were purified by centrifugation through 2.4 M sucrose, 1 mM with respect to $MgCl_2$, pH 7.4. At least three animals were used in all experiments. The tissues from individual animals were pooled. The yield of nuclei based on the recovery of DNA was about 60% in most experiments. Most of the losses arose from incomplete breakage of cells.

ZONAL CENTRIFUGATION OF NUCLEI

All runs were carried out at 5°C. A gradient of sucrose (1000ml.) ranging from 20 to 50 (w/w), 1 mM withrespect to $MgCl_2$, pH 7.4, and linear with respect to volume, was introduced into the MSE A-XII rotor while it was running at 600 rpm. The pumping speed was 50 ml/min. This was followed by an underlay of 55% (w/w) sucrose. The sample of nuclei examined in the zonal centrifuge was equivalent to approximately 6 g wet weight of tissue in the experiments with liver. The sample (10 to 20 ml in 15% sucrose, 1 mM MgCl$_2$, pH 7.4) was introduced through the core line, followed by an overlay of 10% (w/w) sucrose until the leading edge of the sample was 6 cm from the axis of rotation.

A stroboscopic lamp was used to observe the positions of zones against a scale marked on the underside of the rotor. This was illuminated from below by a lamp mounted under a translucent Perspex screen, fitted into the bottom of the bowl of the centrifuge. The stroboscopic lamp permitted an exact determination of rotor speed. Most runs were 1 hr in duration. The contents of the rotor were displaced by pumping in more underlay. Displacement lasted for 26 min. The effluent was passed through a flow cell of 1 cm path length in a Gilford 2000 extinction meter and monitored at 600 nm. The cell compartment was cooled to 7 to 8°C. The effluent was collected either manually in large fractions selected according to extinction, or in 100 of 13 ml fractions in a Chromofrac fraction cutter (Baird and Tatlock Ltd., Chadwell Health, Essex, U.K.) modified to give a 15-sec interval between changes. The tubes used for collecting fractions were cooled in ice-cold water.

Results and Discussion

The spectrophotometric chart from zonal centrifugation of nuclei is shown in Fig. 6.15a. The data are replotted on a linear scale of distance from the axis of rotation, instead of effluent volume scale, in Fig. 6.15b. The slope of the extinction scale takes into account of the variation of zonal volume with distance from the axis of rotation. This method of presenting the information obtained from the spectrophotometric traces gives a clearer and more realistic indication of the migration and separation of zones. The peak at 6 cm marks the position of the sample band. A small amount of material, apparently mostly membranous, remains at this position. Part of the change in extinction represents the passage through the flow cell of the interface between the overlay and the gradient. The end of the gradient is also marked by a small peak. Two main zones of nuclei are seen. The fractions corresponding to each zone were combined. The nuclei were recovered and the nuclear volumes measured. The nuclei from zones shown in Fig. 6.15 were characterized both by microscopic examination and chemical analyses, details of which can be found in Ref. 82.

These data demonstrate that the two well-separated zones contain predominantly diploid and tetraploid nuclei, respectively. Prolonging the duration of the centrifugation, though it increased the separation of the apices of each zone, resulted in broader peaks without appreciable decrease in the contamination of the tetraploid zone. The nuclei from the diploid and tetraploid zones appear in satisfactory condition when examined by phase contrast

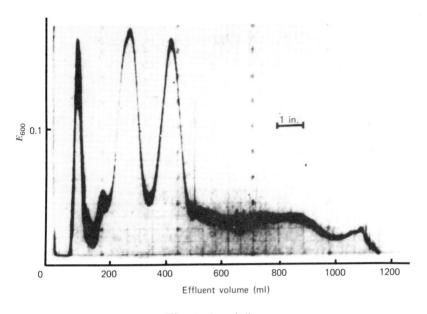

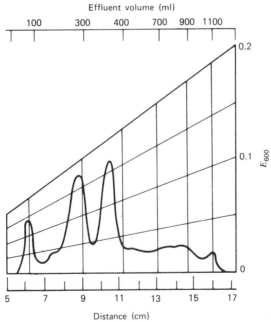

Fig. 6.15 Typical spectrophotogram of nuclei separation by a MSE A-XII zonal centrifuge rotor (*above*). (*a*) Absorbance vs. volume. (*b*) Absorbance vs. radial distance (*below*). *Source:* Reproduced from Ref. 82 with the permission of the Biochemical Society of Great Britain and the authors.

microscopy. Zonal centrifugation removes virtually all the cytoplasmic contaminants present in the original nuclear pellet.

SEDIMENTATION PROPERTIES OF LIVER NUCLEI

An advantage of the A-type rotor is that it permits continuous direct observation of the position of the particle zones and enables an investigation to be made of their sedimentation behavior. Figure 6.16a shows the position of the tetraploid and diploid zones as a function of time at a rotor speed of 600 rpm. The rate of sedimentation at twice the rpm, and therefore a fourfold increase in gravitational field, is shown in Fig. 6.16b.

Estimation of Particle Size and Sedimentation Coefficient

One will see from the discussion below that the plot of $\ln r$ versus $\omega^2 t$ or $\ln r$ versus t will be more advantageous than that of the plot of r versus t in ordinary rectangular coordinates. However, with the information from Fig. 6.16, the size and the observed sedimentation coefficient of diploid and tetraploid nuclei can be estimated roughly. Using (5.2.50) through (5.2.54), the quantity D_p^2 $(\rho_p - \rho_m)\omega^2/18\mu_m$ is seen to be a plot of the slope of $\ln r$ versus t. The quantity D_p is the diameter of sedimenting particle; ρ_p is the solvated density of the particle; ρ_m and μ_m are respectively the density and viscosity of the density gradient medium, which are both functions of rotor position; and ω is the angular velocity of spinning of the centrifuge.

The density of the nuclei varies with the density of the gradient medium presumably because of their permeability to sucrose or other gradient solute solutions. Johnston et al. [82] used a gradient of 5 to 17% (w/w) sucrose at 600 rpm for 40 min and 50 to 69% (w/w) sucrose at 2500 rpm for 6.5 hr to study the density behavior of sedimenting nuclei. They obtained $(\rho_p - \rho_m)$-versus-ρ_m data, and by extrapolation they concluded that the isopycnic $(\rho_p - \rho_m)$ densities for both nuclei are equal and are 1.35 g/cm³ (i.e., $\rho_p = 1.35$ g/cm³).

Now, the following data are obtained from Figs. 6.16a and 6.16b at 1200 rpm:

	Distance Traveled (cm)	Time (min)	Gradient at Both Ends (%, w/w)
Diploid nuclei	6 → 8.6	20	20 → 25
Tetraploid nuclei	6 → 10.45	20	20 → 30

The first approximation to the estimation of the size and the sedimentation coefficent of two nuclei is shown below. The experiments were run at 5°C. The ρ_m and μ_m of sucrose solution at 5°C are listed in Appendix A. Thus one finds the following values at 5°C for sucrose solutions:

Sucrose Solution (%, w/w)	ρ_m (g/cm^3)	μ_m (centipoises)
20	1.0843	3.1354
25	1.1075	4.0425
30	1.1315	5.4216

If the arithmetic average is used as the first approximation to the various quantities in the estimation, one obtains the quantities in (5.2.54) as

	ρ_m	$\rho_p - \rho_m$	$\mu_m \times 10^2$	$d \ln r/dt$
Diploid	1.0959	0.2541	3.5890	3.0×10^{-4}
Tetraploid	1.1079	0.2421	4.2785	4.6×10^{-4}

The angular velocity at 1200 rpm is

$$\omega = \frac{1200 \times 2\pi}{600} = 125.66 \text{ sec}^{-1}$$

$$\omega^2 = 15791.37 \text{ sec}^{-2}$$

Thus, from (5.2.54) one obtains the following relationship:

$$\frac{D_p^{\,2}(\rho_p - \rho_m)\omega^2}{18\mu_m} = \frac{d \ln r}{dt}$$

By rearranging the above equation and substituting the respective values into the equation, one obtains the following diameters of the nuclei:

Diploid $D_p = 2.20 \times 10^{-4}$ cm

Tetraploid $D_p = 3.04 \times 10^{-4}$ cm

As shown in (5.1.54), a plot of $\ln r$ versus $\omega^2 t$ has a slope of $[D_p^{\,2}(\rho_p - \rho_m)]/18\mu_m$, which is equivalent to the (instantaneous) sedimentation coefficient. Therefore, the observed sedimentation coefficient can be obtained by the following:

$$S = \frac{1}{\omega^2} \frac{d \ln r}{dt}$$

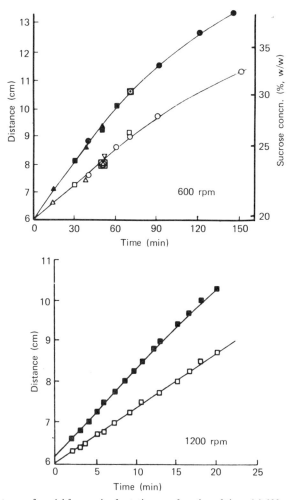

Fig. 6.16   Distance of nuclei from axis of rotation as a function of time. (*a*) 600 rpm (*above*). (*b*) 1200 rpm (*below*). *Source:* Reproduced from Ref. 82 with permission of the Biochemical Society of Great Britain and the authors.

Thus, one obtains

Diploid $S = 1.90 \times 10^5$ Svedberg units

Tetraploid $S = 2.91 \times 10^5$ Svedberg units

If higher-order accuracies are required, an exact gradient profile and the method described in Chapter IV, (4.2.10) through (4.2.23), have to be used.

The calculations become more complicated and more tiedous and time consuming. Programmed calculation by electronic computer is essential.

Proteins

Proteins in general have low bouyant densities in ionic buoyant density-gradient media. Not much use of the technique has been made for fractionation of proteins, despite the natural variation in the amino acid composition of proteins, which gives rise to differences in densities of up to 0.1 g/cm^3 [83]. The classical approach to the isolation of plasma lipoproteins is flotation through a gradient from a serum sample whose density has been increased by addition of salt. Flotation of lipoproteins through a gradient of NaBr is a useful method for examining various protein components of pathological sera. The general methods for the separation have employed multiple centrifugations in angle-head rotors with adjustments of the density of the medium between centrifugations [84–86]. As noted previously, angle-head rotors have the disadvantages of convective disturbance and adherence of large lipoprotein molecules to the walls of the centrifugal tube. These effects are much reduced in swinging-bucket rotors.

Recently, Redgrave, Roberts, and West [87] have used a simple discontinuous density gradient in high-speed swinging-bucket rotors to separate the individual plasma lipoprotein species in a single ultracentrifugation. Their procedures are presented here as an example.

Materials and Methods

Blood was collected from subjects into tubes containing ethylenediaminetetraacetic acid (EDTA), final concentration 4 mM, pH 7.4, either at 9.0 A.M. after an overnight fast or before the midday meal in unfasted subjects. Plasma was separated by centrifugation at 300 × g for 15 min at 4°C. No attempt was made to remove chylomicrons from the midday samples because the results were not obviously different from the samples taken at 9.0 A.M. Lipoprotein samples were usually prepared on the same day or within 72 hr of collection.

Plasma cholesterol concentration was determined in these samples prior to ultracentrifugation by the saponification method of Mann [88] and assayed using the o-phthaldialdehyde color reagent of Zlatkis and Zak [89].

Three types of 6 × 14-ml titanium swinging-bucket rotors were used in their respective centrifuges: the SW 41 rotor in the Beckman L3-50 ultracentrifuge (Beckman Inc., Palo Alto, California); the M.S.E. rotor (Cat. No. 43127-111) in the M.S.E. superspeed 65 Mk II ultracentrifuge (Measuring and Scientific Equipment Ltd., London, U.K.); and the SB-283 rotor in the I.E.C. B-60 ultracentrifuge (International Equipment Co., Needham Heights, Massachusetts).

Samples of plasma were adjusted to $\rho = 1.21$ with solid potassium bromide (0.325 g/ml plasma), and 4.0-ml aliquots were pipetted into 13.5-ml cellulose nitrate centrifuge tubes ($\%_{16}$-in. diam. $\times$ 3½-in. length). Samples of less than 4.0 ml were adjusted with salt solutions and solid potassium bromide to $\rho = 1.21$ in a volume of 4.0 ml. A discontinuous gradient was formed by carefully layering 3.0 ml salt solution, $\rho = 1.063$, above the plasma, followed by 3.0 ml salt solution, $\rho = 1.019$. Finally, the tube was filled with 2.5 to 3.0 ml of $\rho = 1.006$ salt solution. To minimize mixing at the density junctions, the salt solutions were allowed to gravity feed down the side of the centrifuge tube through a hypodermic needle (22 gauge) attached to a glass syringe with the barrel removed. All salt solutions contained EDTA (0.1 mg/ml) and were prepared from potassium bromide and sodium chloride [85].

The samples were centrifuged for 24 hr at 20°C at either 41,000 rpm (SW 41, 286,000 g; SB-283, 283,200 g), or at 40,000 rpm (M.S.E., 284,000 g). After centrifugation, chylomicrons and very low-density lipoprotein (VLDL) were present at the top of the tube, low-density lipoprotein (LDL) at the junctions of the next two steps of the density gradient, and high-density lipoprotein (HDL) at $\rho = 1.21$ and $\rho = 1.063$ junction. In human samples LDL was clearly visible as an orange-yellow band. The density of the fractions after centrifugation was measured with a digital precision density meter (Model DMA 02C, Anton Pear KG, A-8054 Graz, Austria).

The bands, together with part of the salt solutions above and below each band, were carefully transferred by aspiration into tubes, saponified, and assayed for cholesterol as described.

Results and Discussion

Table 6.6 shows the distribution of lipoprotein cholesterol in 19 normal male subjects. Total recovery of loaded lipoprotein cholesterol is 89.3 $\pm$ 1.31%

Table 6.6 Distribution of Lipoprotein Cholestral in 19 Normal Male Subjects

| Band | Density | Designation | Cholesterol distribution[a] | |
			% of Total	mg/100cm³
I	1.006	VLDL and chylomicrons	7.3 $\pm$ 0.81	15.4 $\pm$ 1.78
II	1.006–1.019	LDL	11.4 $\pm$ 2.22	23.9 $\pm$ 4.36
III	1.019–1.063	LDL	50.1 $\pm$ 2.52	109.2 $\pm$ 8.78
IV	1.063–1.21	HDL	24.2 $\pm$ 1.68	49.8 $\pm$ 3.01
	> 1.21	—	7.0 $\pm$ 0.57	14.7 $\pm$ 1.17

[a] Figures are mean $\pm$ SEM.

(SEM), the standard error of the mean. In these subjects, the plasma cholestrol concentrations range from 135 to 287 mg/100 cm^3, with a SEM of 213 ± 9.6 mg/100 cm^3. As expected, most of the plasma cholesterol is recovered in the bands corresponding to LDL (bands II + III = 61.5 ± 2.01%), with the majority in band III at 1.019 < ρ < 1.063. In band IV (1.063 < ρ < 1.21), 24.2 ± 1.68% is recovered, corresponding to HDL. A further 7.0 ± 0.57% is recovered in the fraction ρ > 1.21.

During the course of the centrifugal procedure, there is a smoothing of the discontinuous gradient, as shown in Fig. 6.17. However, the general shape of the gradient is maintained, and the designated fractions are recovered. The technique could be adapted if necessary to a more extensive subfractionation of plasma lipoporteins because of this smoothing of the gradient during the centrifugation. The fractions separated have been verified by cellulose acetate electrophoresis and immunoelectrophoresis.

The method provides a convenient and rapid ultracentrifugal analysis of the plasma lipoproteins. The discontinuous gradient is simple to prepare and inherently stable if mechanical disturbance is carefully avoided. Separation is dependent only on the densities of the lipoportein species, so separations are quite distinct. Analytical results after longer periods of centrifugation are unchanged.

The subfractionation of LDL into two bands is useful in assessing the metabolic significance of the LDL subgroups as illustrated in Table 6.6 by considerable variation in the amount of LDL in bands II and the relative high SEM.

Because both chylomicrons and VLDL have densities less than 1.006, the described method does not separate these species. Chylomicrons can be separated from VLDL only be a preliminary rate centrifugation method.

Remarks

The sedimentation rate of a particle, its banding density, or both of these properties may be altered in a density gradient in a manner useful for achieving separations. The factors involved are [90]:

1. Precipitation (increase in effective particle size) produced by diffusion of a precipitant into a sample zone.

2. Resolubilization in a gradient negative with respect to the precipitating agent.

3. Alteration in volume and density by change in osmotic pressure or biochemical environment.

4. Specific binding to ions or substances, altering sedimentation rate or banding density.

5. Sequential dissection by sedimentation through zones of immobilized reagents.

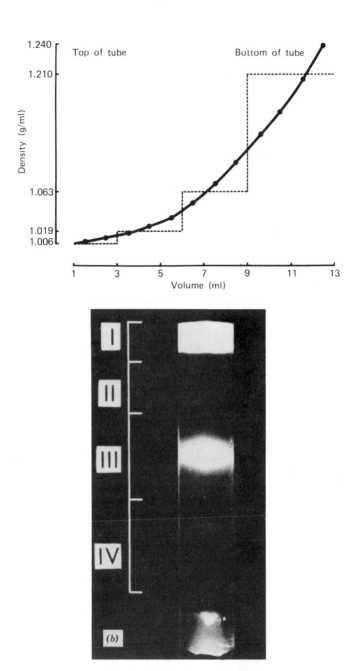

Fig. 6.17 Profile of the gradient after centrifugation (dotted line shows the initial discontinuous step gradient). Right photograph is the gradient in centrifuge tube after separation of human plasma lipoproteins with marked fractions of the gradient that comprise bands I through IV. *Source:* Reproduced from Ref. 87 with permission of Academic Press and the authors.

374

These techniques, singly or in combination, open up many new, interesting avenues of research.

Precipitation by a Diffusion Precipitant

Soluble sample materials of high or low molecular weight may be precipitated by the diffusion of a suitable precipitant into the sample zone. Inorganic salts in aqueous solution may be caused to precipitate, often in the form of small crystals, as an organic solvent miscible with water (and in which the salts are not soluble) is allowed to diffuse in. Similarly, organic substances insoluble in water may be precipitated out of alcohol or acetone by water diffusing into the sample zone. High-molecular-weight substances such as proteins and nucleic acids may be precipitated by water-miscible organic solvents or heavy metal salts. The precipitating agent may be either in the overlay, in the underlying gradient, or in both. In all instances care must be taken to ensure that the proper density increments are present to ensure stability from a gravitational viewpoint. When ethyl alcohol is used as solvent, its density may be adjusted with 2-chloroethanol, for example.

The unique feature about precipitation in a centrifugal field is that precipitated particles sediment out of the sample zone as they increase in size. Since precipitation occurs in response to a diffusion gradient, different substances may precipitate at different times and levels in the diffusion gradient, tending to minimize cross-contamination in the particles formed. Particle size will be a function of the centrifugal field used, and with reasonably high forces only very tiny particles will be formed. If the particles are not soluble at any level as they sediment through the gradient, they may be fractionated by isopycnic banding. This method of preparative particle separation has not been widely applied to precipitates, although the crystal densities of a large number of substances are known.

Gradient Resolubilization

In as much as precipitated particles are sedimented through a gradient, the composition of the gradient may vary in such a way that the particles go back into solution. One of the advantages of this method is that the time during which a substance is precipitated is rather short, especially where a high centrifugal field is employed. The rate at which a sedimenting particle will dissolve increases as the size of the particle increases. Interest in these processes centers chiefly around the possibility of adapting them to continuous on-stream protein fractionation.

Alteration in Volume and Density by Change in Osmotic Pressure or Biochemical Environment

Gradients may be prepared having similar density slopes, but having osmotic pressure gradients which may vary and, in the case of a specially prepared sucrose–dextran gradient, may approach zero. Where particles differ in their response to an osmotic pressure gradient, the gradient may be adjusted to maximize the separation. Mitochondria appear to shrink appreciably more than do peroxisomes in sucrose gradients. The mitochondria initially sediment ahead of the peroxisomes; however, the latter overtake them and band at a slightly denser level. For optimal rate separations, it appears advantageous not to employ isotonic gradients but rather to let the mitochondria shrink and sediment more rapidly ahead of the lysosomes. Sedimentation rates may also be affected by the biochemical environment quite apart from osmotic effects. The volume changes produced in isolated nuclei by small amounts of divalent cations and in mitochondrial volume by ATP or thyroxine are examples of this phenomenon.

Specific Binding to Ions or Substances Altering Sedimentation Rate or Banding Density

The banding density of smooth and rough endoplasmic reticulum fragments, and subfamilies of these, may be altered by small changes in the ionic environment. The use of uranium- or ferritin-labeled antibodies to specifically alter the banding density of an organelle or fragment does not appear to have been explored. A simple and frequently used method for increasing the sedimentability of hemagglutinating viruses is to allow them to attach to red cells, which are easily sedimented. As the viruses desorb, they are easily separated from the red cells by centrifugation.

Sequential Dissection by Sedimentation Through Zones of Immobilized Reagents

Reagents having low sedimentation coefficients may be incorporated at different levels in a density gradient to attach larger particles moving through the gradient. The method was employed in early studies on the extraction of histones from nuclei sedimenting through gradients of increasing acidity; and more recently, detergent gradients have been used to solubilize microsomes. As more interest develops in the dissection of membranes and ribosomes, these methods will probably be more widely employed [14].

References

1. N. G. Anderson, *Anal. Biochem.*, **23**, 72 (1968).
2. DS-494 A, Beckman Instruments, Inc., Spinco Division, Palo Aito, Calif., 1975.
3. H. Noll, "Polysomes: Analysis of Structure and Function" in *Techniques in Protein Biosynthesis*, Vol. 2, P. N. Campbell and J. R. Sargent, Eds., Academic, New York, 1969.
4. N. G. Anderson, *Rev. Sci. Instrum.*, **26**, 891 (1955).
5. N. G. Anderson, "Techniques for the Mass Isolation of Cellular Components," in

Physical Techniques in Biological Research, Vol. III, *Cells and Tissues*, G. Oster and A. W. Pollister, Eds., Academic, New York, 1956.

6. H. Holter, M. Ottensen, and R. Weber, *Experimentia*, **9**, 346 (1953).

7. R. Weber, "Separation of Cytoplasmic Particles by Density-Gradient Centrifugation," in *Eighth International Congress for Cell Biology*, North-Holland, Leyden, 1954.

8. N. G. Anderson, *Biochim. Biophys. Acta*, **25**, 428 (1957).

9. G. A. Bray, *Anal. Biochem.*, **1**, 279 (1960).

10. J. B. Bancroft and P. Kaesberg, *Biochim. Biophys. Acta*, **39**, 519 (1960).

11. R. L. Sinsheimer, *J. Mol. Biol.*, **1**, 37 (1959).

12. R. L. Sinsheimer, *J. Mol. Biol.*, **1**, 43 (1959).

13. A. W. Kozinski and W. Szybalski, *Virology*, **9**, 260 (1959).

14. N. G. Anderson, *The Development of Zonal Centrifuges*, N. G. Anderson, Ed., *Natl. Cancer Inst. Monogr. No. 21*, U.S. Government Printing Office, Washington, D.C., 1966.

15. P. A. Charlwood, *Anal. Biochem.*, **5**, 226 (1963).

16. F. Vedel and M. J. D'Aoust, *Anal. Biochem.*, **35**, 54 (1970)

17. M. Castaneda, R. Sanchez, and R. Santiago, *Anal. Biochem.*, **44**, 381 (1971)

18. M. Meselson, F. W. Stahl, and J. Vinograd, *Proc. Natl. Acad. Sci.*, **43**, 581 (1957).

19. W. G. Flamm, M. L. Birnstiel, and P. M. B. Walker, in *Subcellular Components: Preparation and Fractionation*, 2nd ed., G. D. Birnie, Ed., Butterworth, London, 1972, p. 279.

20. J. B. Ifft, W. R. Martin, and K. Kinzie, *Biopolymers*, **9**, 597 (1970).

21. E. Reid, in *Subcellular Components: Preparation and Fractionation*, 2nd ed., G. D. Birnie, Ed., Butterworth London, 1972, p. 93.

22. R. H. Hinton, M. L. E. Burge, and G. C. Hartman, *Anal. Biochem.*, **29**, 248 (1969).

23. G. C. Hartman, N. Black, R. Sinclair, and R. H. Hinton, in *Methodological Developments in Biochemistry*, Vol. 4, E. Reid, Ed. Longman, London 1974, p. 93.

24. G. Zimmer, A. D. Keith, and L. Packer, *Arch. Biochem. Biophys.*, **152**, 105 (1972).

25. M. Kurokawa, T. Sakamoto, and M. Kato, *Biochem. J.*, **97**, 833 (1965).

26. A. Raynaud and H. H. Ohlenbusch, *J. Mol. Biol.*, **63**, 523 (1972).

27. E. M. Boyd, D. R. Thomas, B. F. Harton, and T. H. J. Huisman, *Clin. Chim. Acta*, **16**, 333 (1967).

28. C. W. Boone, C. S. Harell, and H. E. Bond, *J. Cell Bio.*, **36**, 369 (1968).

29. J. A. Sykes, J. Whitecarrer, L. Briggs, and L. H. Anson, *J. Natl. Cancer Inst.*, **44**, 855 (1975).

30. M. Castagna and J. Chauveau, *Exp. Cell Res.*, **57**, 211 (1969).

31. T. G. Pretlow and E. E. Williams, *Anal. Biochem.*, **55**, 114 (1973).

32. C. deDuve, *Harvey Lect.* Harvey Society, New York, **59**, 49 (1964).

33. S. I. Honda, T. Hongladarom, and G. G. Laties, *J. Exp. Bot.*, **17**, 460 (1966).

34. P. A. Ketchum and S. C. Holt, *Biochim. Biophys. Acta*, **196**, 141 (1970).

35. II. Beaufay, P. Jacques, P. Baudhuin, O. Z. Sellinger, J. Berthet, and C. deDuve, *Biochem. J.*, **92**, (1964).
36. A. Boyum, *Scand. J. Clin. Lab. Invest.*., **21** (Suppl. 9.7), (1968).
37. G. D. Birnie, D. Rickwood, and A. Hell, *Biochim. Biophys. Acta*, **331**, 283 (1973).
38. R. T. L. Chan and I. E. Scheffler, *J. Cell. Biol.*, **61**, 780 (1974).
39. D. Doenecke and B. J. McCarthy, *Biochemistry*, **14**, 1366 (1975).
40. P. Serwer, *J. Mol. Biol.*, **92**, 433 (1975).
41. G. D. Birnie, E. MacPhail, and D. Rickwood, *Nucleic Acids Res.*, **1**, 919 (1974).
42. D. Rickwood, in *Biological Separations in Iodinated Density-Gradient Media*, D. Rickwood, Ed., Information Retrieval Ltd., London, 1976, p. 27.
43. D. Rickwood, A. Hell, G. D. Birnie, and C. C. Gulhuus-Moe, *Biochim. Biophys. Acta*, **342**, 367 (1974).
44. A. Huttermann and G. Wendlberger, in *Biological Separations in Iodinated Denisty-Gradient Media*, D. Rickwood, Ed., Information Retrieval Ltd., London, 1976, p. 25.
45. G. D. Birnie and D. Rickwood, *Centrifugal Separations in Molecular and Cell Biology*, Butterworth, London, 1978.
46. V. Schumaker and A. Rees, in *Principles and Techniques in Plant Virology*, C. I. Kado and H. O. Agrawal, Eds., Van Nostrand–Reinhold, New York, 1972, p. 336.
47. L. Levintow and J. E. Darnell, Jr., *J. Biol. Chem.*, **235**, 70 (1960).
48. H. H. Gschwender, M. Brummund, and F. Lehamnn-Grube, *J. Virol.*, **15**, 1317 (1975).
49. W. H. Wunner, R. M. L. Buller, and C. R. Pringle, in *Biological Separations in Iodinated Density-Gradient Media*, D. Rickwood, Ed., Information Retrieval Ltd., London, 1976, p. 159.
50. D. A. R. Vanden Berghe, G. Van der Groen, and S. R. Pattyn, in *Biological Separations in Iodinated Density-Gradient Media*, D. Rickwood, Ed., Information Retrieval Ltd., London, 1976, p. 175.
51. C. B. Reimer, R. S. Baker, R. M. vanFrank, T. E. Newlin, G. B. Cline, and N. G. Anderson, *Am. Soc. Microbiol.*, **6**, 1207 (1967).
52. N. G. Anderson, *Exp. Cell Res.*, **8**, 91 (1955).
53. N. G. Anderson, A. E. Lansing, I. Lieberman, C. T. Rankin, and H. Elrod, *Wistar Inst. Symp. Monog.*, **8**, 23 (1968).
54. K. S. Kirby, *Progr. Nucleic Acid Res. Mol. Biol.*, **3**, 1 (1964).
55. G. D. Birnie, *FEBS Lett.*, **27**, 19 (1972).
56. W. Szybalski, in *Methods in Enzymology*, Vol. 12B, S. P. Colowick and N. O. Kaplan, Eds., Academic, New York, 1968, p. 330.
57. E. H. Davidson and B. R. Hough, *J. Mol. Biol.*, **56**, 491, (1971).
58. A. Hell, G. D. Birnie, T. K. Slimming, and J. Paul, *Anal. Biochem.*, **48**, 369 (1972).
59. J. O. Bishop, *Biochim. Biophys. Acta*, **174**, 636 (1969).
60. R. R. Burgess, *J. Biol. Chem.*, **244**, 6160 (1969).
61. S. Penman, *J. Mol. Biol.* **17**, 117 (1966).
62. S. Linn and I. R. Lehman, *J. Biol. Chem.*, **240**, 1287 (1965).
63. M. Collot, S. Wattiaux-de Coninck, and R. Wattiaux, *Eur. J. Biochem.*, **51**, 603 (1975).

64. G. H. Hogebuom, W. C. Schneider, and M. J. Striebich, *Cancer Res.*, **13**, 617 (1953).
65. C. deDuve, B. C. Pressman, R. Gianetto, R. Wattiaux, and F. Applemans, *Biochem. J.*, **60**, 604 (1955).
66. C. deDuve, in *Lysosomes in Biology and Pathology*, Vol. 1, J. T. Dingle and H. B. Fell, Eds., North-Holland, Amsterdam, 1969, p. 3.
67. D. B. Rooydn, in *Enzyme Cytology*, D. B. Rooydn, Ed. Academic, London, 1967, p. 103.
68. C. DeDuve and P. Baudhuin, *Physiol. Rev.*, **46**, 323 (1966).
69. H. Schuel and N. G. Anderson, *J. Cell Biol.* **21**, 309 (1964).
70. H. Schuel, S. R. Tipton, and N. G. Anderson, *J. Cell Biol.*, **22**, 317 (1964).
71. N. G. Anderson and C. L. Burger, *Science*, **136**, 464 (1962).
72. L. T. Skeggs, *Am. J. Clin. Pathol.*, **28**, 311 (1957).
73. P. R. N. Kind and E. J. King, *J. Clin. Pathol.*, **7**, 322 (1954).
74. M. E. A. Powell and M. J. H. Smith, *J. Clin. Pathol.*, **7**, 245 (1954).
75. C. deDuve, R. Wattiaux, and P. Baudhuin, *Adv. Enzymol.*, **24**, 291 (1962).
76. J. L. Gamble and A. L. Lehninger, *J. Biol. Chem.*, **223**, 921 (1956).
77. S. J. Cooperstein and A. A. Lazarow, *J. Biol. Chem.*, **189**, 665 (1951).
78. L. Smith, *Methods Biochem. Anal.*, **2**, 427 (1955).
79. H. Beaufay and J. Berthet, *Biochem. Soc. Symp.*, **23**, 66 (1964).
80. C. deDuve, J. Berthet, and H. Beaufay, *Prog. Biophys. Biophys. Chem.*, **9**, 326 (1959).
81. J. F. Thompson and E. T. Mikuta, *Arch. Biochem. Biophys.*, **51**, 487 (1954).
82. I. R. Johnson, A. P. Mathias, F. Pennington, and D. Ridge, *Biochem. J.*, **109**, 127 (1968).
83. C. C. Widnell and J. R. Tate, *Biochem. Biophys. Acta.*, **123**, 478 (1966).
84. O. F. deLalla and J. W. Gofman, *Methods Biochem. Anal.*, **1**, 459 (1954).
85. R. J. Havel, H. A. Eder, and J. H. Bragdon, *J. Clin. Invest.*, **34**, 1345 (1955).
86. F. T. Lindgren, H. A. Elliott, and J. W. Gofman, *J. Phys. Colloid Chem.*, **55**, 80 (1955).
87. T. G. Redgrave, D. C. K. Roberts, and C. E. West, *Anal. Biochem.*, **65**, 42 (1975).
88. G. V. Mann., *Clin. Chem.*, **7**, 275 (1961).
89. A. Zlatkis and B. Zak, *Anal. Biochem.*, **29**, 143 (1969).
90. N. G. Anderson, *Q. Rev. Biophys.*, **1** (3), 217 (1968).

Chapter **VII**

MECHANICAL SEPARATIONS
BY CENTRIFUGES

Mechanical separations by centrifuges are applicable to heterogeneous mixtures and are used commercially to (1) separate immiscible liquids, (2) remove and recover solids from dispersion in liquids, (3) remove excess liquid from solids, and (4) any combination of the first three. The techniques are based on physical differences between each phase, such as the particle size, shape, and density or density of each phase. Centrifuges offer many advantages to processes where high tonnages of materials are involved for separating components from compositions ranging in characteristics from slimy sludges to fast-draining solids.

Two basically different types of centrifuges are employed to perform these functions. In the first, which may be described as a "settling machine," the liquid, or one of the liquid phases, is continuous and the dispersed particles of solids or of the other liquid phase are caused to sediment through it by the acceleration of centrifugal force. The second basic type of centrifuge may be described as a "centrifugal filter." In it, the solid phase is supported on a permeable surface, such as a screen, through which the liquid phase is free to pass under the acceleration of centrifugal force.

1 BASIC THEORY OF CENTRIFUGAL SETTLING

A given particle in a given fluid settles under gravitational force at a fixed maximum rate. To increase the settling rate, the force of gravity acting on the particle may be replaced by a much stronger centrifugal force. Centrifugal separators have to a considerable extent replaced gravity separators in production operations because of their greater effectiveness with fine drops and particles and their much smaller size for a given capacity. The schematic diagrams of various centrifugal settling machines are shown in Fig. 1.1.

The sedimentation of a spherical particle in an incompressible fluid in a centrifugal force field may be written from the equation of motion given in (2.9.9) as

$$\frac{d\mathbf{v}}{dt} + \frac{18\mu}{\rho_p D^2_p}\mathbf{v} = \frac{\omega^2 r}{\rho_p}[\rho_p - \rho_m] \qquad (7.1.1)$$

380

where ρ_p is the density, D_p is the diameter of a particle, and ρ_m is the density of the fluid. Since the sedimentation (settling) process is a very slow process and is taking place in the Stokes region ($10^{-4} < Re < 2$), one may neglect the acceleration term or the velocity at steady state (terminal velocity). Thus, the settling velocity of a particle may be written as

$$\mathbf{v}_c = \frac{(\rho_p - \rho_m)D_p^2\omega^2 r}{18\mu} \tag{7.1.2}$$

The subscript c indicates the settling velocity under a centrifugal force field. If one incorporates the term defined in (2.1), the centrifuge effect Z, the settling velocity of a particle in a centrifugal force field , $\mathbf{v}_c$, becomes

$$\mathbf{v}_c = Z\mathbf{v}_g \tag{7.1.3}$$

where $\mathbf{v}_g$ is the settling velocity of a particle under a gravitational field.

If the size of settling particle is large and the terminal velocity is no longer in the Stokes region, the above analysis is not valid. One has to analyze the settling problems by applications of the mechanics of particle movement through a fluid. For steady flow of a fluid passing a solid particle, boundary layers are established, and a force is exerted on the particle by the fluid. This force is a combination of boundary-layer drag and form drag, and it can be expressed in terms of a drag coefficient, C_D. The drag coefficient C_D is defined as

$$\mathbf{F}_D = C_D A \frac{\rho_p \mathbf{v}^2}{2} \tag{7.1.4}$$

where $\mathbf{F}_D$ is the force acting on the solid particle, the drag force (boundary-layer drag or skin drag and form drag), $\mathbf{v}$ is the free stream velocity relative to the particle (for the settling, the fluid is stationary, thus $\mathbf{v}$ is the terminal velocity), and A is the cross-sectional area of the particle normal to the flow. The forces acting on the falling particle are the external force $\mathbf{F}_E$, a buoyant force $\mathbf{F}_B$, and the drag force $\mathbf{F}_D$ due to fluid friction in the direction of the velocity of fluid relative to the particle. Then, the force balance on a falling particle is

$$(\mathbf{F}_E - \mathbf{F}_D - \mathbf{F}_B) = m\frac{d\mathbf{v}}{dt} \tag{7.1.5}$$

where m is the mass of the particle. The external force $\mathbf{F}_E$ may be expressed by Newton's law as

$$\mathbf{F}_E = m\mathbf{a}_E \tag{7.1.6}$$

where a_E is the acceleration of the particle resulting from the external force. Archimedes' principle yields the buoyant force. The mass of the fluid displaced by the solid particle is $(m/\rho_p)\rho_m$, where ρ_p and ρ_m are densities of the solid particle and fluid, respectively. Therefore,

$$F_B = \left(\frac{m}{\rho_p}\right)\rho_m a_E \qquad (7.1.7)$$

Substituting (7.1.6), (7.1.4), and (7.1.7) into (7.1.5) gives

$$\frac{d\mathbf{v}}{dt} = a_E - \frac{\rho_m a_E}{\rho_p} - \frac{C_D \mathbf{v}^2 \rho_m A}{2m} \qquad (7.1.8)$$

Equation 7.1.8 is a general equation for the total force acting on a particle in any force field. If the external force is gravity, a_E is equal to the acceleration of gravity $\mathbf{g}$; and if it is a centrifugal force, a_E is equal to the centrifugal acceleration $\omega^2 r$. Thus, (7.1.8) under the centrifugal force field becomes

$$\frac{d\mathbf{v}_c}{dt} = \left(1 - \frac{\rho_m}{\rho_p}\right)\omega^2 r - \frac{C_D \mathbf{v}_c^2 \rho_m A}{2m} \qquad (7.1.8a)$$

The solution of (7.1.8a) requires knowledge of the drag coefficient C_D for a given flow condition.

For a spherical particle in various flow regions, the drag coefficients are [1]

$$\mathbf{C_D} = \frac{24}{\mathbf{Re}} \qquad \mathbf{Re} < 0.4 \qquad \text{(Stokes region)} \qquad (7.1.9a)$$

$$\mathbf{C_D} = \left(\frac{10}{\mathbf{Re}}\right)^{1/2} \qquad 0.4 < \mathbf{Re} < 500 \qquad \text{(Allen region)} \qquad (7.1.9b)$$

$$\mathbf{C_D} = 0.44 \qquad 500 < \mathbf{Re} < 2 \times 10^5 \qquad \text{(Newton region)} \qquad (7.1.9c)$$

where $\mathbf{Re}(= D_p \rho_p \mathbf{v}/\mu)$ is the Reynolds number. If one substitutes (7.1.9) and $A = \pi D_p^2/4$ into (7.1.4), the drag forces of a settling particle at various flow regions become

$$F_D = 3\pi D_p \mathbf{v}\mu \qquad \mathbf{Re} < 0.4 \qquad (7.1.10a)$$

$$F_D = \frac{5}{4}\pi\sqrt{\upsilon\rho_m}(D_p \mathbf{v})^{3/2} \qquad 0.4 < \mathbf{Re} < 500 \qquad (7.1.10b)$$

$$\mathbf{F}_D = 0.055\pi\rho_m(D_p\mathbf{v})^2 \qquad 500 < \mathbf{Re} < 2 \times 10^5 \qquad (7.1.10c)$$

Equation 7.1.10a is exactly the same as (2.9.4), except for the sign. The negative sign in (2.9.4) is for the case when a fluid is flowing through a stationary particle, while for the settling process, the fluid is stationary and the particle is moving. Hence, the reverse sign is apparent.

Then, substituting (7.1.9) and $A = \pi D_p^2/4$ into (7.1.8a) and letting $d\mathbf{v}/dt = 0$ (steady state), one obtains the terminal velocity of a particle at various flow regions:

$$\mathbf{v}_c = \frac{(\rho_p - \rho_m)D_p^2\omega^2 r}{18\mu} \qquad \mathbf{Re} < 0.4 \qquad (7.1.11a)$$

$$\mathbf{v}_c = \left[\frac{4}{225}\frac{(\rho_p - \rho_m)^2\omega^4 r^2}{\rho_p\mu}\right]^{1/3} D_p \qquad 0.4 < \mathbf{Re} < 500 \qquad (7.1.11b)$$

$$\mathbf{v}_c = \left[3\omega^2 r(\rho_p - \rho_m)\frac{D_p}{\rho_m}\right]^{1/2} \qquad 500 < \mathbf{Re} < 2 \times 10^5 \qquad (7.1.11c)$$

Again, (7.1.11a) is exactly the same as (7.1.2). Thus, the terminal velocity under centrifugal force and gravitational force can be related by use of the centrifuge effect Z in such a way that

$$\mathbf{v}_c = Z^{1/3}\mathbf{v}_g \qquad 0.4 < \mathbf{Re} < 500 \qquad (7.1.12a)$$

$$\mathbf{v}_c = Z^{1/2}\mathbf{v}_g \qquad 500 < \mathbf{Re} < 2 \times 10^5 \qquad (7.1.12b)$$

Comparing (7.1.3) with (7.1.12), it is easily seen that the centrifuge effect becomes more dominant over the particle settling velocity as the size of the particle becomes smaller.

2 TUBULAR CENTRIFUGAL SETTLING MACHINES

Tubular centrifuge settling machines have various geometries depending on the system to be separated. In general, the settling machine for a liquid–liquid system has a tall, narrow bowl, while the machine for a solid–liquid system has a broad, short bowl.

Liquid–Liquid System

A tubular settling centrifuge for a liquid–liquid separation is schematically shown in Fig. 1.1a. The bowl turns in a stationary casing at about 15,000 rpm. Feed enters from a stationary nozzle inserted through an opening in the bottom of the bowl. It separates into two concentric layers of liquid inside the bowl. The inner, or lighter, layer spills over a weir at the top of the bowl; it is thrown outward into a stationary discharge cover and from there to a spout. Heavy liquid flows over another weir into a separate cover and discharge spout. The weir over which the heavy liquid flows is removable and may be replaced with another having an opening of a different size. The position of the liquid-liquid interface (the neutral zone) is maintained by a mechanical equilibrium between two hydraulic layers. Assume that the heavy liquid, of density ρ_H, overflows the weir at radius r_H and the light liquid, of density ρ_L, leaves through ports at radius r_L. Then, if a mechanical equilibrium is established inside the tubular centrifuge, one has

$$p_i - p_L = p_i - p_H \qquad (7.2.1)$$

where p is the centrifugal pressure in the radial direction, and the subscripts i, L, and H indicate the radial position at the liquid-liquid interface at the free surface of the light liquid at r_L and at the free surface of the heavy liquid at r_H, respectively. From (2.9.8),

$$p_i - p_L = \frac{\rho_L \omega^2 (r_i^2 - r_L^2)}{2g_c} \quad \text{and} \quad p_i - p_H = \frac{\rho_H \omega^2 (r_i^2 - r_H^2)}{2g_c} \qquad (7.2.2)$$

Equating these pressure drops, simplifying, and solving for r_i gives

$$r_i = \left[\frac{r_H^2 - (\rho_L/\rho_H) r_L^2}{1 - \rho_L/\rho_H} \right]^{1/2} \qquad (7.2.3)$$

Equation 7.2.3 shows that r_i, the radius of the neutral zone, is sensitive to the density ratio, especially when the density is near unity [2]. It also shows that if r_L is held constant and r_H, the radius of the discharge lip for the heavier liquid, increases, the neutral zone is shifted toward the wall of the bowl. If r_H is decreased, the zone is shifted toward the axis. An increase in r_L, at constant r_H, also shifts the neutral zone toward the axis, and a decrease in r_L cause a shift toward the wall. The position of the neutral zone is practically important in the design of operation.

Solid-Liquid System

In a settling centrifuge, a particle of given size is removed from the liquid if sufficient time is available for the particle to reach the wall of the separator bowl. If it is assured that the particle is at all times moving radially at its terminal velocity, the diameter of the smallest particle that should just be removed can be calculated. A schematic of a particle settling centrifugal machine is presented in Fig. 7.1. The feed point is at the bottom, and the liq-

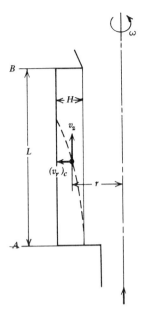

Fig. 7.1   Schematic diagram of particle-settling centrifugal machine.

uid discharge is at the top. Assume that all the liquid moves upward through the bowl at a constant velocity $\mathbf{v}_z$, carrying solid particles with it. A given particle, as shown in the figure, begins to settle at the bottom of the bowl at some position in the liquid, say, at a distance r_A from the axis of rotation; r_1 is the liquid–air interface and can be determined by (7.2.3). The particle's settling time is limited by the residence time of the liquid in the bowl. At the end of this time, let the particle be at a distance r_B from the axis of rotation. If $r_B <$ r_2, the particle leaves the bowl with the liquid; if $r_B = r_2$, it is deposited on the bowl wall and removed from the liquid.

In a steady-state operation, for complete separation of particles in a settling centrifuge, the residence time of the liquid in the bowl has to be longer than the time required for the particles to travel in a radial direction to reach the bowl wall. This condition can be written mathematically as

$$\frac{H}{(\mathbf{v}_r)_c} \leq \frac{L}{\mathbf{v}_z} = \frac{V}{Q} \tag{7.2.4}$$

where V is the volume of the centrifuge, Q is the volumetric flow rate of liquid through the centrifuge bowl, L is the length of the bowl, $(\mathbf{v}_r)_c$ is the radial settling velocity of particles in a centrifugal force field, H is the thickness of fluid layer, and v_z is the axial velocity of upward flow of liquid. Rearranging (7.2.4), one has a criterion of complete separation for a volumetric flow rate:

$$Q \leq (v_r)_c \frac{V}{H} = v_g \frac{ZV}{H} \tag{7.2.5}$$

or

$$Q \leq v_g \Sigma_c \tag{7.2.5a}$$

where v_g is the particle's settling velocity in a gravitational field and Σ_c is a factor, called a centrifugal settling area, defined by $\Sigma_c = ZV/H$ and has a dimension in units of area. The centrifugal settling area Σ_c depends on the size and geometry of the centrifugal machine and on the centrifugal force field strength, or the capacity of the machine and the operating conditions.

The Σ_c factor can then be used as a means of comparing centrifuges [2]. It is the cross-sectional area that will remove particles down to the same diameter as those separated in the centrifuge when its volumetric feed rate equals that of the centrifuge. If two centrifuges are to perform the same function,

$$\frac{Q_1}{\Sigma_{c_1}} = \frac{Q_2}{\Sigma_{c_2}} = \cdots = \frac{Q_i}{\Sigma_{c_i}} \tag{7.2.5b}$$

The quantity Σ_c can be determined for commercial centrifuges, although in some cases the determination requires approximation methods.

Equation 7.2.5a may be written with an equal sign by introducing a dimensionless efficiency factor ϵ_c, the settling separation efficiency factor:

$$Q = \epsilon_c v_g \Sigma_c \qquad (\epsilon_c \leq 1) \tag{7.2.6}$$

It can be seen that the settling separation efficiency factor depends on Q, v_g, and Σ_c; and ϵ_c may be summarized as being dependent on (1) the size and geometry of the centrifugal machine, (2) the operating conditions, and (3) the physical properties of the separation system, such as viscosity of the liquid, densities of liquid and particles, sizes of particles, and so on.

The residence time of a particle may be evaluated for the three flow regions from (7.1.11a, b, c). Since $(v_r)_c = dr/dt$ and the boundary conditions are $r = r_A$ at $t = 0$ and $r = r_B$ at $t = t_f$, rearranging of (7.1.11) gives

A. In the Stokes region (**Re** < 2):

$$t_f = \frac{18\mu}{(\rho_p - \rho_m)\omega^2 D_p^2} \ln \frac{r_B}{r_A} \tag{7.2.7a}$$

B. In the Allen region (2 < **Re** < 500):

$$t_f = 0.783 \left[\frac{\rho_m \mu}{(\rho_p - \rho_m)\omega^4} \right]^{1/3} \frac{(r_B^{1/3} - r_A^{1/3})}{D_p} \qquad (7.2.7b)$$

C.   In the Newton region ($500 < \mathrm{Re} < 2 \times 10^5$):

$$t_f = 0.155 \left[\frac{\rho_m}{[\rho_p - \rho_m]\omega^2 D_p} \right]^{1/2} (r_B^{1/2} - r_A^{1/2}) \qquad (7.2.7c)$$

The volume of the settling machine, V, equals $\pi L(r_2^2 - r_1^2)$. If one divides V by the volumetric flow rate Q, a residence time of the liquid can be obtained. If the particle settles in the Stokes region and the settling separation efficiency factor is unity, the volumetric flow rate gives

$$Q \le \frac{\pi L(\rho_p - \rho_m)\omega^2 D_p^2}{18\mu} \frac{(r_2^2 - r_1^2)}{\ln (r_B/r_A)} \qquad (7.2.8)$$

In the other two flow regions, the volumetric flow rate can be obtained in an exactly similar manner.

Ambler [2] has defined a "cutoff point" as the diameter of the particle which just reaches one half the distance between r_1 and r_2. If $(D_p)_c$ is the cutoff diameter, a particle of this size moves a distance $H/2 = (r_2 - r_1)/2$ during the settling time allowed. If a particle of diameter $(D_p)_c$ is to be removed, it must reach the bowl wall in the available time. Thus, $r_B = r_2$ and $r_A = (r_1 + r_2)/2$. Equating 7.2.8 then becomes

$$Q_c \le \frac{\pi L(\rho_p - \rho_m)\omega^2(D_p)_c^2}{18\mu} \frac{(r_2^2 - r_1^2)}{\ln [2r_2/(r_1 + r_2)]} \qquad (7.2.9)$$

where Q_c is the volumetric flow rate corresponding to the cutoff diameter. At this flow rate, most particles greater in size than $(D_p)_c$ will be largely removed, and those smaller in size will tend to remain in suspension. The critical diameter $(D_p)_c$ or the cutoff diameter of particles may be obtained as

$$(D_p)_c \le \left\{ \frac{18 \, Q\mu \ln [2r_2/(r_1 + r_2)]}{\pi L(\rho_p - \rho_m)\omega^2(r_2^2 - r_1^2)} \right\}^{1/2} \qquad (7.2.10)$$

If the thickness of the liquid layer is thin compared to the radius of the bowl, $r_1 \approx r_2$, and (7.2.9) and (7.2.10) become indeterminate. Under these conditions, however, the settling velocity may be considered constant and given by (7.1.11a). The thickness of the liquid layer is H, and the settling distance for particles of cutoff diameter $(D_p)_c$ is $H/2$. Then,

$$\mathbf{v}_c t_f = \frac{H}{2} \tag{7.2.11}$$

where t_f is the residence time. For a thin layer of liquid, t_f is given by

$$t_f = \frac{V}{Q_c} = \frac{2\pi r_2 HL}{Q_c} \tag{7.2.12}$$

Using (7.1.11a), (7.2.11), and (7.2.12) and solving for Q_c, one obtains

$$Q_c = \frac{2\pi(\rho_p - \rho_m)(D_p)_c^2 r_2^2 \omega^2 L}{9\mu} \tag{7.2.13}$$

This analysis is somewhat oversimplified, since the pattern of liquid flow in a centrifuge bowl is more complicated than the assumed plug flow. For rigorous treatment of the liquid flow pattern, the method used in Section 4, Chapter III, may be followed.

3 DISK CENTRIFUGES

For some liquid–liquid separations, the disk-type centrifuge illustrated in Fig. 1.1b is highly effective. A short, wide bowl turns on a vertical axis. The bowl has a flat bottom and a conical top. Feed enters from above through a stationary pipe set into the neck of the bowl. Two liquid layers are formed as in a tubular centrifuge; they flow over adjustable dams into separate discharge spouts. Inside the bowl and rotating with it are closely spread "disks," which are actually cones of sheet metal set one above the other. Matching holes in the disks about halfway between the axis and the wall of the bowl form channels through which the liquids pass. Feed liquid from the tube, as usually enters the bowl at the bottom, flows into the channels and then upward past the disks. Heavier liquid is thrown outward, displacing lighter liquid toward the center of the bowl. In its travel, the heavy liquid very soon strikes the underside of a disk and flows beneath it to the periphery of the bowl without encountering any more light liquid. Light liquid similarly flows inward and upward over the upper surfaces of the disks. Since the disks are closely spaced, the distance that a drop of either liquid must travel to escape from the other phase is short, much shorter than in the comparatively thick liquid layers in a tubular centrifuge. In addition, in a disk centrifuge there is considerable shearing at the liquid–liquid interface, as one phase flows in one direction and the other phase in the opposite direction [3]. This shearing helps break certain types of emulsions. Disk centrifuges are particularly valuable where the purpose of the centrifuging is not for complete separation but for the concentration of one fluid phase, as in the separation of cream from milk and in the concentration of rubber latex.

Analyses of conical disk centrifuges have been made by Jury and Locke [4]. Their analyses and model are presented in the following: A schematic diagram of the stack of cones contained in a conical disk-type centrifuge is shown in Fig. 7.2a. The N cones of inner radius R_1 and outer radius R_2 form an angle θ with the vertical and enclose $N-1$ spaces of thickness s. Feed slurry is fed as indicated, and during the course of its progress up through $N-1$ spaces, it is centrifuged with angular velocity ω into two cuts, the dilute top cut and the concentrated bottom cut. In this model of centrifugation, it is assumed that the feed slurry contains fluid of density ρ_m and viscosity μ and particles of diameter D_p and density ρ_p, where $\rho_p > \rho_m$. The feed slurry approaches its angular velocity ω as it enters the $N-1$ spaces and moves at a flow rate Q or linear velocity $\mathbf{v}$ up through each space.

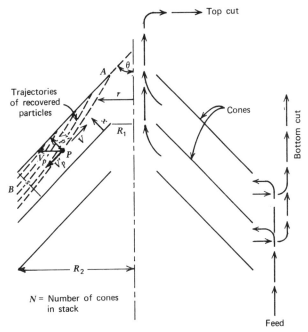

Fig. 7.2 Schematic diagram of cones in a disk centrifuge. (*a*) Schematic drawing of cones (*above*). (*b*) Cross area of the annular space (*see p. 390*). *Source:* Reproduced from Ref. 4 by permission of AIChE and the author.

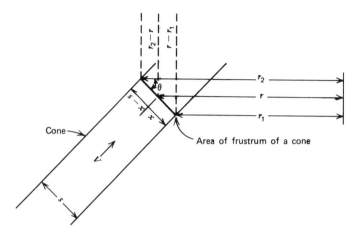

Fig. 7.2   Continued

A given particle, say, P, moves laterally with a velocity $\mathbf{v}_p$ relative to the fluid. The components of this velocity are $\mathbf{v}_p{}'$ and $\mathbf{v}_p{}''$ (as shown in Fig. F.2a), wherein the contribution of gravity is neglected because of its negligible effect in strong centrifugal fields. It is immediately apparent that whether the particle P moves up or down through a space is determined by whether $\mathbf{v} - \mathbf{v}_p{}'$ is positive or negative. Negative values of this quantity are associated with low capacities and are of no particular interest here. On the assumption that the particle P moves up through a space, its trajectory involves an increase along the x-coordinate as its position along the r-coordinate decreases in value. Any particle P entering at B will appear in the bottom cut if its trajectory does not cross a vertical line drawn through point A; otherwise, the particle will appear in the top cut.

If the trajectories are relatively flat, as shown in Fig. 7.2a, there will be a limiting trajectory passing through point A. If the trajectories have much more curvature than that shown in Fig. 7.2a, there will be a limiting trajectory which is tangent to a vertical line drawn through A. In any event, the limiting trajectory also passes through the entrance B at some point x, say, x_0. The fraction of solid feed particles recovered in the bottom cut is $(s - x_0)/s$, and this fraction can be calculated if the mathematical expression for the trajectory under consideration is known..

Fractional Recovery of Particle Size D_p by Approximate Analytical Solution

The particle velocity relative to the cone wall along x may be defined as

$$\frac{dx}{dt} = \mathbf{v}_p{}'' \tag{7.3.1}$$

and along r is defined as

$$\frac{dr}{dt} = \mathbf{v}_p - \mathbf{v}\sin\theta \tag{7.3.2}$$

The ratio of (7.3.1) to (7.3.2) gives

$$\frac{dx}{dr} = \frac{\mathbf{v}_p{''}}{\mathbf{v}_p - \mathbf{v}\sin\theta} \tag{7.3.3}$$

The problem is to express $\mathbf{v}_p{''}$, $\mathbf{v}_p$, and $\mathbf{v}$ as functions of x and r so that (7.3.3) may be integrated. To simplify integration, $\mathbf{v}$ may be approximated by the ratio of Q to the cross-sectional area through which the slurry flows. The cross-sectional area under consideration is indicated in Fig. 7.2b. The area is

$$\pi(r_1 + r_2)s \tag{7.3.4}$$

but

$$\frac{r_2 - r}{s - x} = \cos\theta \tag{7.3.5}$$

$$\frac{-r + r_1}{x} = -\cos\theta \tag{7.3.6}$$

Adding (7.3.5) and (7.3.6) shows that

$$r_2 + r_1 = 2r + (s - 2x)\cos\theta \tag{7.3.7}$$

Therefore, one obtains the fluid velocity in terms of the flow rate through a single passage, $\mathbf{q}$, to yield

$$\mathbf{v} = \frac{\mathbf{q}}{\pi[2r + (s - 2x)\cos\theta]s} \tag{7.3.8}$$

and since $2r \gg (s - 2x)\cos\theta$, one finds that

$$\mathbf{v} \approx \frac{\mathbf{q}}{2\pi rs} \tag{7.3.9}$$

Assuming that particles are in the Stokes free settling region, the velocity of particles is given in (7.1.11a). Thus,

$$\mathbf{v}_p = \frac{(\rho_p - \rho_m)D_p{}^2\omega^2 r}{18\mu} \tag{7.3.10}$$

Also,

$$\mathbf{v}_p{}'' = \frac{(\rho_p - \rho_m)D_p{}^2\omega^2 r}{18\mu}\cos\theta \tag{7.3.11}$$

The differential equation (7.3.3) becomes

$$\int_{x_0}^{s} dx = -\cos\theta \int_{R_2}^{R_1} \frac{r^2 dr}{b^2 - r^2} \tag{7.3.12}$$

in which

$$b = \left[\frac{9\mathbf{q}\mu \sin\theta}{s\pi(\rho_p - \rho_m)D_p{}^2\omega^2}\right]^{1/2} \tag{7.3.13}$$

The solution is

$$\frac{s - x_0}{s} = \frac{\cos\theta}{s}\left\{\frac{b}{2}\ln\left[\left(\frac{b + R_2}{b - R_2}\right)\left(\frac{b - R_1}{b + R_1}\right)\right] - (R_2 - R_1)\right\} \tag{7.3.14}$$

The dimensionless groups $\cos\theta$, b/x, R_2/s, R_1/s, b/R_2, and b/R_1 determine the recovery efficiency of the centrifugation. Equation 7.3.14 is plotted in Fig. 7.3 for the case of a Merco centrifuge with specifications as follows: $N = 24$, $\theta = 45°$, $s = 1.6838$ cm, $R_1 = 5.12$ cm, $R_2 = 7.62$ cm, and $\omega = 859.85$ radians/sec (8210 rpm). Of course, θ, s, R_1, and R_2 are the only variables that need to be known for plotting. It is obvious from Fig. 7.3 that until b reaches a value of approximately 21.7, all particles are recovered. When b is increased beyond this value, losses set in, until the recovery finally approaches zero as b approaches infinity.

Sludge Film Thickness for Particle Size D_p

Once the particles arrive at the cone surface, they migrate back down the upper cone surface to appear in the bottom cut. At the two extremities of s, the value of fluid velocity $\mathbf{v}$ is zero, while toward the center of s, the value of $\mathbf{v}$ reaches a maximum. Thus, a particle of diameter D_p migrates up through a space until $\mathbf{v} - \mathbf{v}_p{}'$ goes through zero and becomes negative, when it begins to migrate downward to appear as the bottom cut.

The definition of Newtonian viscosity is

$$g_c \frac{\mathbf{F}}{A} = \mu \frac{d\mathbf{v}}{dx} \tag{7.3.15}$$

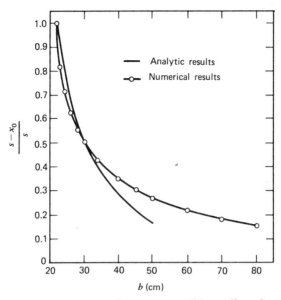

Fig. 7.3 Solutions to the equations governing a cone-type disk centrifuge. *Source:* Reproduced from Ref. 4 by permission of AIChE and the author.

If the linear distance along the surface of a cone normal to x is designated as ϕ, then over a differential elemental section $d\phi$, there will be a corresponding pressure drop dp. If s is small,

$$g_c \mathbf{F} \simeq \frac{dp}{2}(s - 2x)2\pi r \tag{7.3.16}$$

and

$$A = 2\pi r\, d\phi \tag{7.3.17}$$

Substituting (7.3.16) and (7.3.17) into (7.3.15), one has

$$g_c \frac{1}{2}\frac{dp}{d\phi}\Big|_0^x (s - 2x)dx = \mu \int_0^{\mathbf{v}} d\mathbf{v} \tag{7.3.18}$$

or

$$g_c \frac{1}{2}\frac{dp}{d\phi}[sx - x^2] = \mu\mathbf{v} \tag{7.3.19}$$

For small s, $\mathbf{v}$ rises to its maximum value at $x = s/2$. From (7.3.19), then, one has .

$$g_c \frac{1}{2} \frac{dp}{d\phi} \frac{s^2}{4} = \mu \mathbf{v}_{max} \tag{7.3.20}$$

The ratio of (7.3.19) to (7.3.20) gives the result

$$\mathbf{v} = 4\mathbf{v}_{max} \frac{sx - x^2}{s^2} \tag{7.3.21}$$

For small s, the flow rate through a single passage can be expressed as

$$\mathbf{q} = 4\pi r \int_0^{s/2} \mathbf{v} \, dx \tag{7.3.22}$$

Substituting (7.3.21) into (7.3.22) and integrating gives

$$\mathbf{q} = \mathbf{v}_{max} \frac{4\pi r}{3} s \tag{7.3.23}$$

Eliminating $\mathbf{v}_{max}$ between (7.3.21) and (7.3.23), the velocity distribution in the annulus is

$$\mathbf{v} = \frac{3\mathbf{q}}{\pi r} \frac{sx - x^2}{s^3} \tag{7.3.24}$$

The free settling velocity of particles relative to fluid, $\mathbf{v}_p$, is

$$\mathbf{v}_p = \frac{(\rho_p - \rho_m)D_p^2\omega^2 r \sin \theta}{18\mu} \tag{7.3.25}$$

At some distance $x = y$ (sludge film thickness), the velocities $\mathbf{v}$ and $\mathbf{v}_p$ become equal, and (7.3.24) and (7.3.25) may be combined to show that

$$\frac{y}{s} = \frac{1}{2}\left\{1 - \left[1 - \frac{2}{3}\left(\frac{r}{b}\sin \theta\right)^2\right]^{1/2}\right\} \tag{7.3.26}$$

The quantity y is the thickness of the sludge film flowing down the cone wall, which is determined by the dimensionless groups $\sin \theta$ and r/b. While this treatment is valid only for small values of y, it is valuable in that it shows, for example, that in order to keep y small for a given centrifuge and particle size, one must operate with large b. Also, as one passes from R_2 to R_1, the film thickness decreases for a given particle size.

Fractional Recovery of Particle Size D_p by Numerical Solution

If (7.3.10), (7.3.11), and (7.3.24) are used to substitute the velocities v_p, v_p'' and v in (7.3.3), one obtains

$$\frac{dx}{dr} = \frac{\cos \theta}{1 - 6\left(\frac{b}{r}\right)^2 \left(\frac{x}{s} - \frac{x^2}{s^2}\right)} \qquad (7.3.27)$$

This equation is difficult to solve in closed analytical form. Jury and Locke [4] have solved it by numerical methods by writing (7.3.27) in a finite difference form. Thus,

$$\frac{x_{n+1} - x_n}{r_{n+1} - r_n} = \frac{\cos \theta}{1 - 6\left(\frac{b}{r_n}\right)^2 \left(\frac{x_n}{s} - \frac{x_n^2}{s^2}\right)} \qquad (7.3.28)$$

It will be noted from (7.3.27) that if $dr/dx = 0$ at $r = R_1$, then

$$1 - 6\left(\frac{b}{R_1}\right)^2 \left(\frac{x}{s} - \frac{x^2}{s^2}\right) = 0 \qquad (7.3.29)$$

or, in the finite difference case,

$$\frac{x_n}{s} = \frac{1}{2} \left\{ 1 + \left[1 - \frac{2}{3} \left(\frac{R_1}{b}\right)^2 \right]^{1/2} \right\} \qquad (7.3.30)$$

Equation 7.3.30 was the starting point for numerical calculation of a trajectory. By substituting s, R_1, and b, Jury and Locke [4] calculated the point x_1 on the vertical line drawn through A of Fig. 7.2a. This is the point of tangency of the limiting trajectory. If one substitutes x_1, R_1, b, s, and θ in (7.3.28) and selects an interval for Δx, then r_2 can be calculated. This process is simply repeated to obtain r_3, r_4, and so on. For $\Delta x = -0.09935$, $b = 80$, the fractional recovery was 0.1802. When Δx was cut in half, the recovery became 0.1728. When halved three more times, recoveries became 0.1660, 0.1615, and 0.1610, respectively. The latter calculation for this and the other values of b lead to the data which form the basis for the plot in Fig. 7.3. It is obvious from Fig. 7.3 that the analytical solution is a fairly good approximation to the numerical solution, particularly at the smaller values of b. A similar study has been made by Inoue and Kojima [6]. They used fly-ash particles to study separating characteristics of disk centrifuges for solid–liquid systems. Results similar to Fig. 7.3 were obtained experimentally.

Centrifugal Settling Area Σ_c

The centrifugal settling area for a cone-type disk centrifuge may be obtained by using (7.3.3) together with (7.3.9), (7.3.10), and (7.3.11). This gives

$$\frac{dr}{dx} = \sec\theta - \frac{9\mu Q \tan\theta}{N\pi(\rho_p - \rho_m)D_p^2\omega^2 r^2 s} \tag{7.3.31}$$

This equation is also difficult to solve in a closed analytical form. Kuwai and Inoue [5] have reported that the first term at the right-hand side of (7.3.31), $\sec\theta$, is negligible. Moreover, elimination of the term provides an allowance for the safety factor in the design and will give a better result. The quantity Q is the total slurry feed rate to the centrifuge, and N is the total number of cones in the centrifuge. Rearranging (7.3.31) by dropping the $\sec\theta$ term, one has

$$\int_{R_2}^{R_1} r^2 dr = \frac{9\mu Q \tan\theta}{N\pi(\rho_p - \rho_m)D_p^2\omega^2 s} \int_{x_0}^{s} dx \tag{7.3.32}$$

After integration, the total slurry feed rate Q can be expressed in the form of (7.2.6), $Q = \epsilon_c v_g \Sigma_c$. Then, the centrifugal settling area Σ_c becomes

$$\Sigma_c = \frac{2\pi N(R_2^3 - R_1^3)\omega^2}{3g\tan\theta} \frac{s}{s - x_0} \tag{7.3.33}$$

Equation 7.3.33 shows that the capacity of a cone-type disk centrifuge is proportional to N, $(R_2^3 - R_1^3)$, and the square of ω and is inversely proportional to $\tan\theta$ and the fractional recovery of particle size $(s - x_0)/s$.

4 DISCHARGE COEFFICIENTS IN CENTRIFUGAL FIELD

When the feed liquid contains more than a few percent of solids, means must be provided for discharging the solids automatically. A modified cone-type disk centrifuge easily provides this type of separation. In the periphery of the bowl at its maximum diameter is a set of small holes or nozzles perhaps 3 to 4 mm in diameter. The central part of the bowl operates in the same way as the usual disk centrifuge, overflowing either one or two streams of clarified liquid. Solids are thrown to the periphery of the bowl and escape continuously through the nozzles, together with considerable liquid. Thus, an understanding of flow rates through small holes or nozzles is important in the design of this type of centrifuge.

The discharge coefficient C_Q is the factor that accounts for losses through the flow passages due to the changes in cross-sectional area of flow, frictional losses, and other obstructions, and so on. It is defined as follows:

$$C_Q = \frac{Q_{\text{actual}}}{Q_{\text{ideal}}} \qquad (7.4.1)$$

Studies on the discharge coefficient in a centrifugal field have been made by Inoue [7] for a nozzle in a tubular centrifuge and in a cone-type disk centrifuge and by Takamatsu and his co-workers [8-10] for small holes for cylindrical geometries.

Nozzle in Tubular Centrifuges

Given a tubular centrifuge revolving at a specified rotational speed ω (radians/sec) with the liquid-free surface in the centrifuge maintained at r_f from the axis of rotation. The diameter d of a nozzle attached to the centrifuge is so small compared with the diameter of the centrifuge bowl that the flow of fluid within the nozzle can be assumed to be one-dimensional and in the radial direction only. The schematic diagram depicting the nozzle in a tubular centrifuge is shown in Fig. 7.4. A force balance for a differential distance dr in the nozzle is

$$-dp = -\frac{\rho\omega^2 r}{g_c} dr + dp_f \qquad (7.4.2)$$

where dp is the pressure gradient in the differential distance dr, ρ is the density of the liquid, r is the radial position from the axis of rotation, and dp_f is

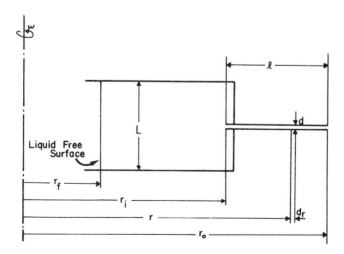

Fig. 7.4 Schematic diagram of a nozzle in a tubular centrifuge.

the frictional and other losses in the interval dr. Integrating (7.4.2) from the nozzle inlet r_i to the outlet r_o, one obtains

$$-p_i = \frac{\rho\omega^2}{2g_c}(r_o{}^2 - r_i{}^2) + \Sigma p_f \qquad (7.4.3)$$

where p_i is the pressure at the nozzle inlet r_i and Σp_f is the total losses in the nozzle consisting of the pressure drop in the entrance region due to the contraction in the cross-sectional area, of the frictional losses due to the wall of the nozzle, and of the losses due to the inside surface condition of the nozzle. If the frictional losses can be expressed by Fanning's equation and the other two losses are expressed in terms of dynamic pressure [11], the total losses may be

$$\Sigma p_f = \left[\left(\frac{1}{\alpha} - 1\right)^2 + \frac{4fl}{d} + \xi\right]\frac{\rho v^2}{2g_c} \qquad (7.4.4)$$

in which f is the Fanning friction factor, l is the length of the nozzle, α is the contraction ratio, v is the average velocity of the fluid in the nozzle, and ξ is the relative roughness of the nozzle surface. The conservation of energy at the inlet of the nozzle r_i is

$$p_i = \frac{\rho\omega^2}{2g_c}(r_i{}^2 - r_f{}^2) + \left[\left(\frac{d}{L}\right)^4 - 1\right]\frac{\rho v^2}{2g_c} \qquad (7.4.5)$$

where r_f is the radial position of the free surface of the liquid in steady rotation and L is the height of the centrifuge. Equating (7.4.3) with (7.4.5), one has

$$\frac{\rho\omega^2(r_o - r_f{}^2)}{2g_c} = \left[\left(\frac{1}{\alpha} - 1\right)^2 + \frac{4fl}{d} + \xi - 1 - \left(\frac{d}{L}\right)^4\right]\frac{\rho v^2}{2g_c} \qquad (7.4.6)$$

The volumetric flow rate of liquid through the nozzle is $Q_{ideal} = Av$. Then, Q_{actual} may be written as

$$Q_{actual} = C_Q A \omega \sqrt{r_o{}^2 - r_f{}^2} \qquad (7.4.7a)$$

and

$$C_Q = \left[\left(\frac{1}{\alpha} - 1\right)^2 + \frac{4fl}{d} + \xi - 1 - \left(\frac{d}{L}\right)^4\right]^{-1/2} \qquad (7.4.7b)$$

where C_Q is the volumetric flow discharge coefficient ($C_Q \leq 1$) and A is the cross-sectional area of the nozzle ($A = \pi d^2/4$). If (7.4.7a) is written in terms of rpm $n(= 60\omega/2\pi)$, one obtains

$$Q_{\text{actual}} = \frac{(\pi d)^2}{120} C_Q n \sqrt{r_o^2 - r_f^2} \qquad (7.4.8)$$

In a gravitational field, the volumetric flow rate may be written as

$$Q_{\text{actual}} = C_Q A \sqrt{2g_c \frac{\Delta p}{\rho}} \qquad (7.4.9)$$

where Δp is the pressure head applied to the liquid at the nozzle inlet.

Inoue [7] used stainless steel syringe material as nozzles with an experimental apparatus for the centrifugal field with $r_i = 8.10$ cm and $L = 1.0$ cm to obtain the discharge coefficients. The nozzle dimensions are presented in Table 7.1. The variation of discharge, the actual volumetric flow rate Q_{actual}, with respect to rpm (n) in (7.4.8) is linear. Experimental verification of this fact was made and presented in Fig. 7.5. The variation of the discharge coefficient C_Q

Table 7.1 Dimensions of Nozzle Used by Inoue [7]

Nozzle Symbol	Diameter d (mm)	Length l (mm)	l/d
A_1	0.620	12.5	20.2
A_2	0.620	10.0	16.1
A_3	0.618	7.5	12.2
A_4	0.624	5.0	8.1
B_1	0.464	12.5	27.0
B_2	0.460	10.0	21.8
B_3	0.458	7.5	16.4
B_4	0.449	5.0	11.2
C_1	0.318	12.5	39.4
C_2	0.319	10.0	31.4
C_3	0.318	7.5	23.6
C_4	0.321	5.0	15.5
D_1	0.211	12.5	50.2
D_2	0.215	10.0	46.6
D_3	0.210	7.5	35.7
D_4	0.215	5.0	23.2

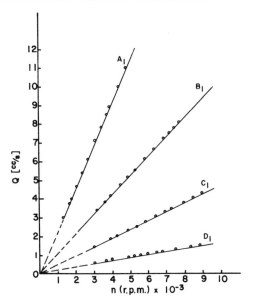

Fig. 7.5   Discharge rate Q as function of n number of revolutions per minute, rpm. *Source:* Reproduced from Ref. 7.

with respect to Reynolds number and the length–diameter ratio of various nozzles used are presented in Fig. 7.6. The results were summarized as follows: (1) The discharge coefficient in a centrifugal field is identical with that in a gravitational field. (2) The discharge coefficient is nearly a constant value if **Re** > 3000.

Discharge in Cone-Type Disk Centrifuges

Discharges from a cone-type disk centrifuge were also investigated by Inoue [7]. The schematic of nozzles in a cone-type disk centrifuge and a coordinate system used in his analysis are reproduced in Fig. 7.7.

A force balance from the nozzle inlet to the outlet may be written in a similar manner as (7.4.7a) to yield

$$\frac{\rho}{2g_c} \left(\frac{Q}{C_Q A} \right)^2 = \Delta p + \frac{\rho \omega_s^2}{2g_c} (r_o^2 - r_i^2) \qquad (7.4.10)$$

where Δp is the pressure difference between the rotating body and the nozzle outlet and ω_s is the angular velocity of the apparatus (in radians/sec).

In evaluating Δp in (7.4.10), the frictional losses created by the disk may have to be taken into consideration. To simplify the mathematical analysis, the

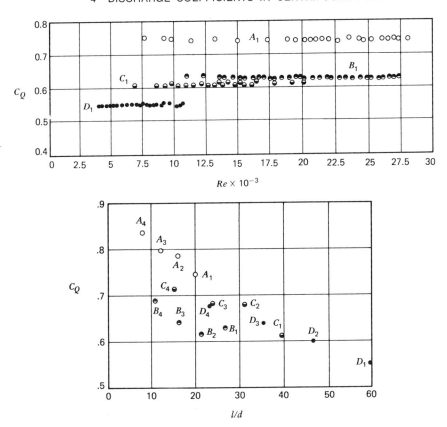

Fig. 7.6   Discharge coefficient C_Q obtained experimentally by Inoue. (*a*) Discharge coefficient as function of Reynolds number (*above*). (*b*) Variation of l/d ratio to discharge coefficient (*below*). *Source:* Reproduced from Ref. 7.

volumetric flow rate or the discharge between two cones for a single passage was studied with assumptions that (1) liquid is flowing uniformly in the axial direction, (2) the flow normal to the disk is negligible, and (3) the liquid flow parallel to the disk surface is a laminar flow with a parabolic velocity profile and the disk surface is very smooth.

If one uses the new coordinate system (α, β, γ) as shown in Fig. 7.8 for a cone, the transformations of the coordinates are

$$x = (\alpha \cos \theta + \beta \sin \theta)\cos \gamma \qquad (7.4.11a)$$

$$y = (\alpha \cos \theta + \beta \sin \theta)\sin \gamma \qquad (7.4.11b)$$

$$z = \alpha \sin \theta - \beta \cos \theta \qquad (7.4.11c)$$

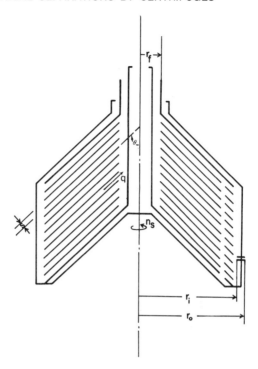

Fig. 7.7   Schematic diagram of a nozzle in a cone-type disk centrifuge. Left-hand side for separating liquids, right-hand side for separating solid from liquid.

The pressure drop between the two cone disks shown in Fig. 7.8 between AB and CD may be obtained from the equation of motion in the β-direction [12]:

$$\mathbf{u}\frac{\partial \mathbf{v}}{\partial a} + \mathbf{v}\frac{\partial \mathbf{v}}{\partial \beta} + \frac{\mathbf{w}}{r}\frac{\partial \mathbf{v}}{\partial r} - \frac{\mathbf{w}^2 \sin \theta}{r} = -\frac{g_c}{\rho}\frac{\partial p}{\partial \beta} + \nu \left(\frac{\partial^2 \mathbf{v}}{\partial \alpha^2} \right.$$

$$+ \frac{\partial^2 \mathbf{v}}{\partial \beta^2} + \frac{1}{r^2}\frac{\partial^2 \mathbf{v}}{\partial r^2} + \frac{\cos \theta}{r}\frac{\partial \mathbf{v}}{\partial \alpha} + \frac{\sin \theta}{r}\frac{\partial \mathbf{v}}{\partial \beta} - \frac{2 \sin \theta}{r}\frac{\partial \mathbf{w}}{\partial \gamma}$$

$$\left. - \frac{\mathbf{u} \cos \theta \sin \theta + \mathbf{v} \sin^2 \theta}{r^2} \right) \qquad (7.4.12)$$

where α, β, and γ, are the new orthogonal coordinate system; $\mathbf{u}$, $\mathbf{v}$, and $\mathbf{w}$ are the velocity components in the α-, β-, and γ-directions, respectively; ν is the kinematic viscosity; θ is the angle between the cone disk and the rotational

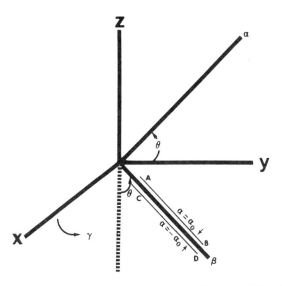

Fig. 7.8 Coordinate system used by Inoue. *Source:* Reproduced from Ref. 7.

axis; and r is a radius given by $r = \alpha \cos \theta + \beta \sin \theta$ (however, $\alpha \ll \beta$, $r \eqsim \beta \sin \theta$).

The velocity distribution between two cone disks may be written as

$$\mathbf{v} = \pm \frac{3(\alpha_0{}^2 - \alpha^2)\mathbf{q}}{\pi s^3 \beta \sin \theta} = \pm \frac{3(\alpha_0{}^2 - \alpha^2)\mathbf{q}}{\pi s^3 r} \qquad (7.4.13)$$

in which α_0 is the position of the disk in the α-coordinate, s is the distance between cones, and $\mathbf{q}$ is the volumetric flow rate between cones. A positive sign indicates the flow is outward and a negative one is inward flow. Using the equation of continuity, the velocity distribution given in (7.4.13), and replacing β by r, (7.4.12) may be simplified to

$$\frac{g_c}{\rho}\frac{dp}{dr} = \frac{\mathbf{w}^2}{r} \pm \frac{9(\alpha_0{}^2 - \alpha^2)^2\mathbf{q}^2}{\pi^2 s^6 r^3} \mp \frac{6\nu\mathbf{q}}{\pi s^3 r \sin \theta}\left(1 + \frac{\alpha \cos \theta}{r}\right) \qquad (7.4.14)$$

Integrating between r_1 and r_2, one obtains the pressure difference between these distances:

$$\Delta p = \frac{\rho}{g_c}\int_{r_1}^{r_2}\frac{\mathbf{w}^2}{r}\,dr \pm \frac{3\rho\mathbf{q}^2}{20\pi^2 g_c s^2}\left(\frac{1}{r_1{}^2} - \frac{1}{r_2{}^2}\right) \mp \frac{6\nu\mathbf{q}}{\pi g_c s^3 \sin \theta}\ln\frac{r_2}{r_1} \qquad (7.4.15)$$

The terms on the right-hand side of (7.4.15) represent centrifugal effect, the inertia effect, and the viscous effect. In practice, the orders of magnitude for the second and the third terms are about 10^{-3} to that of the first term, therefore they are negligible. Substituting (7.4.15) into (7.4.10) and changing the limit of integration from r_f to r_i, one obtains

$$\frac{\rho}{g_c} \left(\frac{Q}{C_Q A} \right)^2 = \frac{\rho}{g_c} \int_{r_f}^{r_i} \frac{w^2}{r} \, dr + \frac{\rho \omega_s^2}{2g_c} (r_o^2 - r_i^2) \qquad (7.4.16)$$

Then changing the angular velocities to revolutions per minute, (7.4.16) becomes

$$\left(\frac{Q}{C_Q A} \right)^2 = \left(\frac{\pi}{30} \right)^2 \left[2 \int_{r_f}^{r_i} n^2 r \, dr + n_s^2 (r_o^2 - r_i^2) \right] \qquad (7.4.17)$$

where n and n_s are the revolutions per minute (rpm's) of liquid and centrifugal machine, respectively. If $n = n_s$, then there is no slip between the liquid and the machine during the rotation.

Inoue [7] used (7.4.17) to measure the angular velocity of liquid between the disks using a nozzle with a known discharge coefficient and found that n and n_s are approximately equal. The volumetric flow rate Q is linear with respect to n_s, therefore n^2 in the integration must be a constant. Thus, $n \approx n_s$.

Discharge Coefficient of Small Holes in the Centrifugal Field

The discharge coefficients of small holes drilled through a rotating cylinder wall were investigated analytically and experimentally by Takamatsu and his co-workers [8–10]. They considered various geometries of holes, namely, a perfect orifice, a quasi-orifice, a nozzle, and a critical-state geometry (a capillary side hole along the hole wall), and derived expressions of discharge coefficients for those hole geometries. They concluded that no difference was recognized whether a hole is in a centrifugal field or in a gravitational field. This result was the same as that obtained by Inoue [7]. In addition, Takamatsu's group investigated the effect of liquid surface tension σ and found that there is no surface tension effect if the Weber number We ($= 2\Delta p d / \sigma$) is greater than 200. If the Weber number is less than 200, the effects, represented in Fig. 7.9, are given by the following formulas:

1. A perfect orifice:

$$\frac{C_Q}{C_\sigma^{(1)}} = \sqrt{1 + \frac{4}{We \sqrt{\alpha_c}}} \qquad (7.4.18a)$$

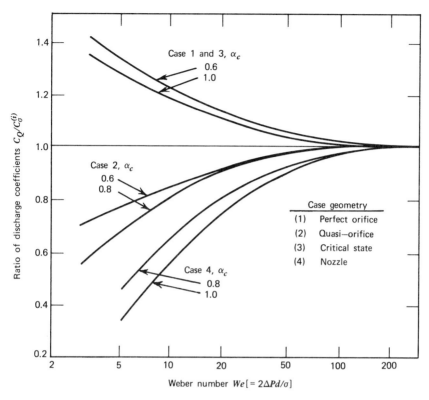

Fig. 7.9 Discharge coefficients of various small-hole geometries as function of the Weber number. *Source:* Reproduced from Ref. 9.

2. A quasi-orifice:

$$\frac{C_Q}{C_\sigma^{(2)}} = \sqrt{1 + \frac{4}{We\sqrt{\alpha_c}} - \frac{8}{We}} \qquad (7.4.18b)$$

3. A critical state:

$$\frac{C_Q}{C_\sigma^{(3)}} = \sqrt{1 + \frac{4}{We\sqrt{\alpha_c}}} \qquad (7.4.18c)$$

4. A nozzle:

$$\frac{C_Q}{C_\sigma^{(4)}} = \sqrt{1 + \frac{4}{We\sqrt{\alpha_c}} - \frac{4\sqrt{3}}{We}} \qquad (7.4.18d)$$

where d is the diameter of the hole, σ is the liquid surface tension, α_c is the contraction coefficient and $C_o^{(i)}$ are the discharge coefficients when the surface tension is taken into consideration.

5 DECANTER CENTRIFUGES

The decanter-type centrifuges are especially suited for high-tonnage applications with very different characteristics, from slimy sludges down to the 5-μ range or coarse, fast-draining crystals. Continuous mechanical removal of solids from a sedimentation centrifuge is done by one or more helical conveyors revolving in the same direction as the bowl, usually at a slightly lower speed. The moderate relative speed of the screw in the bowl moves the separated solids toward their exit as shown in Fig. 1.1c. The clarified effluent leaves the bowl at the opposite end, usually through a number of exit holes, the radial location of which essentially determines the free surface radius r_f of the rotating liquid volume in the bowl.

The decanter-type centrifuges are not especially good clarifiers. However, they process feeds containing large amounts of solids of a wide range of particle sizes and deliver drier solids than other types of sedimentation machines, since they provide for drainage of the solids after they are removed from the liquid. Thus, most applications in this type of centrifuge are in problems involving removal of liquid from solids, not in clarification of liquids.

As shown in Fig. 1.1c, the bowl is conical and rotates about a horizontal axis. Feed enters through a stationary central inlet pipe and is sprayed into the revolving bowl. The liquid, under the influence of the centrifugal force forms a layer—the pool—at the large end of the bowl. The pool depth is set by liquid overflow ports in the bowl end plate. Clarified liquid discharges through these ports into the casing, from which it is removed by the outlet pipe. Solids settle outward to the inner surface of the bowl and are simultaneously moved toward the small end of the bowl by the conveyor. They are conveyed out of the pool up the beach and delivered to the discharge slots through which they fly into the casing and drop through the large sludge discharge line. Wash liquid may be sprayed from wash nozzles on the solids as they are conveyed up the beach; this liquid flows toward the large end of the bowl and mixes with the clarified liquor.

A major problem in a decanter-type centrifuge is satisfactory clarification of the liquor. For the solids to convey properly, there must be just the correct balance between the frictional force of the solids on the conveyor and the frictional force on the bowl well. If the solids slide too easily on the bowl wall, they turn freely with the conveyor, building up the helix without moving toward the discharge. When the solids are under the liquid surface, in the pool, the buoyancy of the liquid may greatly reduce the radial force, so that the frictional force is small. When the solids have been lifted out of the liquid onto the

beach, the buoyant force no longer acts on them, and even though the centrifugal force is less than where the diameter is large, the net radial force on the
solids may be greatly increased.

Thus, solids which convey well under the liquid may not convey up the
beach, and those which move readily on the dry beach may not convey well in
the pool. In extreme cases, the solids are so buoyant that they float on the liquid or at the interface between two liquid layers of differing densities.
Sometimes, strips about 3 to 4 mm thick are attached to the bowl wall to increase the frictional resistance and permit the helix to convey the solids. In
more difficult situations the solution may be found in a machine with a bowl
which is a cylindrical-conical combination and in which the liquid layer ex-
tends almost from one end of the bowl to the other. Nearly all of the conveying,
therefore, is done on submerged solids in a cylindrical portion. Near the solids
discharge, the diameter of the helix is reduced, so that deposited solids form a
layer increasing in thickness toward the discharge ports. This stationary layer
of solids becomes a steep beach, up which even recalcitrant solids can often be
conveyed.

Centrifugal Settling Area Σ_c

The centrifugal settling area for decanter-type centrifuges may be obtained
in a similar manner as the other types. In practice, there will be an effect of
conveyors, but for simplicity this effect is neglected in the derivation. Both
conical and cylindrical-conical combination types for the derivation are
depicted in Fig. 7.10.

The centrifugal settling velocity and the transport velocity of liquid from inlet to outlet in the centrifuges are

$$\frac{dr}{dt} = \mathbf{v}_g \frac{\omega^2 r}{g} \tag{7.5.1}$$

$$\frac{dl}{dt} = \frac{Q}{\pi \left[(r_2 - r_1)\dfrac{r}{L} + r_1 \right]^2 - r_1^2} \tag{7.5.2}$$

Eliminating dt from (7.4.9) and (7.5.2) and integrating from $r = r_1$ to $r = r_2$ and from $l = 0$ to $l = L$, one has

$$Q = \mathbf{v}_g \frac{\pi L \omega^2}{g} \frac{r_2^2 + r_2 r_1 - 2r_1^2}{3 \ln \dfrac{r_2}{r_1}}$$

$$= \mathbf{v}_g \frac{\pi L \omega^2}{g} \frac{2r_1 + r_2}{3} r_{1m} \tag{7.5.3}$$

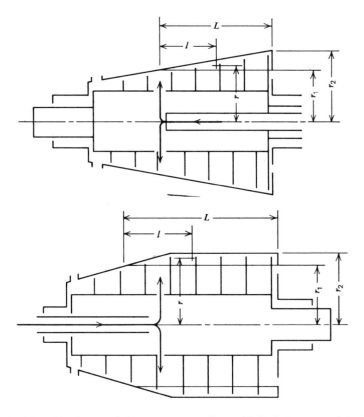

Fig. 7.10   Schematic diagram of decanter-type centrifuges. (*a*) Conical geometry (*above*). (*b*) Cylindrical–conical combination geometry (*below*).

Therefore, from (7.4.9), the centrifugal settling area for the decanter-type centrifuge becomes

$$\Sigma_c = \frac{\pi L \omega^2}{g} \frac{2r_1 + r_2}{3} r_{1m} \tag{7.5.4}$$

where r_{1m} is the logarithmic mean radius defined by $r_{1m} = (r_2 - r_1)/\ln (r_2/r_1)$, and the radii r_1 and r_2 and the length L are as specified in Fig. 7.10.

In the usual operational case for the decanter-type centrifuges, the solid-particle population is high, and the surrounding particles interfere with the motion of other individual particles. The influence of the neighboring particles affects the velocity gradient surrounding each particle. In the hindered settling of particles, the higher the concentration of solids, the lower is the particle velocity. The mechanism of settling differs in that the particle is settling through a suspension of particles in a fluid rather than through the fluid itself.

The density of the fluid phase becomes effectively the bulk density of the slurry, which is the quotient of total mass of liquid plus solid divided by the total volume. The viscosity of the slurry is considerably higher than that of the fluid because of the interference of boundary layers around interacting solid particles, and because of the increase of form drag caused by the solid particle. The viscosity of such a slurry is frequently a function of the rate of shear, of the previous history as it affects clustering of particles, and of the shape and roughness of the particles insofar as these factors contribute to a centrifuge boundary layer. As would be surmised from the above, a generalized prediction of the viscosity of slurries is impossible. Experimental measurements are necessary for accurate values, and extrapolation of any variable should be made with caution.

A correction factor ϕ was used by Schnittger [13] to account for deviation in the velocity in hindered settling from the Stokes free settling in a decanter-type centrifuge. For the range $0.001 < \text{Re} < 58$ and $0 < c < 42\%$, the correction factor for a hindered settling velocity is [14]

$$\phi = (1 - c)^2 \exp \left[- \frac{4.1c}{1.64 - c} \right] \tag{7.5.5}$$

in which c is the solid concentration in the solid-liquid phase expressed in relative percent. Thus, the terminal velocity in hindered settling in a centrifugal field becomes

$$\mathbf{v} = \mathbf{v}_g \phi Z \tag{7.5.6}$$

The centrifugal settling area for decanter-type centrifuges may be modified to give

$$\Sigma_c = \phi \; \frac{\pi L \omega^2}{g} \; \frac{2r_1 + r_2}{3} \; r_{1m} \tag{7.5.7}$$

Operational Relation Between Flow Rates and Rotational Speed

If the settling separation efficiency factor ϵ_c is unity, one may see from (7.2.8) that the discharge or volumetric flow rate Q and the rotational speed of the machine in rpm (n) have the relation $Q \propto n^2$. However, experience in actual operation shows that a rise in speed would not be as advantageous as the above relationship implies. In other words, the speed exponential would be less than 2. Schnittger [13] conducted a series of experiments in the plant during actual processing conditions and found that the Q-n relation for the speed exponential lies between 1.1 and 1.5. With a constant machine efficiency of 80%, the Q-n relationship shows a marked dependence on solid concentration in the feed. Schnittger's results are reproduced in Fig. 7.11. This reduction in the speed exponent may be accounted for in the settling separation efficiency factor ϵ_c.

6   CENTRIFUGAL FILTERS

A filtering operation can be carried out using a perforated basket centrifuge rather than conventional vacuum or pressure filters. The mechanism of centrifugal filtration is identical to that of pressure filtration, and the same theories apply to both cases if the terms within the theories are properly understood and evaluated. For centrifugal filtration, the driving force is the centrifugal force acting on the fluid and the filter cake rather than the

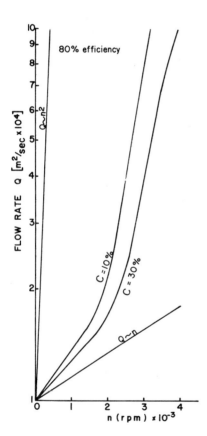

Fig. 7.11   Relation between volumetric flow rate and rotational speed of centrifuges for constant machine efficiency of 80% at $C = 10$ and 30% v/v of inlet feed. *Source:* Reproduced for Ref. 13 by permission of American Chemical Society and the author.

pressure force delivered by the pumping or the vacuum system. Simple substitution of the fluid pressure force $-\Delta p/L$ by the centrifugal force $\rho\omega^2 r/g_c$ seems to be the logical analogy.

Filters using centrifugal force are usually used in the filtration of coarse granular or crystalline solids which form a porous cake and are available for batch or continuous operation. Slurry is fed to a rotating basket or rotating cylinder having a slotted or perforated wall covered with a filter medium such as canvas or metal cloth. Pressure resulting from the centrifugal force field forces the liquor through the filter medium, leaving the solids behind. If the feed to the basket is then shut off and the cake of solids spun for a short time, much of the residual liquid in the cake drains off the particles, leaving the solids much drier than those from a filter press or vacuum filter, because of a higher pressure gradient due to the centrifugal force field. When the filtered material must subsequently be dried by thermal means, considerable savings may result from the use of centrifugal filtration. The rate of cake-thickness growth is perhaps the single most important guide in selecting equipment. Growth rate determines whether pressures, vacuum, or gravity filters or solid or perforated basket centrifuges are indicated. As an approximate guide, initial choice of equipment is suggested in Table 7.2 by Tiller and Crump [15].

Batch Centrifugal Filters

These commonly consist of a basekt with perforated sides rotated around a vertical axis as shown in Fig. 1.2a. An electric motor, located either above or

Table 7.2 Filtration Equipment Selection Guide According to Cake Buildup

Separation Type	Cake Buildup Rate Measure	Equipment Used
Rapid filtering	0.1 to 10 cm/sec	Gravity pans; screens; horizontal belt, or top-feed drum filter; continuous feed centrifuge, pusher type
Medium filtering	0.1 to 10 cm/min	Vacuum drum, disk, horizontal belt, or pan filters; or peeler centrifuges
Slow filtering	0.1 to 10 cm/hr	Pressure filters; disk and tubular centrifuges, sedimenting centrifuges
Clarification	Negligible cake	Cartridges, granular beds, precoat drums, filter aid admix

Source: Reproduced from Ref. 15 by permission of A.I.Ch.E. and the authors.

below the basekt, turns it at rates usually below 4000 rpm. Basket diameters may be as great as 122 cm (48 in.). The slurry is fed into the center of the rotating basket and is forced against the basket sides by centrifugal force. There the liquid passes through the filter medium which is placed around the inside surface of the basket sides and is caught in a shielding vessel, called a curb, within which the basket rotates. The solid phase builds up a filter cake against the filter medium. When this cake is thick enough to retard the filtration to an uneconomical rate or to endanger the balance of the centrifuge, the machine is slowed, and the cake is scraped into a bottom discharge or is scooped out of the centrifuge. In an underdriven centrifuge, the entire cake and filter medium may be removed and a clean filter cloth inserted.

In the automatically discharging batch centrifugal filter, unloading occurs automatically while the centrifuge is rotating, but the filtration cycle is still a batch one. The constant rotation speed of these machines permits lower power requirements for a given amount of filtrate collected as well as lower labor requirements than would be obtained in a batch centrifuge.

Continuous Centrifugal Filters

A multistage push-type centrifugal filter is schematically shown in Fig. 1.2b. In this filter, solids handling is held to a minimum, which permits filtration of fragile solids with the least possible breakage. As with other centrifugal filters, this unit is best suited for handling coarse-granular or coarse-crystalline solids in nonviscous liquids. Units are available with capacities up to 25 tons solids/h when handling this type of material.

In operation, the feed slurry is fed through a tube to the feed funnel. The funnel is attached to the pusher and rotates at the same speed as the centrifuge drum. In the funnel, the feed is accelerated to drum speed and fed into the back end of the filter drum where the filtrate is forced through the filter screen, and the solids form a cake on the screen. Intermittently, the cake is pushed toward the discharge end of the centrifuge by the pusher, which then retreats again leaving an open region for the buildup of new cake. In this way, the cake moves across the face of the filter screen until it is pushed off the end into the solids-collector housing. As it moves, the cake passes through a wash region where wash liquid passes through it into the filtrate housing. The filtrate and wash liquid are kept separate by partition in the wet collector.

Other continuous centrifuge filters are built, such as the solid-bowl centrifuge as depicted in Fig. 1.2c and 1.2d. They, however, would have the bowl wall consisting of a screen through which the filtrate and wash water would drain. Frequently, the cake moves across the face of the filter screen by a screw conveyor or an inverted shape of the filter screen in the centrifuge, permitting the cake to self-discharge, as shown in Figs. 1.2c and 1.2d, respectively.

7 THEORY OF PRESSURE AND CENTRIFUGAL FILTRATION

As pointed out by Tiller and Crump [15], it is customary to divide the mathematical analysis of filter operations into two parts. First, the mechanism of flow within the cake is studied, and second, the external conditions imposed upon the filter cake by the fluid pressure and the filter structure are derived. The external analysis establishes boundary conditions that must be satisfied at the extreme faces of the cake. The cake resistance determines the pressure drop across the cake. In analyzing the internal flow (i.e., within the cake itself), the distribution of hydraulic pressure p_f, porosity ϵ, and internal flow rate $\mathbf{q}$ are determined as functions of distance x through the cake. The filtrate volume V and applied pressure p are developed as functions of time t. Only in simplified cases is it possible to obtain analytical expressions relating these variables.

The flow of filtrate through the filter cake should be described by any of the general equations for flow through packed beds, such as the Blake-Kozeny equation [16]:

$$f_p = \frac{180}{\mathbf{Re}_p} \qquad \text{for } \mathbf{Re}_p < 1 \qquad (7.7.1)$$

in which f_p is the friction factor and $\mathbf{Re}_p$ is the Reynolds number for flow in a packed bed. Both are defined by

$$-\frac{\Delta p}{L} = \frac{\rho_m \mathbf{v}_\infty^2 (1 - \epsilon) f_p}{D_p \epsilon^3} \qquad (7.7.2a)$$

$$\mathbf{Re}_p = \frac{D_p \mathbf{v}_\infty \rho_m}{\mu(1 - \epsilon)} \qquad (7.7.2b)$$

Since they are based on empirical arguments, the numerical coefficients associated with (7.7.2) have been neglected. The quantity ϵ is the void fraction of the bed, and $\mathbf{v}_\infty$ is the superficial velocity at the average of inlet and outlet pressure. Substituting (7.7.2) into (7.7.1), one obtains

$$-\frac{\Delta p}{L} = 180 \frac{\mathbf{v}_\infty \mu (1 - \epsilon)^2}{D_p^2 \epsilon^3} \qquad (7.7.3)$$

Equation (7.7.3) relates the pressure drop through the cake to the flow rate $Q(=\mathbf{v}_\infty A/\epsilon)$, the cake porosity ϵ, the thickness L, and the solid-particle diameter D_p. Some modification of the equation is necessary so that the measurable variables of filtration can be introduced into it.

For nonspherical particles, an effective particle diameter is defined [17] as

$$D_p = \frac{6}{s_0} \tag{7.7.4}$$

where s_0 is the specific area of a particle, in m^2/m^3 of solid volume. Thus, (7.7.3) may be rewritten as

$$-\frac{\Delta p}{L} = 5 \frac{\mathbf{v}_\infty \mu (1 - \epsilon)^2 s_0^2}{\epsilon^3} \tag{7.7.5}$$

Solving (7.7.5) for the velocity of flow yields

$$\mathbf{v}_\infty = \frac{(-\Delta p)\epsilon^3}{5L\mu s_0^2 (1 - \epsilon)^2} = \frac{1}{A}\left(\frac{dV}{dt}\right) \tag{7.7.6}$$

where dV/dt is the rate of filtrate, that is, the volume of filtrate passing through the bed per unit time, and A is the filtration area. As written, the quantities V, t, L, $(-\Delta p)$, s_0, and ϵ may all vary. The cake thickness L may be related to the volume of filtrate by a material balance, since the thickness will be proportional to the volume of feed delivered to the filter:

$$LA(1 - \epsilon)\rho_s = w(V + \epsilon LA) \tag{7.7.7}$$

in which ρ_s is the density of the solids in the cake and w is the weight of solids in the feed slurry per volume of liquid in this slurry. The term ϵLA represents the volume of filtrate held in the filter cake. It is normally infinitesimal compared to V, the filtrate which has passed through the filter cake. Assuming this term to be negligible and combining (7.7.7) with (7.7.6) to eliminate L yields

$$\frac{1}{A}\frac{dV}{dt} = \frac{(-\Delta p)\epsilon^3}{5\frac{wV}{A\rho_s}\mu(1 - \epsilon)s_0^2} = \frac{(-\Delta p)}{\alpha\frac{\mu w V}{A}} \tag{7.7.8}$$

where α is the specific cake resistance, defined as

$$\alpha = \frac{5(1 - \epsilon)s_0^2}{\rho_s\epsilon^3} \tag{7.7.9}$$

The collection of all terms involving the filter cake properties into the specific cake resistance does not imply that the resistance α will be constant for a given feed slurry, regardless of filtering pressure drop or of filter type or size. The void fraction ϵ usually varies with variation of the compacting stress applied to the filter cake. This stress will be directly proportional to $-(\Delta p/L)$; and, since L varies throughout the process, ϵ may also vary. Both ϵ and s_0 are sensitive to the degree of flocculation of the precipitate in the feed.

Clear understanding of porosity characteristics is essential to an understanding of cake filtration in general. To a very large degree, the final liquor content of a cake is determined by the average porosity. Quantitative calculations indicate that substantial deliquoring can be accomplished by control of the hydraulic pressure patterns during washing. A valuable quantitative understanding of the mechanisms involved can be achieved even where quantitative solutions for the pertinent differential equations cannot be obtained [15].

Equation 7.7.8 is expressed in the familiar form of a flux proportional to a driving force divided by a resistance where both driving force and resistance apply to the filter cake alone. However, practically any total pressure drop $(-\Delta p_T)$ measured will at least include the pressure drop across the filter medium and will probably include the pressure drop of various flow channels before and after the actual filtering area. Since the flow resistances including these additional parts of the apparatus are arranged in series, (7.7.8) may be modified to

$$\frac{dV}{dt} = \frac{(-\Delta p_T)}{\mu\left(\dfrac{\alpha w V}{A^2} + \dfrac{R_M}{A}\right)} \tag{7.7.10}$$

where R_M has the units of (length)$^{-1}$ and represents the resistance of the filter medium and piping to filtrate flow. For convenience in analyzing filtration performance data, the resistance of the filter cloth and flow channels is usually expressed in terms of an equivalent volume of filtrate, V_e. Thus, (7.7.10) may be written as

$$\frac{dV}{dt} = \frac{(-\Delta p_T)A^2}{\mu \alpha W(V + V_e)} = Q \tag{7.7.11}$$

Here, the equivalent volume of filtrate, V_e, is the volume of filtrate necessary to build up a fictitious filter cake, the resistance of which is equal to the resistance of the filter medium and the piping between the pressure taps used to measure $-\Delta p_T$. The filter medium resistance of significance here is the resistance of the medium with the pores partially blanked with filter cake and with the initial layer of filter cake, on which the bulk of the filter cake will be built, in place.

Incompressible Cakes

Darcy's law demonstrates that, if porosity ϵ and specific resistance α are constant, the hydraulic pressure gradient is linear through the cake [16]. If such is the case, the cake is called "incompressible." For operation at constant $-\Delta p_T$, one may integrate (7.7.11) to yield

$$t = \frac{\mu\alpha w}{A^2(-\Delta p_T)} \left(\frac{V^2}{2} + V_e V \right) \tag{7.7.12}$$

from which the time necessary to pass any given volume of filtrate can be calculated.

The solution of (7.7.12) requires evaluation of the two constants α and V_e. The specific cake resistance α may be evaluated from the properties of the filter cake if ϵ and s_0 are known for the particular filtration condition from (7.7.9). However, the volume of filtrate equivalent to the filtration medium and piping flow resistances V_e must be determined from experimental filtration data. For this reason, it is usual practice to evaluate both α and V_e from a pilot filtration run using the actual slurry to be filtered under conditions as close to those to be employed in the plant as possible. To permit evaluation of these constants from experimental data, (7.7.11) is inverted to give [17]

$$\frac{1}{Q} = \frac{\mu\alpha w}{A^2(-\Delta p_T)} (V + V_e) \tag{7.7.13}$$

This equation is a straight line between $1/Q$ versus V if $-\Delta p_T$ is constant. Thus, from actual constant-pressure filtration data, $1/Q$ may be plotted as the ordinate as a function of V as the abscissa. The slope of the line is $\mu\alpha w/A^2(-\Delta p_T)$ and the intercept is $[\mu\alpha w/A^2(-\Delta p_T)]V_e$. Thus, the specific cake resistance α may be determined from the slope and the volume V_e of filtrate equivalent obtained by dividing the intercept by the slope.

Compressible Cakes

As mentioned before, most chemical precipitates form compressible filter cakes, in which higher compressive force deforms the solid particles, breaks up flocculent aggregates, and forms the particles closer together. Characteristics of porosity and specific resistance are that porosity decreases and specific resistance increases rapidly with increasing pressure in proportion to compressibility of the cake. The relationship between fluxes (flow rate per unit area) and pressure drop across the cake with respect to compressibility of the cake is schematically shown by Tiller and Crump [15], (Fig. 7.12).

In the development of filtration theory, porosity plays a fundamental role in its relation to flow rate, pressure drop, and other parameters involved in the differential equations of flow through compressible, porous media. Porosity variation determines the average porosity and liquid content of the filter cake in commercial operation. Since a dry cake is frequently desired, it is important for design purposes to know how the average porosity varies with total pressure. Because of the nonanalytical functions involved in dealing with compressible materials, it is generally necessary to relate the specific resistance α and the porosity ϵ to the pressure drop across the cake with empirical relationships.

For practical purposes, Tiller and Cooper [18] have found empirically that it is possible to represent both porosity ϵ and local specific resistance by power functions of the compressible pressure p_s and ϵ and α are assumed to be constant values. If p_s is above some low pressure designated as $p_0 (p_s > p_0)$, then the following model can be used:

$$\alpha = ap_s{}^n \tag{7.7.14a}$$

$$\epsilon = Ep_s{}^{-\lambda} \tag{7.7.14b}$$

$$1 - \epsilon = Bp_s{}^\beta \tag{7.7.14c}$$

If $p_s < p_0$, then the model becomes

$$\alpha = \alpha_0 = ap_0{}^n \tag{7.7.15a}$$

$$\epsilon = \epsilon_0 = Ep_0{}^n \tag{7.7.15b}$$

$$1 - \epsilon = 1 - \epsilon_0 = Bp_0{}^\beta \tag{7.7.15c}$$

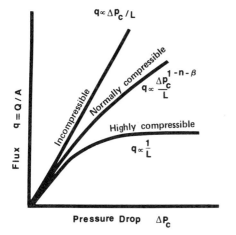

Fig. 7.12   Schematic diagram of flux vs. pressure drop across the filter cake; n and β are constants defined in (7.2.14) and (7.2.15). *Source:* Reproduced from Ref. 15 by permission of AIChE and the authors.

where p_0 is a low pressure in the range of 6.89×10^2 to $6.89 \times 10^3 \, \text{N/m}^2$ (0.1 to $1.0 \, \text{lb/in.}^2$). The approximations represented by these equations result from the linearity of logarithmic plots of α and ϵ versus p_s in the region up to $6.8 \times 10^5 \, \text{N/m}^2$ and for values of n less than roughly 0.5 to 0.7. The quantities α_0, ϵ_0, B, n, and β are empirical constants. While it may seem peculiar to represent both ϵ and $1 - \epsilon$ by different power functions, the relatively limited changes in porosity permit the data to be accurately rectified by (7.7.14c) and (7.7.15c).

The integration of (7.7.10) or (7.7.11) requires the writing of $-\Delta p_T$ as a differential function of p, which gives

$$\frac{dV}{dt} = \frac{A^2}{\mu w V} \int_0^{\Delta p_T} \frac{dp}{\alpha} \tag{7.9.16}$$

If one writes w in a differential form for per unit area, one obtains from (7.7.7)

$$dw = (1 - \epsilon)\rho_s \, dL \tag{7.9.17}$$

For relatively dilute slurries and long filtration cycles, $dV/dt = Q$ may be considered as approximately constant. Under these conditions, combining (7.7.16) and (7.7.17) will yield

$$\int_0^L dL = L = \frac{A^2}{\mu \rho_s Q} \int_0^{\Delta p_T} \frac{dp}{\alpha(1 - \epsilon)} \tag{7.7.18}$$

The practicality of (7.7.14) and (7.7.15) is related to two integrals in (7.7.17) and (7.7.18) which occur with regularity in filtration theory. The integral in (7.7.17) can be obtained as follows:

$$
\begin{aligned}
I_1 &= \int_0^{\Delta p_T} \frac{dp_s}{\alpha} = \int_0^{p_i} \frac{dp_s}{\alpha_0} + \int_{p_i}^{\Delta p_T} \frac{dp_s}{\alpha p_s{}^n} \\[2mm]
&= \frac{p_0{}^{1-n}}{\alpha_0} + \frac{\Delta p_T{}^{1-n} - p_0{}^{1-n}}{\alpha_0 p_0{}^n} \\[2mm]
&= \frac{\Delta p_T{}^{1-n} - np_i{}^{1-n}}{\alpha_0(1 - n)}
\end{aligned}
\tag{7.7.19}
$$

If n is 0.6 or less, the term in p_i can be neglected for total pressure above $6.89 \times 10^4 \, \text{N/m}^2$ ($10 \, \text{lb./in.}^2$). However, if n is as large as 0.7, the p_i term must be included. When n becomes large, the power function approximation is less accurate. When p_i is neglected, (7.7.19) reduces to

$$I_1 = \frac{\Delta p_T^{1-n}}{\alpha_0(1-n)} \tag{7.7.19a}$$

A precisely equivalent procedure for the integral in (7.7.18) yields

$$I_2 = \int_0^{\Delta p_T} \frac{dp_s}{\alpha(1-\epsilon)} = \frac{1}{\alpha_0 B} \frac{\Delta p_T^{1-n-\beta} - (n+\beta)p_0^{1-n-\beta}}{1-n-\beta} \tag{7.7.20}$$

which reduces to

$$I_2 = \frac{\Delta p_T^{1-n-\beta}}{\alpha_0 B(1-n-\beta)} \tag{7.7.20a}$$

In the evaluation of I_1 and I_2, the pressure loss at the interface of the supporting medium and the cake has been neglected. These integrals can be used for deriving many practical results involving cake filtration. Other empirical forms which have been found for most of these precipitates are

$$\alpha = \alpha_0 + a p_s^n \tag{7.7.21}$$

or

$$\alpha = \alpha_0 \left(\frac{1+p_s}{p_0}\right)^n \tag{7.7.21a}$$

Equation 7.7.21 has the disadvantage of the difficulty of integrating the form dp_s/α. Equation 7.7.21a suffers from difficulty in obtaining the parameters as compared with (7.7.14) and (7.7.15) used by Tiller and Cooper [18].

The relations between p and α, ϵ, and s_0 can be experimentally determined by "compressibility-permeability" experiments. The slurry to be tested is enclosed in a cylinder with a porous bottom, and a filter cake is built up on this porous surface by letting the filtrate drain through it. A piston is then applied with an increasing series of weights. At each piston loading, the porosity is determined by noting the piston position. Filtrate is fed to the filter cake and the value of α determined by solving (7.7.16) assuming α to be constant at any given loading. Then, s_0 can be calculated from the values of α and ϵ already determined. Figure 7.13 shows typical compression-permeability data of Grace [19], giving α, ϵ, s_0, and K_1 [$= \Delta V/(V \Delta p)$], cake compressibility, as functions of Δp. Grace has shown that the results of such experiments can be used under industrial filtration conditions.

In process calculations, the average liquid content is of more interest than

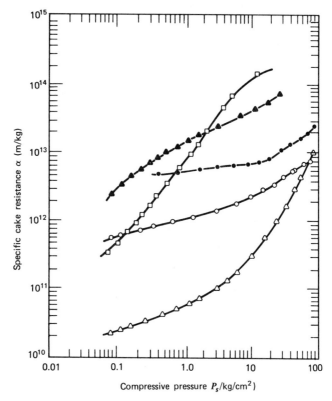

Fig. 7.13 Compression-permeability data (recalculated from Ref. 19 to cgs units). (*a*) Specific resistance (*above*). (*b*) Porosity as a function of specific surface (*right*). (*c*) Cake porosity (*see p. 422*). (○) E. & A. Kaolin, 50 g/liter 0.01 *M* Al$_2$[SO$_1$)$_2$. (●) E. & A. Kaolin, 50 g/liter 0.01 *M* Na$_1$P$_2$O$_7$. (△) G-60 Darco, 50 g/liter 0.01 *M* Na$_1$P$_2$O$_7$. (▲) Iron blue pigment, plant filter feed.

the distribution or profile of porosity with respect to distance. The average porosity can be obtained by

$$\epsilon_{avg} = \frac{1}{L} \int_0^L \epsilon \, dL \tag{7.7.22}$$

or the equation developed by Tiller and Cooper [20] and Shirato and Okamura [21, 22, 23]:

$$\epsilon_{avg} = 1 - \frac{\displaystyle\int_0^{\Delta p_T} dp_s/\epsilon}{\displaystyle\int_0^{\Delta p_T} dp_s/[\alpha(1 - \epsilon)]} \tag{7.7.23}$$

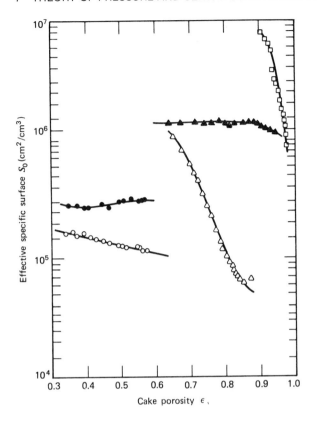

Fig. 7.13 (*b*).

Substituting (7.7.19) and (7.7.20) into (7.7.23), one obtains

$$\epsilon_{avg} = 1 - B\left(\frac{1-n-\beta}{1-n}\right)\left[\frac{p^{1/n} - np_0^{1-n}}{p^{1-n-\beta} - (n+\beta)p_0^{1-n-\beta}}\right] \quad (7.7.24)$$

or

$$1 - \epsilon_{avg} = Bp^{\beta}\left(\frac{1-n-\beta}{1-n}\right)\left[1 + \frac{\beta}{(p/p_i)^{1-n} - (n+\beta)(p/p_i)^{\beta}}\right] \quad (7.7.25)$$

As the pressure increases, in many cases the factor containing p_i can be neglected, and (7.7.25) reduces to

$$1 - \epsilon_{avg} = B\left(\frac{1-n-\beta}{1-n}\right)p^{\beta} \quad (7.7.25a)$$

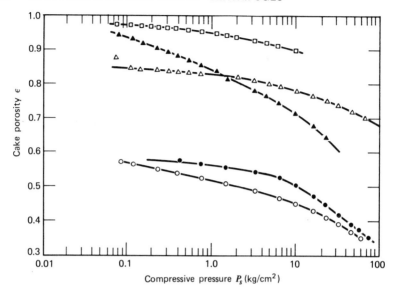

Fig. 7.13 (c).

Using (7.7.14c), one can write (7.7.25a) as

$$1 - \epsilon_{avg} = \left(\frac{1 - n - \beta}{1 - n} \right)(1 - \epsilon) = \text{const}\,(1 - \epsilon) \qquad (7.7.25b)$$

where ϵ_{avg} and ϵ are compared at equal pressures. In accordance with (7.7.25b), ϵ_{avg} will be given approximately as a linear function of ϵ, or $(1 - \epsilon_{avg})$ will be linear in $(1 - \epsilon)$. Logarithmic plots of $(1 - \epsilon_{avg})$ and $(1 - \epsilon)$ versus pressure will be parallel and separated by the logarithm of the constant in (7.7.25b). It should be noted that ϵ is the local porosity of a given pressure and ϵ_{avg} is the average porosity of a cake at the same pressure.

Various effects on porosity in filtration have been studied by many investigators. Tiller and his associates have investigated the effects of side-wall friction [24], cake nonuniformity effect [25], and skin effect with highly compressible materials [26].

Centrifugal Filtration

A number of factors exist in both theory and practice which show the differences between a perforated centrifugal filter and conventional vacuum or pressure filter [27]. Primarily, the geometry of the centrifugal bowl results in an effective filtration area which decreases as cake thickness increases, and this results in a change in superficial fluid velocity through the cake. The driving force causing filtration not only is the pressure gradient across the cake but also includes the centrifugal force on the liquid as it flows through the cake. The centrifugal field is not uniform as in gravity filtration but varies considerably with bowl radius. The compressive pressure acting on cake solids at any point in the cake results not only from friction loss of the fluid flowing, but also from action of the centrifugal field on the mass of cake solids and from kinetic energy change in the fluid flowing through the cake.

In a typical basket filter, the cake forms on the inside vertical walls of the filter. The filter area decreases as the cake thickens, and the kinetic energy of the flowing filtrate also changes. These differences are illustrated in Fig. 7.14

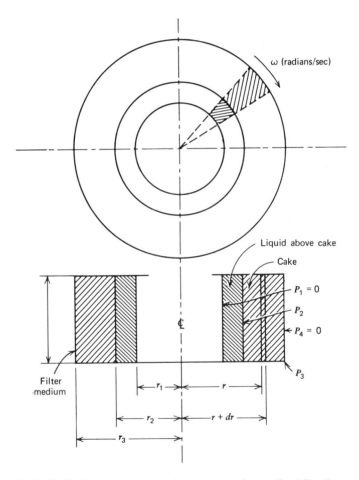

Fig. 7.14 Physical arrangement and nomenclature in centrifugal filtration.

showing the physical situation and the nomenclature to be used in the following discussion [17, 27].

A differential pressure balance about a very thin layer of filter cakes gives

$$-dp_T = -dp_c - dp_v - dp_f \qquad (7.7.26)$$

where $-dp_T$ is the total effective pressure drop, $-dp_c$ is the hydraulic-pressure gradient developed by fluid in the centrifugal field flowing through the cake, $-dp_v$ is the hydraulic pressure gradient as a result of kinetic energy changes in passing through the fluid, and $-dp_f$ is the hydraulic pressure gradient as a result of frictional drag. For this differential cake, the area through which flow passes is

$$A = 2\pi L r \qquad (7.7.27)$$

A solid in a differential thickness dr of the filter cake is

$$dW = w\,dV = \rho_s (1 - \epsilon)\, 2\pi L r\, dr \qquad (7.7.28)$$

Where dW is the differential weight of the solids, w is the weight of the solids per unit volume of original slurry, and dV is the increase in filtrate volume in a differential time span dt. Then the linear radial velocity of filtration flowing through the differential cake thickness dr is obtained by differentiating the radial velocity component of the filtrate $v_r = Q/(2\pi L r)$, which yields

$$d\mathbf{v}_r = -\frac{Q}{2\pi L}\frac{dr}{r^2} = -\left(\frac{dV}{dt}\right)\frac{dr}{2\pi L r^2} \qquad (7.7.29)$$

The various terms of (7.7.26) may be written in terms of particular nomenclature and physical arrangement of centrifugal filtration as depicted in Fig. 7.14. The pressure gradient resulting from kinetic energy variation must be

$$dp_v = \frac{\rho_m V\,dV}{g_c} = -\frac{\rho_m}{g_c}\left(\frac{Q}{2\pi L}\right)^2\frac{dr}{r^3} \qquad (7.7.30)$$

The pressure gradient resulting from hydraulic head is $dp_g = (g/g_c)dz$. Since $\mathbf{g} = \omega^2 r$ and the centrifugal acceleration acts on the liquid in the direction of flow, dz is replaced by dr. As a result, one has for the centrifugal filtration

$$dp_c = \frac{\rho_m \omega^2 r\, dr}{g_c} \qquad (7.7.31)$$

The pressure drop resulting from drag and skin friction can be obtained by appropriately replacing flow area, volume, and weight of the solids of (7.7.8) by corresponding expressions for a differential cake thickness dr:

$$-dp_f = \frac{\rho_s(1 - \epsilon)\alpha\mu}{g_c 2\pi L}\left(\frac{dV}{dt}\right)\frac{dr}{r} = \frac{\mu\rho_s Q\alpha(1 - \epsilon)dr}{g_c 2\pi Lr} \qquad (7.7.32)$$

Substituting (7.7.30), (7.7.31), and (7.7.32) into (7.7.26) gives

$$-\int_{p_2}^{p_3} dp = \frac{-\rho_m\omega^2}{g_c}\int_{r_2}^{r_3} r\, dr + \frac{\rho_m}{g_c}\left(\frac{Q}{2\pi L}\right)^2\int_{r_2}^{r_3}\frac{dr}{r^3}$$

$$+ \frac{\mu\rho_s Q}{g_c 2\pi L}\int_{r_2}^{r_3}\frac{\alpha(1 - \epsilon)}{r}\, dr \qquad (7.7.33)$$

where p_2 and p_3 are pressures at the filter cake surface and at the filter cloth surface, respectively; and r_2 and r_3 are radii from the rotational axis to the filter cake surface and to the filter cloth surface, respectively.

Equation 7.7.33 is the basic rate equation for centrifugal filtration and gives the pressure drop across the filter cake. The pressure at the filter cloth surface p_3 can be related to the normally measured downstream pressure p_4 in terms of a filter medium resistance as was done for pressure filtration:

$$p_3 - p_4 = \frac{\left(\dfrac{dV}{dt}\right)\mu R_M{}'}{2\pi Lr_3 g_c} \qquad (7.7.34)$$

where p_4 is pressure at the low-pressure side of the filter cloth and $R_M{}'$ is the resistance of the filter medium, analogous to the medium resistance given in (7.9.10). The pressure at the cake surface p_2 is

$$p_2 - p_1 = \frac{\rho_m\omega^2(r_2^2 - r_1^2)}{2g_c} \qquad (7.7.35)$$

where r_1 is the radius to the surface of the liquid over the filter cake. Thus, replacing the limits of integration from p_2 to p_3 by p_1 to p_4, one obtains the basic general rate equation of centrifugal filtration. If a gauge pressure is used, p_1 and p_4 are atmospheric pressures. Thus, $p_1 = p_4 = 0$.

For the centrifugal filtration case, $-dp \neq dp_s$, since compressive pressures other than the pressure drop act on the cake solids. In this instance the mechanical compressive pressure acting on the cake solids in the direction of flow results from both frictional drag dp_f and from a direct mechanical compressive pressure dp_w due to the weight of cake solids suspended in liquid under the centrifugal field. Thus, the mechanical compressive pressure acting on the cake solids is

$$dp_s = dp_w + dp_f \qquad (7.7.36)$$

The quantity dp_w resulting from centrifugal accleration $\omega^2 r$ acting on cake solids is

$$dp_w = \frac{(\rho_s - \rho_f)(1 - \epsilon)\omega^2 r\, dr}{\rho_m g_c} \qquad (7.7.37)$$

Using dp_f from (7.7.26) as $dp_f = -dp_T + dp_c + dp_v$, (7.7.36) may be rewritten as

$$p_s = \int_0^{p_s} dp_s = \frac{(\rho_s - \rho_m)\omega^2}{g_c} \int_{r_2}^{r_3} (1 - \epsilon) r\, dr - \int_{p_2}^{p_3} dp$$

$$+ \frac{\rho_m \omega^2}{g_c} \int_{r_2}^{r_3} r\, dr + \frac{\rho_m}{g_c} \left(\frac{Q}{2\pi L}\right)^2 \int_{r_2}^{r_3} \frac{dr}{r^3} \qquad (7.7.38)$$

Substituting (7.7.32) and (7.7.37) into (7.7.36) and rearranging give

$$\int_0^{p_s} \frac{dp_s}{1 - \epsilon} = \frac{(\rho_s - \rho_m)\omega^2}{g_c} \int_{r_2}^{r_3} r\, dr - \frac{\mu \rho_s Q}{g_c 2\pi L} \int_{r_2}^{r_3} \alpha \frac{dr}{r} \qquad (7.7.39)$$

The limit of the mechanical compressive pressure acting on the cake solids in (7.7.39) could be obtained from (7.7.38). Integration of (7.7.38) is not possible without a knowledge of variation of α and ϵ with r. Thus, solution in the general case is not possible from compression permeability data alone.

Special Case: Nearly Incompressible Cake

For the special case of a nearly incompressible cake, solution of (7.7.33) is possible. For this simplified case, $dp_w \ll dp_f$, as might result with low ρ_s solids or thin cakes and considerable liquid head carried above the cake. Then one substitutes (7.7.34) and (7.7.35) into (7.7.33), integrates with $p_1 = p_4 = 0$, and rearranges the result to obtain

$$Q = \frac{dV}{dt} = \frac{\pi L \rho_m \omega^2 (r_3{}^2 - r_1{}^2) - \dfrac{\rho(dV/dt)^2}{4\pi L}\left(\dfrac{1}{r_3{}^2} - \dfrac{1}{r_1{}^2}\right)}{\mu \left[\alpha \rho_s (1 - \epsilon) \ln \dfrac{r_3}{r_2} + \dfrac{R_M{}'}{r_3}\right]} \qquad (7.7.40)$$

The second term in the numerator derives from the changes in fluid kinetic energy. It is almost always very small and can be neglected. Noting this and that $\omega = \pi n/30$ (n is rpm), one can write the final rate equation for this simplified case as

$$Q = \frac{\pi^3 n^2 \rho_m L (r_3{}^2 - r_1{}^2)}{(30)^2 \mu \left[\alpha \rho_s (1 - \epsilon) \ln \dfrac{r_3}{r_2} + \dfrac{R_M{}'}{r_3} \right]} \tag{7.7.40a}$$

In order to compare this simplified centrifugal filtration rate equation with the standard filtration rate equation (7.7.10), one can consider the total hydraulic driving force Δp_T to replace Δp, where

$$\Delta p_T = \frac{\rho_m \omega^2 (r_3{}^2 - r_1{}^2)}{2g_c} \tag{7.7.41}$$

Noting from (7.7.7) and neglecting the volume of filtrate held in the filter cake term that

$$\rho_s (1 - \epsilon) = \frac{w}{\pi L (r_3{}^2 - r_2{}^2)} \tag{7.7.42}$$

and substituting (7.7.41) and (7.7.42) into (7.7.40a) gives

$$Q = \frac{(\Delta p_T) g_c}{\mu \left[\dfrac{\alpha w}{[2\pi L (r_3 + r_2)/2][2\pi L (r_3 - r_2)/\ln(r_3/r_2)]} + \dfrac{R_M{}'}{2\pi L r_3} \right]} \tag{7.7.43}$$

which can be written as

$$Q = \frac{(\Delta p_T) g_c}{\mu \left(\dfrac{\alpha w}{A_m A_{lm}} + \dfrac{R_M{}'}{A_c} \right)} \tag{7.7.44}$$

Thus, it appears that the A^2 term in (7.7.10) for the cake resistance term should be made up of $A_m \times A_{lm}$ while the area of the filter cloth A_c is correct as a replacement for A when dealing with the resistance medium. The quantities A_m and A_{lm} are defined as

$$A_m = 2\pi L \left(\frac{r_3 + r_2}{2} \right) = \text{area based on arithmetic radius} \qquad (7.7.44a)$$

$$A_{lm} = 2\pi L \left(\frac{r_3 - r_2}{\ln \frac{r_3}{r_2}} \right) = \text{area based on logarithmic radius} \qquad (7.7.44b)$$

SYMBOLS

a	Empirical constant
$\mathbf{a}_E$	Acceleration due to external force
A	Area
b	Parameter as defined in (7.3.13)
B	Empirical constant
$C_\sigma^{(i)}$	Discharge coefficient with surface tension effect for the ith geometry
C_i	Concentration at interface
$\mathbf{C_D}$	Drug coefficient
C_Q	Volumetric flow discharge coefficient as defined in (7.4.1)
d	Diameter of nozzle
D_p	Diameter of particle
E	Empirical constant
f	Fanning friction factor
f_p	Fanning friction factor for packed bed
$\mathbf{F}_B$	Buoyant force
$\mathbf{F}_D$	Drag force
$\mathbf{F}_E$	External force
g_c	Unit conversion factor
H	Thickness of fluid layer
I_1	Quantity as defined in (7.7.19)
I_2	Quantity as defined in (7.7.20)
l	Length of nozzle
L	Length or height of centrifuge
m	Mass
n	Empirical constant
n	Revolutions per minute

N	Total number of cones in centrifuge
p_i	Centrifugal pressure in the radial direction at interface
p_L	Centrifugal pressure in the radial direction at free surface of light liquid at r_L
p_H	Centrifugal pressure in the radial direction at free surface of heavy liquid at r_H
p_0	A reference low pressure
p_s	Compressible pressure
$\mathbf{q}$	Volumetric flow rate through a single passage in disk centrifuge
Q	Volumetric flow rate
r	Radial variable in cylindrical coordinate
r_i	Radial position of interface
r_L	Radial position of light liquid-free surface
r_H	Radial position of heavy liquid-free surface
r_f	Radial position of liquid-free surface
r_A	Radial position at $t = 0$
r_B	Radial position at $t = t_f$
R	Radius of top or bottom of rotor
$\mathbf{Re}$	Reynolds number $(=D_p\, v\rho/\mu)$
s	Distance between cones
s_0	Specific area of particle
t	Time
T	Temperature
$\mathbf{v}$	Velocity
$\mathbf{v}_c$	Velocity due to centrifugal force
$\mathbf{v}_g$	Velocity due to gravitational force
$\mathbf{v}_p$	Velocity of particle
$\mathbf{v}_p{}',\ \mathbf{v}_p{}''$	Velocity component of particle as shown in Fig. 7.2a
$\mathbf{v}_r$	Velocity in radial direction
$\mathbf{v}_z$	Velocity in axial direction
V	Volume
V_e	Fictitious equivalent volume of filtrate
w	Weight of solid
$\mathbf{We}$	Weber number $(=2\,\Delta pd/\sigma)$
x	Linear distance normal to and measured from top side of a cone

y	Sludge film thickness
Z	Centrifugal effect as defined in (2.1)

Greek Symbols

α	Contraction ratio
α	Specific filter cake resistance
α_c	Contraction coefficient
α, β, γ	Transformed orthogonal coordinate system as shown in Fig. 7.8
β	Empirical constant
ϵ	Void fraction of solid, porosity
θ	Cone angle
λ	Empirical constant
μ	Dynamic viscosity
ξ	Relative roughness of nozzle surface
ρ_L, ρ_H	Density of light and heavy liquid, respectively
ρ_m	Density of fluid medium
ρ_p	Density of particle
ρ_s	Density of solid cake
σ	Surface tension
Σ_c	Centrifugal settling area $(=ZV/H)$
ϕ	Linear distance measured along cone
ω	Angular velocity
ϕ	Correction factor for a hindered settling velocity as defined in (7.5.5)

Subscripts

A, B	Indication of positions at $t = 0$ and $t = t_f$, respectively
c	Quantity due to centrifugation, cutoff, or hydraulic loss through cake
f	Final time or quantity due to frictional losses
g	Quantity due to gravitational force
H	Heavy liquid
i	Interface
L	Light liquid
m	Mass
n	Independent variable in finite difference

T	Total quantity
v	Quantity due to kinetic energy loss
1	Top end of cone or the first in a series
2	Bottom end of cone or the second in a series
1, 2, 3, 4	Indication of position at respective interface
0	Reference state

Superscripts

n	Empirical constant
β	Empirical constant
λ	Empirical constant

References

1. D. Kunii and O. Levenspiel, *Fluidization Engineering,* Wiley, New York, 1969, p. 76.
2. C. M. Ambler, *Chem. Eng. Progr.*, **48**, 150 (1952).
3. W. L. McCabe and J. C. Smith, *Unit Operations of Chemical Engineering,* 3rd ed., McGraw-Hill, New York, 1976, p. 967.
4. S. H. Jury and W. L. Locke, *A.I.Ch.E. J.*, **3** (4), 480 (1957).
5. H. Kuwai and I. Inoue, *Chem. Machinery* (Japan), **14**, 90 (1950).
6. I. Inoue and H. Kojima, *Kagaku Kogaku* (Japan), **22** (6), 357 (1958).
7. I. Inoue, *Kagaku Kogaku* (Japan), **22** (3), 143 (1958).
8. T. Takamatsu, T. Takahashi, and H. Shoji, *Kagaku Kogaku* (Japan) **22** (9), 555 (1958).
9. T. Takamatsu, T. Takahashi, S. Shiga, and H. Shoji, *Kagaku Kogaku* (Japan), **22** (9), 569 (1958).
10. T. Takamatsu, T. Takahashi, and K. Tanaka, *Kagaku Kogaku* (Japan) **32** (9), 890 (1968).
11. R. M. Olson, *Engineering Fluid Mechanics,* International Text Book Co., Scranton, Penn., 1967.
12. S. Goldstein, *Modern Developments in Fluid Dynamics,* Vol. 1, Oxford Univ. Press, 1938, pp. 101, 297.
13. J. R. Schnittger, *Ind. Eng. Chem. Process Des. Develop.*, **9** (3), 407 (1970).
14. F. A. Zens and D. F. Othmer, *Fluidization and Fluid Particle System, Chemical Engineering Series,* Rheinhold, New York, 1963, p. 237.
15. F. M. Tiller and J. R. Crump, *Chem. Eng. Progr.*, **73** (10), 65 (1977).
16. R. A. Greenkorn and D. P. Kessler, *Transfer Operations,* McGraw-Hill, New York, 1972, p. 257.
17. A. S. Foust, L. A. Wenzel, C. W. Clump, L. Mans, and L. B. Anderson, *Principles of Unit Operations,* Wiley, New York, 1962, p. 473.
18. F. M. Tiller and H. Cooper, *A.I.Ch.E. J.*, **8** (4), 445 (1962).
19. H. P. Grace, *Chem. Eng. Prog.*, **49** (6), 303, 367 (1953).

20. F. M. Tiller and H. Cooper, *A.I.Ch.E. J.*, **6** (4), 595 (1960).
21. M. Shirato and S. Okamura, *Kagaku Kogaku* (Japan), **19**, 104, 111 (1955).
22. M. Shirato and S. Okamura, *Kagaku Kogaku* (Japan), **20**, 98, 678 (1956).
23. M. Shirato and S. Okamura, *Kagaku Kogaku* (Japan), **23**, 11, 226 (1959).
24. F. M. Tiller, S. Haynes, Jr., and W.-M. Lu, *A.I.Ch.E. J.*, **18** (1), 13 (1972).
25. F. M. Tiller and W.-M. Lu, *A.I.Ch.E. J.*, **18** (3), 569 (1972).
26. F. M. Tiller and T. C. Green, *A.I.Ch.E. J.*, **19** (6), 1266 (1973).
27. H. P. Grace, *Chem. Eng. Progr.*, **49** (8), 427 (1953).

TABLES OF DENSITY AND VISCOSITY OF SUCROSE AQUEOUS SOLUTIONS

Sucrose (%)	Density^a (g/ml)	Viscosity^b (cp)	Sucrose (%)	Density^a (g/ml)	Viscosity^b (cp)
Temperature 30.0°C					
0	0.9963	0.7978	28	1.1136	2.143
1	1.0000	0.8176	29	1.1183	2.255
2	1.0038	0.8384	30	1.1230	2.376
3	1.0075	0.8601	31	1.1278	2.506
4	1.0113	0.8830	32	1.1326	2.648
5	1.0152	0.9069	33	1.1374	2.802
6	1.0191	0.9322	34	1.1423	2.970
7	1.0230	0.9588	35	1.1472	3.153
8	1.0270	0.9868	36	1.1522	3.353
9	1.0310	1.016	37	1.1572	3.572
10	1.0350	1.048	38	1.1622	3.813
11	1.0391	1.081	39	1.1672	4.079
12	1.0432	1.116	40	1.1723	4.372
13	1.0473	1.154	41	1.1775	4.697
14	1.0515	1.193	42	1.1826	5.058
15	1.0557	1.235	43	1.1878	5.461
16	1.0599	1.280	44	1.1931	5.912
17	1.0642	1.328	45	1.1983	6.418
18	1.0685	1.380	46	1.2036	6.988
19	1.0728	1.434	47	1.2090	7.632
20	1.0772	1.493	48	1.2144	8.344
21	1.0816	1.555	49	1.2198	9.168
22	1.0861	1.622	50	1.2252	10.10
23	1.0906	1.694	51	1.2307	11.18
24	1.0951	1.771	52	1.2362	12.42
25	1.0997	1.854	53	1.2418	13.84
26	1.1043	1.943	54	1.2474	15.50
27	1.1089	2.039	55	1.2530	17.43

Sucrose (%)	Density^a (g/ml)	Viscosity^b (cp)	Sucrose (%)	Density^a (g/ml)	Viscosity^b (cp)
56	1.2586	19.69	64	1.3052	64.53
57	1.2643	22.36	65	1.3112	77.35
58	1.2701	25.52	66	1.3172	93.54
59	1.2758	29.30	67	1.3232	114.2
60	1.2816	33.84	68	1.3293	140.9
61	1.2875	39.34	69	1.3354	175.8
62	1.2933	46.05	70	1.3416	222.0
63	1.2992	54.30			

Temperature 25.0 °C

1	1.0014	0.9139	33	1.1395	3.2493
2	1.0052	0.9376	34	1.1444	3.4512
3	1.0090	0.9625	35	1.1493	3.6721
4	1.0128	0.9886	36	1.1543	3.9143
5	1.0167	1.0162	37	1.1593	4.1808
6	1.0206	1.0452	38	1.1643	4.4746
7	1.0245	1.0758	39	1.1694	4.7994
8	1.0285	1.1082	40	1.1745	5.1596
9	1.0325	1.1424	41	1.1797	5.5604
10	1.0366	1.1787	42	1.1848	6.0079
11	1.0407	1.2171	43	1.1901	6.5088
12	1.0448	1.2579	44	1.1953	7.0715
13	1.0489	1.3013	45	1.2006	7.7057
14	1.0531	1.3474	46	1.2059	8.4226
15	1.0574	1.3965	47	1.2113	9.2354
16	1.0616	1.4488	48	1.2167	10.1386
17	1.0659	1.5047	49	1.2221	11.1884
18	1.0702	1.5644	50	1.2276	12.3906
19	1.0746	1.6282	51	1.2331	13.7732
20	1.0790	1.6965	52	1.2386	15.3705
21	1.0835	1.7697	53	1.2442	17.2251
22	1.0879	1.8483	54	1.2498	19.3894
23	1.0924	1.9326	55	1.2554	21.9288
24	1.0970	2.0233	56	1.2611	24.9251
25	1.1016	2.1209	57	1.2668	28.4828
26	1.1062	2.2260	58	1.2726	32.7334
27	1.1108	2.3395	59	1.2784	37.8474
28	1.1155	2.4621	60	1.2842	44.0442
29	1.1202	2.5948	61	1.2901	51.6111
30	1.1250	2.7386	62	1.2959	60.9275
31	1.1298	2.8946	63	1.3019	72.4955
32	1.1346	3.0643	64	1.3078	86.9939

Sucrose (%)	Density[a] (g/ml)	Viscosity[b] (cp)	Sucrose (%)	Density[a] (g/ml)	Viscosity[b] (cp)
65	1.3138	105.3420	68	1.3320	198.8241
66	1.3199	128.8052	69	1.3382	251.3983
67	1.3259	159.1388	70	1.3444	321.9836

Temperature 20.0 °C

1	1.0026	1.0296	36	1.1563	4.6205
2	1.0064	1.0569	37	1.1613	4.9485
3	1.0102	1.0856	38	1.1663	5.3115
4	1.0140	1.1159	39	1.1714	5.7144
5	1.0179	1.1478	40	1.1766	6.1630
6	1.0219	1.1815	41	1.1817	6.6641
7	1.0258	1.2171	42	1.1870	7.2259
8	1.0298	1.2548	43	1.1922	7.8574
9	1.0339	1.2947	44	1.1975	8.5698
10	1.0380	1.3371	45	1.2028	9.3760
11	1.0421	1.3820	46	1.2081	10.2914
12	1.0462	1.4297	47	1.2135	11.3338
13	1.0504	1.4806	48	1.2189	12.4979
14	1.0546	1.5347	49	1.2244	13.8568
15	1.0588	1.5923	50	1.2299	15.4209
16	1.0631	1.6539	51	1.2354	17.2296
17	1.0675	1.7196	52	1.2409	19.3311
18	1.0718	1.7899	53	1.2465	21.7855
19	1.0762	1.8652	54	1.2522	24.6681
20	1.0806	1.9459	55	1.2578	28.0727
21	1.0851	2.0325	56	1.2635	32.1183
22	1.0896	2.1255	57	1.2693	36.9575
23	1.0941	2.2255	58	1.2750	42.7847
24	1.0987	2.3332	59	1.2808	49.8539
25	1.1033	2.4493	60	1.2867	58.4957
26	1.1079	2.5747	61	1.2926	69.1469
27	1.1126	2.7103	62	1.2985	82.3910
28	1.1173	2.8571	63	1.3044	99.0098
29	1.1221	3.0164	64	1.3104	120.0721
30	1.1268	3.1894	65	1.3164	147.0460
31	1.1316	3.3777	66	1.3225	181.9787
32	1.1365	3.5831	67	1.3286	227.7530
33	1.1414	3.8076	68	1.3347	288.5073
34	1.1463	4.0535	69	1.3408	370.2395
35	1.1513	4.3234	70	1.3470	481.7917

Sucrose (%)	Density[a] (g/ml)	Viscosity[b] (cp)	Sucrose (%)	Density[a] (g/ml)	Viscosity[b] (cp)
Temperature 15.0°C					
1	1.0034	1.1702	36	1.1581	5.5221
2	1.0073	1.2019	37	1.1631	5.9315
3	1.0111	1.2354	38	1.1682	6.3864
4	1.0150	1.2708	39	1.1734	6.8934
5	1.0189	1.3082	40	1.1785	7.4604
6	1.0229	1.3477	41	1.1837	8.0965
7	1.0269	1.3895	42	1.1889	8.8126
8	1.0309	1.4339	43	1.1942	9.6213
9	1.0350	1.4809	44	1.1995	10.5377
10	1.0391	1.5309	45	1.2048	11.5795
11	1.0432	1.5840	46	1.2102	12.7679
12	1.0474	1.6405	47	1.2156	14.1276
13	1.0516	1.7007	48	1.2211	15.6545
14	1.0558	1.7648	49	1.2265	17.4452
15	1.0601	1.8333	50	1.2320	19.5179
16	1.0644	1.9064	51	1.2376	21.9286
17	1.0688	1.9847	52	1.2432	24.7470
18	1.0732	2.0685	53	1.2488	28.0603
19	1.0776	2.1582	54	1.2544	31.9782
20	1.0820	2.2546	55	1.2601	36.6392
21	1.0865	2.3582	56	1.2658	42.2200
22	1.0911	2.4696	57	1.2716	48.9497
23	1.0956	2.5896	58	1.2774	57.1228
24	1.1002	2.7191	59	1.2832	67.1281
25	1.1048	2.8590	60	1.2891	79.4768
26	1.1095	3.0103	61	1.2950	94.8528
27	1.1142	3.1743	62	1.3009	114.1797
28	1.1189	3.3523	63	1.3069	138.7120
29	1.1237	3.5458	64	1.3129	170.1869
30	1.1285	3.7567	65	1.3189	211.0244
31	1.1334	3.9869	66	1.3250	264.6538
32	1.1382	4.2387	67	1.3311	335.9805
33	1.1431	4.5148	68	1.3373	432.1655
34	1.1481	4.8183	69	1.3434	563.7854
35	1.1531	5.1527	70	1.3497	746.7412
Temperature 10.0°C					
1	1.0040	1.3430	4	1.0157	1.4617
2	1.0079	1.3804	5	1.0196	1.5059
3	1.0118	1.4199	6	1.0236	1.5528

Sucrose (%)	Density[a] (g/ml)	Viscosity[b] (cp)	Sucrose (%)	Density[a] (g/ml)	Viscosity[b] (cp)
7	1.0277	1.6025	39	1.1752	8.4383
8	1.0317	1.6553	40	1.1803	9.1667
9	1.0358	1.7114	41	1.1856	9.9878
10	1.0400	1.7711	42	1.1908	10.9164
11	1.0441	1.8345	43	1.1961	11.9701
12	1.0484	1.9022	44	1.2014	13.1700
13	1.0526	1.9743	45	1.2068	14.5409
14	1.0569	2.0513	46	1.2122	16.1127
15	1.0612	2.1336	47	1.2176	17.9204
16	1.0655	2.2216	48	1.2231	19.9623
17	1.0699	2.3159	49	1.2286	22.3695
18	1.0743	2.4170	50	1.2341	25.1728
19	1.0788	2.5254	51	1.2397	28.4541
20	1.0833	2.6420	52	1.2453	32.3162
21	1.0878	2.7675	53	1.2509	36.8887
22	1.0923	2.9027	54	1.2566	42.3360
23	1.0969	3.0486	55	1.2623	48.8679
24	1.1016	3.2063	56	1.2681	56.7541
25	1.1062	3.3771	57	1.2739	66.3478
26	1.1109	3.5622	58	1.2797	78.1086
27	1.1157	3.7632	59	1.2855	92.6489
28	1.1204	3.9820	60	1.2914	110.7842
29	1.1252	4.2206	61	1.2973	133.6190
30	1.1301	4.4812	62	1.3033	162.6641
31	1.1349	4.7666	63	1.3093	200.0011
32	1.1398	5.0798	64	1.3153	248.5555
33	1.1448	5.4245	65	1.3214	312.4668
34	1.1498	5.8047	66	1.3275	397.6987
35	1.1548	6.2252	67	1.3336	512.9407
36	1.1598	6.6916	68	1.3398	671.1133
37	1.1649	7.2107	69	1.3460	891.7009
38	1.1700	7.7899	70	1.3522	1204.6399

Temperature 5.0 °C

Sucrose (%)	Density[a] (g/ml)	Viscosity[b] (cp)	Sucrose (%)	Density[a] (g/ml)	Viscosity[b] (cp)
1	1.0043	1.5583	8	1.0323	1.9329
2	1.0082	1.6028	9	1.0365	2.0006
3	1.0121	1.6500	10	1.0406	2.0728
4	1.0161	1.7000	11	1.0448	2.1497
5	1.0201	1.7530	12	1.0491	2.2318
6	1.0241	1.8094	13	1.0534	2.3194
7	1.0282	1.8692	14	1.0577	2.4131

Appendix A *(Continued)*

Sucrose (%)	Density[a] (g/ml)	Viscosity[b] (cp)	Sucrose (%)	Density[a] (g/ml)	Viscosity[b] (cp)
15	1.0620	2.5133	43	1.1979	15.1605
16	1.0664	2.6207	44	1.2033	16.7632
17	1.0708	2.7358	45	1.2087	18.6042
18	1.0753	2.8594	46	1.2141	20.7267
19	1.0798	2.9923	47	1.2195	23.1819
20	1.0843	3.1354	48	1.2250	25.9727
21	1.0889	3.2896	49	1.2306	29.2818
22	1.0935	3.4561	50	1.2361	33.1611
23	1.0981	3.6361	51	1.2417	37.7337
24	1.1028	3.8310	52	1.2474	43.1553
25	1.1075	4.0425	53	1.2530	49.6241
26	1.1122	4.2723	54	1.2587	57.3936
27	1.1169	4.5226	55	1.2645	66.7908
28	1.1217	4.7957	56	1.2702	78.2404
29	1.1266	5.0943	57	1.2760	92.3043
30	1.1315	5.4216	58	1.2819	109.7226
31	1.1364	5.7811	59	1.2877	131.4927
32	1.1413	6.1771	60	1.2937	158.9610
33	1.1463	6.6144	61	1.2996	193.9751
34	1.1513	7.0987	62	1.3056	239.0989
35	1.1563	7.6365	63	1.3116	297.9216
36	1.1614	8.2356	64	1.3176	375.5669
37	1.1665	8.9053	65	1.3237	479.4153
38	1.1717	9.6560	66	1.3298	620.2959
39	1.1768	10.5003	67	1.3360	814.3062
40	1.1821	11.4535	68	1.3422	1085.8899
41	1.1873	12.5331	69	1.3484	1472.7625
42	1.1926	13.7605	70	1.3546	2034.3164

Temperature 0.0°C

1	1.0043	1.8306	12	1.0496	2.6533
2	1.0082	1.8843	13	1.0539	2.7614
3	1.0122	1.9414	14	1.0583	2.8771
4	1.0162	2.0021	15	1.0627	3.0010
5	1.0203	2.0666	16	1.0671	3.1340
6	1.0244	2.1352	17	1.0716	3.2768
7	1.0285	2.2083	18	1.0761	3.4304
8	1.0326	2.2862	19	1.0806	3.5958
9	1.0368	2.3692	20	1.0852	3.7742
10	1.0411	2.4578	21	1.0898	3.9668
11	1.0453	2.5523	22	1.0944	4.1751

Appendix A   *(Continued)*

Sucrose (%)	Density[a] (g/ml)	Viscosity[b] (cp)	Sucrose (%)	Density[a] (g/ml)	Viscosity[b] (cp)
23	1.0991	4.4008	47	1.2214	30.6598
24	1.1038	4.6457	48	1.2269	34.5693
25	1.1085	4.9121	49	1.2324	39.2342
26	1.1133	5.2023	50	1.2380	44.7427
27	1.1181	5.5192	51	1.2436	51.2860
28	1.1229	5.8660	52	1.2493	59.1069
29	1.1278	6.2464	53	1.2550	68.5183
30	1.1327	6.6646	54	1.2607	79.9240
31	1.1376	7.1257	55	1.2665	93.8504
32	1.1426	7.6355	56	1.2723	110.9897
33	1.1476	8.2006	57	1.2781	132.2682
34	1.1527	8.8291	58	1.2840	158.9211
35	1.1578	9.5301	59	1.2899	192.6349
36	1.1629	10.3146	60	1.2958	235.7207
37	1.1680	11.1955	61	1.3018	291.3948
38	1.1732	12.1880	62	1.3078	364.1926
39	1.1784	13.3100	63	1.3138	460.5754
40	1.1837	14.5831	64	1.3199	589.9290
41	1.1889	16.0331	65	1.3260	766.0437
42	1.1943	17.6905	66	1.3321	1009.5676
43	1.1996	19.5917	67	1.3383	1351.8850
44	1.2050	21.7808	68	1.3445	1841.7961
45	1.2104	24.3104	69	1.3507	2556.5417
46	1.2159	27.2445	70	1.3570	3621.1423

Source: Calculations were programmed by Barbara S. Bishop based on equations developed by E. J. Barber in *J. Natl. Cancer Inst. Monograph 21*, 1966, p. 219, Ref. 49 in Chapter IV; equations (4.7.26) through (4.7.29). Reproduced from "Useful Data for Zonal Centrifugation," C. T. Rankin, Jr., and L. T. Elrod, The Molecular Anatomy Program, Oak Ridge National Laboratory, January 1970.
[a]Original data were stated to a precision of about 1 part in 10,000. Maximum deviation from original data is 7 parts in 10,000.
[b]Precision of original data was between 1 part in 1000 and 1 part in 10,000. Maximum deviation from original data is 4 parts in 1000 in the range covered in this set of tables.

DENSITY AT 25°C OF CsCl SOLUTION AS A FUNCTION OF REFRACTIVE INDEX

Refractive Index (Sodium D Line, 25°C)	Density (g/cm³)	Refractive Index (Sodium D Line, 25°C)	Density (g/cm³)	Refractive Index (Sodium D Line, 25°C)	Density (g/cm³)
1.34400	1.09857	1.35020	1.16591	1.35640	1.23324
1.34410	1.09966	1.35030	1.16690	1.35650	1.23433
1.34420	1.10075	1.35040	1.16808	1.35660	1.23541
1.34430	1.10183	1.35050	1.16917	1.35670	1.23650
1.34440	1.10292	1.35060	1.17025	1.35680	1.23758
1.34450	1.10400	1.35070	1.17134	1.35690	1.23867
1.34460	1.10509	1.35080	1.17242	1.35700	1.23976
1.34470	1.10618	1.35090	1.17351	1.35710	1.24084
1.34480	1.10726	1.35100	1.17460	1.35720	1.24193
1.34490	1.10835	1.35110	1.17568	1.35730	1.24301
1.34500	1.10943	1.35120	1.17677	1.35740	1.24410
1.34510	1.11052	1.35130	1.17785	1.35750	1.24519
1.34520	1.11161	1.35140	1.17894	1.35760	1.24627
1.34530	1.11269	1.35150	1.18003	1.35770	1.24736
1.34540	1.11378	1.35160	1.18111	1.35780	1.24844
1.34550	1.11486	1.35170	1.18220	1.35790	1.24953
1.34560	1.11595	1.35180	1.18328	1.35800	1.25062
1.34570	1.11704	1.35190	1.18437	1.35810	1.25170
1.34580	1.11812	1.35200	1.18546	1.35820	1.25279
1.34590	1.11921	1.35210	1.18654	1.35830	1.25387
1.34600	1.12029	1.35220	1.18763	1.35840	1.25496
1.34610	1.12138	1.35230	1.18871	1.35850	1.25605
1.34620	1.12247	1.35240	1.18980	1.35860	1.25713
1.34630	1.12355	1.35250	1.19089	1.35870	1.25822
1.34640	1.12464	1.35260	1.19197	1.35880	1.25930
1.34650	1.12572	1.35270	1.19306	1.35890	1.26039

Refractive Index (*Sodium* D Line, 25°C)	Density (g/cm³)	Refractive Index (*Sodium* D Line, 25°C)	Density (g/cm³)	Refractive Index (*Sodium* D Line, 25°C)	Density (g/cm³)
1.34660	1.12681	1.35280	1.19414	1.35900	1.26148
1.34670	1.12790	1.35290	1.19523	1.35910	1.26256
1.34680	1.12898	1.35300	1.19632	1.35920	1.26365
1.34690	1.13007	1.35310	1.19740	1.35930	1.26473
1.34700	1.13115	1.35320	1.19849	1.35940	1.26582
1.34710	1.13224	1.35330	1.19957	1.35950	1.26691
1.34720	1.13333	1.35340	1.20066	1.35960	1.26799
1.24730	1.13441	1.35350	1.20175	1.35970	1.26908
1.34740	1.13550	1.35360	1.20283	1.35980	1.27016
1.34750	1.13658	1.35370	1.20392	1.35990	1.27125
1.34760	1.13767	1.35380	1.20500	1.36000	1.27234
1.34770	1.13876	1.35390	1.20609	1.36010	1.27342
1.34780	1.13984	1.35400	1.20718	1.36020	1.27451
1.34790	1.14093	1.35410	1.20826	1.36030	1.27559
1.34800	1.14201	1.35420	1.20935	1.36040	1.27668
1.34810	1.14310	1.35430	1.21043	1.36050	1.27777
1.34820	1.14419	1.35440	1.21152	1.36060	1.27885
1.34830	1.14527	1.35450	1.21261	1.36070	1.27994
1.34840	1.14636	1.35460	1.21369	1.36080	1.28102
1.34850	1.14744	1.35470	1.21478	1.36090	1.28211
1.34860	1.14853	1.35480	1.21586	1.36100	1.28320
1.34870	1.14962	1.35490	1.21695	1.36110	1.28428
1.34880	1.15070	1.35500	1.21804	1.36120	1.28537
1.34890	1.15179	1.35510	1.21912	1.36130	1.28645
1.34900	1.15287	1.35520	1.22021	1.36140	1.28754
1.34910	1.15396	1.35530	1.22129	1.36150	1.28863
1.34920	1.15505	1.35540	1.22238	1.36160	1.28971
1.34930	1.15613	1.35550	1.22347	1.36170	1.29080
1.34940	1.15722	1.35560	1.22455	1.36180	1.29188
1.34950	1.15830	1.35570	1.22564	1.36190	1.29297
1.34960	1.15939	1.35580	1.22672	1.36200	1.29406
1.34970	1.16048	1.35590	1.22781	1.36210	1.29514
1.34980	1.16156	1.35600	1.22890	1.36220	1.29623
1.34990	1.16265	1.35610	1.22998	1.36230	1.29731
1.35000	1.16373	1.35620	1.23107	1.36240	1.29840
1.35010	1.16482	1.35630	1.23215	1.36250	1.29949
1.36260	1.30057	1.36900	1.37008	1.37550	1.44067
1.36270	1.30166	1.36910	1.37116	1.37560	1.44175
1.36280	1.30274	1.36920	1.37225	1.37570	1.44284

Refractive Index (*Sodium* D *Line,* 25°C)	Density (g/cm³)	Refractive Index (*Sodium* D *Line,* 25°C)	Density (g/cm³)	Refractive Index (*Sodium* D *Line,* 25°C)	Density (g/cm³)
1.36290	1.30383	1.36930	1.37333	1.37580	1.44393
1.36300	1.30492	1.36940	1.37442	1.37590	1.44501
1.36310	1.30600	1.36950	1.37551	1.37600	1.44610
1.36320	1.30709	1.36960	1.37659	1.37610	1.44718
1.36330	1.30817	1.36970	1.37768	1.37620	1.44827
1.36340	1.30926	1.36980	1.37876	1.37630	1.44936
1.36350	1.31035	1.36990	1.37985	1.37640	1.45044
1.36360	1.31143	1.37000	1.38094	1.37650	1.45153
1.36370	1.31252	1.37010	1.38202	1.37660	1.45261
1.36380	1.31360	1.37020	1.38311	1.37670	1.45370
1.36390	1.31469	1.37030	1.38420	1.37680	1.45479
1.36400	1.31578	1.37040	1.38528	1.37690	1.45587
1.36410	1.31686	1.37050	1.38637	1.37700	1.45696
1.36420	1.31795	1.37060	1.38745	1.37710	1.45804
1.36430	1.31903	1.37070	1.38854	1.37720	1.45913
1.36440	1.32012	1.37080	1.38963	1.37730	1.46022
1.36450	1.32121	1.37090	1.39071	1.37740	1.46130
1.36460	1.32229	1.37100	1.39180	1.37750	1.46239
1.36470	1.32338	1.37110	1.39288	1.37760	1.46347
1.36480	1.32446	1.37120	1.39397	1.37770	1.46456
1.36490	1.32555	1.37130	1.39506	1.37780	1.46565
1.36500	1.32664	1.37140	1.39614	1.37790	1.46673
1.36510	1.32772	1.37150	1.39723	1.37800	1.46782
1.36520	1.32881	1.37160	1.39831	1.37810	1.46890
1.36530	1.32989	1.37170	1.39940	1.37820	1.46999
1.36540	1.33098	1.37180	1.40049	1.37830	1.47108
1.36550	1.33207	1.37190	1.40157	1.37840	1.47216
1.36560	1.33315	1.37200	1.40266	1.37850	1.47325
1.36570	1.33424	1.37210	1.40374	1.37860	1.47433
1.36580	1.33532	1.37220	1.40483	1.37870	1.47542
1.36590	1.33641	1.37230	1.40592	1.37880	1.47651
1.36600	1.33750	1.37240	1.40700	1.37890	1.47759
1.36610	1.33858	1.37250	1.40809	1.37900	1.47868
1.36620	1.33967	1.37260	1.40917	1.37910	1.47976
1.36630	1.34075	1.37270	1.41026	1.37920	1.48085
1.36640	1.34184	1.37280	1.41135	1.37930	1.48194
1.36650	1.34293	1.37290	1.41243	1.37940	1.48302
1.36660	1.34401	1.37300	1.41352	1.37950	1.48411
1.36670	1.34510	1.37310	1.41460	1.37960	1.48519

Refractive Index (Sodium D Line, 25°C)	Density (g/cm³)	Refractive Index (Sodium D Line, 25°C)	Density (g/cm³)	Refractive Index (Sodium D Line, 25°C)	Density (g/cm³)
1.36680	1.34618	1.37320	1.41569	1.37970	1.48628
1.36690	1.34727	1.37330	1.41678	1.37980	1.48737
1.36700	1.34836	1.37340	1.41786	1.37990	1.48845
1.36710	1.34944	1.37350	1.41895	1.38000	1.48954
1.36720	1.35053	1.37360	1.42003	1.38010	1.49062
1.36730	1.35161	1.37370	1.42112	1.38020	1.49171
1.36740	1.35270	1.37380	1.42221	1.38030	1.49280
1.36750	1.35379	1.37390	1.42329	1.38040	1.49388
1.36760	1.35487	1.37400	1.42438	1.38050	1.49497
1.36770	1.35596	1.37410	1.42546	1.38060	1.49605
1.36780	1.35704	1.37420	1.42655	1.38070	1.49714
1.36790	1.35813	1.37430	1.42764	1.38080	1.49823
1.36800	1.35922	1.37440	1.42872	1.38090	1.49931
1.36810	1.36030	1.37450	1.42981	1.38100	1.50040
1.36820	1.36139	1.37460	1.43089	1.38110	1.50148
1.36830	1.36247	1.37470	1.43198	1.38120	1.50257
1.36840	1.36356	1.37480	1.43307	1.38130	1.50366
1.36850	1.36465	1.37490	1.43415	1.38140	1.50474
1.36860	1.36573	1.37500	1.43524	1.38150	1.50583
1.36870	1.36682	1.37510	1.43632	1.38160	1.50691
1.36880	1.36790	1.37520	1.43741	1.38170	1.50800
1.36890	1.36899	1.37530	1.43850	1.38180	1.50909
		1.37540	1.43958	1.38190	1.51017
1.38200	1.51126	1.38850	1.58185	1.39500	1.65244
1.38210	1.51234	1.38860	1.58293	1.39510	1.65353
1.38220	1.51343	1.38870	1.58402	1.39520	1.65461
1.38230	1.51452	1.38880	1.58511	1.39530	1.65570
1.38240	1.51560	1.38890	1.58619	1.39540	1.65678
1.38250	1.51669	1.38900	1.58728	1.39550	1.65787
1.38260	1.51777	1.38910	1.58836	1.39560	1.65896
1.38270	1.51886	1.38920	1.58945	1.39570	1.66004
1.38280	1.51995	1.38930	1.59054	1.39580	1.66113
1.38290	1.52103	1.38940	1.59162	1.39590	1.66221
1.38300	1.52212	1.38950	1.59271	1.39600	1.66330
1.38310	1.52320	1.38960	1.59379	1.39610	1.66439
1.38320	1.52429	1.38970	1.59488	1.39620	1.66547
1.38330	1.52538	1.38980	1.59597	1.39630	1.66656
1.38340	1.52646	1.38990	1.59705	1.39640	1.66764
1.38350	1.52755	1.39000	1.59814	1.39650	1.66873

Refractive Index (Sodium D Line, 25°C)	Density (g/cm³)	Refractive Index (Sodium D Line, 25°C)	Density (g/cm³)	Refractive Index (Sodium D Line, 25°C)	Density (g/cm³)
1.38360	1.52863	1.39010	1.59923	1.39660	1.66982
1.38370	1.52972	1.39020	1.60031	1.39670	1.67090
1.38380	1.53081	1.39030	1.60140	1.39680	1.67199
1.38390	1.53189	1.39040	1.60248	1.39690	1.67307
1.38400	1.53298	1.39050	1.60357	1.39700	1.67416
1.38410	1.53406	1.39060	1.60466	1.39710	1.67525
1.38420	1.53515	1.39070	1.60574	1.39720	1.67633
1.38430	1.53624	1.39080	1.60683	1.39730	1.67742
1.38440	1.53732	1.39090	1.60791	1.39740	1.67850
1.38450	1.53841	1.39100	1.60900	1.39750	1.67959
1.38460	1.53949	1.39110	1.61009	1.39760	1.68068
1.38470	1.54058	1.39120	1.61117	1.29770	1.68176
1.38480	1.54167	1.29130	1.61226	1.39780	1.68285
1.38490	1.54275	1.39140	1.61334	1.39790	1.68393
1.38500	1.54384	1.39150	1.61443	1.39800	1.68502
1.38510	1.54492	1.39160	1.61552	1.39810	1.68611
1.38520	1.54601	1.39170	1.61660	1.39820	1.68719
1.38530	1.54710	1.39180	1.61769	1.39830	1.68828
1.38540	1.54818	1.39190	1.61877	1.39840	1.68936
1.38550	1.54927	1.39200	1.61986	1.39850	1.69045
1.38560	1.55035	1.39210	1.62095	1.39860	1.69154
1.38570	1.55144	1.39220	1.62203	1.39870	1.69262
1.38580	1.55253	1.39230	1.62312	1.39880	1.69371
1.38590	1.55361	1.39240	1.62420	1.39890	1.69479
1.38600	1.55470	1.39250	1.62529	1.39900	1.69588
1.38610	1.55578	1.39260	1.62638	1.39910	1.69697
1.38620	1.55687	1.39270	1.62746	1.39920	1.69805
1.38630	1.55796	1.39280	1.62855	1.39930	1.69914
1.38640	1.55904	1.39290	1.62963	1.39940	1.70022
1.38650	1.56013	1.39300	1.63072	1.39950	1.70131
1.38660	1.56121	1.39310	1.63181	1.39960	1.70240
1.38670	1.56230	1.39320	1.63289	1.39970	1.70348
1.38680	1.56339	1.39330	1.63398	1.39980	1.70457
1.38690	1.56447	1.39340	1.63506	1.39990	1.70565
1.38700	1.56556	1.39350	1.63615	1.40000	1.70674
1.38710	1.56664	1.39360	1.63724	1.40010	1.70783
1.38720	1.56773	1.39370	1.63832	1.40020	1.70891
1.38730	1.56882	1.39380	1.63941	1.40030	1.71000
1.38740	1.56990	1.39390	1.64049	1.40040	1.71108

Appendix B *(Continued)*

Refractive Index (Sodium D Line, 25°C)	Density (g/cm³)	Refractive Index (Sodium D Line, 25°C)	Density (g/cm³)	Refractive Index (Sodium D Line, 25°C)	Density (g/cm³)
1.38750	1.57099	1.39400	1.64158	1.40050	1.71217
1.38760	1.57207	1.39410	1.64267	1.40060	1.71326
1.38770	1.57316	1.39420	1.64375	1.40070	1.71434
1.38780	1.57425	1.39430	1.64484	1.40080	1.71543
1.38790	1.57533	1.39440	1.64592	1.40090	1.71651
1.38800	1.57642	1.39450	1.64701	1.40100	1.71760
1.38810	1.57750	1.39460	1.64810	1.40110	1.71869
1.38820	1.57859	1.39470	1.64918	1.40120	1.71977
1.38830	1.57968	1.39480	1.65027	1.40130	1.72086
1.38840	1.58076	1.39490	1.65135	1.40140	1.72194
1.40150	1.72303	1.40800	1.79362	1.41450	1.86421
1.40160	1.72412	1.40810	1.79471	1.41460	1.86530
1.40170	1.72520	1.40820	1.79579	1.41470	1.86638
1.40180	1.72629	1.40830	1.79688	1.41480	1.86747
1.40190	1.72737	1.40840	1.79796	1.41490	1.86856
1.40200	1.72846	1.40850	1.79905	1.41500	1.86964
1.40210	1.72955	1.40860	1.80014	1.41510	1.87073
1.40220	1.73063	1.40870	1.80122	1.41520	1.87181
1.40230	1.73172	1.40880	1.80231	1.41530	1.87290
1.40240	1.73280	1.40890	1.80339	1.41540	1.87399
1.40250	1.73389	1.40900	1.80448	1.41550	1.87507
1.40260	1.73498	1.40910	1.80557	1.41560	1.87616
1.40270	1.73606	1.40920	1.80665	1.41570	1.87724
1.40280	1.73715	1.40930	1.80774	1.41580	1.87833
1.40290	1.73823	1.40940	1.80882	1.41590	1.87942
1.40300	1.73932	1.40950	1.80991	1.41600	1.88050
1.40310	1.74041	1.40960	1.81100	1.41610	1.88159
1.40320	1.74149	1.40970	1.81208	1.41620	1.88267
1.40330	1.74258	1.40980	1.81317	1.41630	1.88376
1.40340	1.74366	1.40990	1.81425	1.41640	1.88485
1.40350	1.74475	1.41000	1.81534	1.41650	1.88593
1.40360	1.74584	1.41010	1.81643	1.41660	1.88702
1.40370	1.74692	1.41020	1.81751	1.41670	1.88810
1.40380	1.74801	1.41030	1.81860	1.41680	1.88919
1.40390	1.74909	1.41040	1.81969	1.41690	1.89028
1.40400	1.75018	1.41050	1.82077	1.41700	1.89136
1.40410	1.75127	1.41060	1.82186	1.41710	1.89245
1.40420	1.75235	1.41070	1.82294	1.41720	1.89353
1.40430	1.75344	1.41080	1.82403	1.41730	1.89462

Refractive Index (Sodium D Line, 25°C)	Density (g/cm³)	Refractive Index (Sodium D Line, 25°C)	Density (g/cm³)	Refractive Index (Sodium D Line, 25°C)	Density (g/cm³)
1.40440	1.75452	1.41090	1.82512	1.41740	1.89571
1.40450	1.75561	1.41100	1.82620	1.41750	1.89679
1.40460	1.75670	1.41110	1.82729	1.41760	1.89788
1.40470	1.75778	1.41120	1.82837	1.41770	1.89896
1.40480	1.75887	1.41130	1.82946	1.41780	1.90005
1.40490	1.75995	1.41140	1.83055	1.41790	1.90114
1.40500	1.76104	1.41150	1.83163	1.41800	1.90222
1.40510	1.76213	1.41160	1.83272	1.41810	1.90331
1.40520	1.76321	1.41170	1.83380	1.41820	1.90439
1.40530	1.76430	1.41180	1.83489	1.41830	1.90548
1.40540	1.76538	1.41190	1.83598	1.41840	1.90657
1.40550	1.76647	1.41200	1.83706	1.41850	1.90765
1.40560	1.76756	1.41210	1.83815	1.41860	1.90874
1.40570	1.76864	1.41220	1.83923	1.41870	1.90982
1.40580	1.76973	1.41230	1.84032	1.41880	1.91091
1.40590	1.77081	1.41240	1.84141	1.41890	1.91200
1.40600	1.77190	1.41250	1.84249	1.41900	1.91308
1.40610	1.77299	1.41260	1.84358	1.41910	1.91417
1.40620	1.77407	1.41270	1.84466	1.41920	1.91525
1.40630	1.77516	1.41280	1.84575	1.41930	1.91634
1.40640	1.77624	1.41290	1.84684	1.41940	1.91743
1.40650	1.77733	1.41300	1.84792	1.41950	1.91851
1.40660	1.77842	1.41310	1.84901	1.41960	1.91960
1.40670	1.77950	1.41320	1.85009	1.41970	1.92068
1.40680	1.78059	1.41330	1.85118	1.41980	1.92177
1.40690	1.78167	1.41340	1.85227	1.41990	1.92286
1.40700	1.78276	1.41350	1.85335	1.42000	1.92394
1.40710	1.78385	1.41360	1.85444	1.42010	1.92503
1.40720	1.78493	1.41370	1.85552	1.42020	1.92611
1.40730	1.78602	1.41380	1.85661	1.42030	1.92720
1.40740	1.78710	1.41390	1.85770	1.42040	1.92829
1.40750	1.78819	1.41400	1.85878	1.42050	1.92937
1.40760	1.78928	1.41410	1.85987	1.42060	1.93046
1.40770	1.79036	1.41420	1.86095	1.42070	1.93154
1.40780	1.79145	1.41430	1.86204	1.42080	1.93263
1.40790	1.79253	1.41440	1.86313	1.42090	1.93372
1.42100	1.93480	1.42120	1.93697	1.42140	1.93915
1.42110	1.93589	1.42130	1.93806		

Source: Compiled by Norman G. Anderson and Norman L. Anderson. Calculated from the equation $\rho^{25.0\,°C} = (10.8601 \times R.I.) - 13.4974$ of J. B. Ifft, D. H. Voet, and J. Vinograd, *J. Phys. Chem.*, *65*, 1138 (1961). *R.I.* is the refractive index (Sodium *D* line. 25°C). Reproduced from "Useful Data for Zonal Centrifugation," C. T. Rankin, Jr., and L. T. Elrod, The Molecular Anatomy Program, Oak Ridge National Laboratory, January 1970.

Appendix **C**

GRADIENT PROPERTIES OF FICOLL

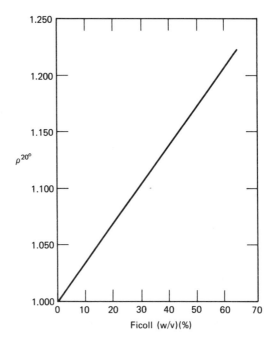

Fig. C.1 Density of Ficoll solutions. Figure obtained from Pharmacia Fine Chemicals, Inc. (Reproduced from "Useful Data for Zonal Centrifugation," by C. T. Rankin, Jr., and L. T. Elrod, The Molecular Anatomy Program, Oak Ridge National Laboratory, January 1970.)

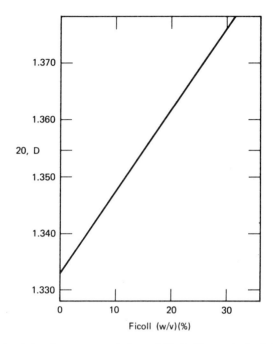

Fig. C.2 Refractive index for aqueous solutions of Ficoll. Figure obtained from Pharmacia Fine Chemicals, Inc.

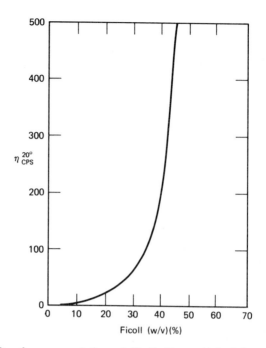

Fig. C.3 Viscosity of aqueous solutions of Ficoll. Figure obtained from Pharmacia Fine Chemicals, Inc. [T. G. Pretlow, C. W. Boone, R. I. Shrager, and G. H. Weiss, *Anal. Biochem.*, **29** (1969).]

Appendix **D**

REFERENCES ON EXPERIMENTAL PROCEDURES OF BIOPARTICLE SEPARATIONS OR PURIFICATIONS

VIRUSES

Anderson, N. G., "Centrifugal Methods for Large-Scale Virus Vaccine Purification," in *Proceedings of the Xth International Congress of the Permanent Section of Microbiological Standardization,* Prague, September 19–23, 1967 (Abstract).

Anderson, N. G., and G. B. Cline, "New Centrifugal Methods for Virus Isolation," In *Methods in Virology,* Vol. II, K. Maramorosch and H. Koprowski, Eds., Academic, New York, 1967, p. 137.

Anderson, N. G., G. B. Cline, W. W. Harris, and J. G. Green, "Isolation of Viral Particles from Large Fluid Volumes," In *Transmission of Viruses by the Water Route,* G. Berg, Ed., Interscience, New York, 1967, p. 75.

Anderson, N. G., D. A. Waters, C. E. Nunley, R. F. Gibson, R. M. Schilling, E. C. Denny, G. B. Cline, E. F. Babelay, and T. E. Perardi, *Anal. Biochem.,* **32,** 460 (1969).

Bond, H. E., and W. T. Hall, *J. Infect. Dis.,* **125,** 263 (1972).

Breillatt, J. P., J. N. Brantley, H. M. Mazzone, M. E. Martignoni, J. E. Franklin, and N. G. Anderson, *Am. Soc. Microbiol.,* **23,** 923 (1972).

Bruening, G., *Virology,* **37,** 577 (1969).

Bruening, G., and H. O. Agrawal, *Virology,* **32,** 306 (1967).

Cline, G. B., H. Coates, N. G. Anderson, R. M. Chanock, and W. W. Harris, *J. Virol.,* **1,** 659 (1967).

Dostal, V., *Z. Immunitätsforsch.,* **135,** 55 (1968).

Gerin, J. L., and N. G. Anderson, *Nature,* **221,** 1255 (1969).

Gerin, J. L., R. M. Faust, and P. V. Holland, *J. Immunol.,* **115,** 100 (1975).

Gerin, J. L., W. R. Richter, J. D. Fenters, and J. C. Holper, *J. Virol.,* **2,** 937 (1968).

Kado, C. I., and C. A. Knight, *J. Mol. Biol.,* **36,** 15 (1968).

Martignoni, M., J. P. Breillatt, and N. G. Anderson, *J. Invert. Pathol.,* **2,** 507 (1968).

Martignoni, M. E., and P. J. Iwai, *J. Invert. Pathol.,* **18,** 219 (1971).

McCombs, R. M., *Appl. Microbiol.,* **17,** 636 (1969).

451

Mizell, M., I. Toplin, and J. J. Isaacs, *Science,* **165,** 1134 (1969).

Murakami, W. T., R. Fine, M. R. Harrington, and Z. Ben Sassan, *J. Mol. Biol.,* **36,** 153 (1968).

Newell, G. R., W. W. Harris, K. O. Bowman, C. W. Boone, and N. G. Anderson, *N. Engl. J. Med.,* **278,** 1185 (1968).

Okuda, K., K. Itoh, K. Miyake, M. Morita, M. Ogonuki, and S. Matsui, *J. Clin. Microbiol.,* **1,** 96 (1975).

Reimer, C. B., R. S. Baker, R. M. van Frank, T. E. Newlin, G. B. Cline, and N. G. Anderson, *J. Virol.,* **1,** 1207 (1967).

Stackpole, C. W., and M. Mizell, *Virology,* **36,** 63 (1968).

Toplin, I., *Appl. Microbiol.,* **15,** 582 (1967).

Toplin, I., R. Boyden, A. De Padova, P. Brandt, and P. Sottong, *Biotechnol. Bioeng.,* **10,** 651 (1968).

Van Frank, R. M., L. F. Ellis, and W. J. Kleinschmidt, *Fed. Proc.,* **28,** 816 (1969).

MEMBRANES

Anderson, N. G., A. I. Lansing, I. Lieberman, C. T. Rankin, and H. Elrod, "Isolation of Rat Liver Cell Membranes by Combined Rate and Isopycnic Zonal Centrifugation," in *Wistar Inst. Symp. Monograph 8,* Philadelphia, 1968, p. 23.

Cotman, C., H. R. Mahler, and N. G. Anderson, *Biochim. Biophys. Acta,* **163,** 272 (1968).

Cotman, C. W., H. R. Mahler, and T. E. Hugli, *Arch. Biochem. Biophys.,* **126,** 821 (1968).

Davis, D. C., H. G. Wilcox, G. Dishman, and M. Heimberg, *Fed. Proc.,* **28,** 848 (1969).

Fleischer, B., S. Fleischer, and H. Ozawa, *J. Cell. Biol.,* **43,** 59 (1969).

Gavard, D., G. deLamirande, and S. Karasaki, *Biochim. Biophys. Acta,* **332,** 145 (1974).

Pfleger, R. C., N. G. Anderson, and F. Snyder, *Biochemistry,* **7,** 2826 (1968).

Quigley, J. W., and S. S. Cohen, *J. Biol. Chem.,* **244,** 2450 (1969).

Spenney, J. G., A. Strych, A. H. Price, H. F. Helander, and G. Sachs, *Biochem. Biophys. Acta,* **311,** 545 (1973).

Vergara, J., F. Zambrano, J. D. Robertson, and H. Elrod, *J. Cell Biol.,* **61,** 83 (1974).

Weaver, R. A., and W. Boyle, *Biochim. Biophys. Acta,* **173,** 377 (1969).

DNA AND RNA

Cech, T. R., G. Wiesehahn, and J. E. Hearst, *Biochemistry,* **15,** 1865 (1976).

Epler, J. L., *Biochemistry,* **8,** 2285 (1969).

Halsall, H. G., and V. N. Schumaker, *Nature,* **221,** 772 (1969).

Jaenisch, R., P. H. Hofschneider, and A. Preuss, *Biochim. Biophys. Acta,* **190,** 88 (1969).

Legerski, R. J., and H. B. Gray, Jr., *Biochem. Biophys. Acta,* **442,** 129 (1976).

Moriyama, Y., J. L. Hodnett, A. W. Prestayko, and H. Bush, *J. Mol. Biol.,* **39,** 335 (1969).

Rudland, P. S., and S. K. Dube, *J. Mol. Biol.,* **43,** 273 (1969).

Wheeler, K. T., J. D. Linn, R. Franklin, and E. L. Pautler, *Anal. Biochem.,* **64,** 329 (1975).

Wiesehahn, G., T. R. Cech, and J. E. Hearst, *Biopolymers,* **15,** 1591 (1976).

Wolf, H., *Anal. Biochem.,* **68,** 505 (1975).

LIPOPROTEINS

Danielsson, B., B. G. Johansson, and B. G. Peterson, *Clin. Chim. Acta,* **47,** 365 (1973).

Heinstein, P. F., and P. K. Stumpf, *J. Biol. Chem.,* **244,** 5374 (1969).

Patsch, W., J. R. Patsch, G. M. Kostner, S. Sailer, and H. Braunsteiner, *J. Biol. Chem.,* **253,** 4911 (1978).

Pfleger, R. C., C. Piantadosi, and F. Synder, *Biochim. Biophys. Acta,* **144,** 633 (1967).

Redgrave, T. G., D. C. K. Roberts, and C. E. West, *Anal. Biochem.,* **65,** 42 (1975).

Viikari, J., E. Haahti, T. T. Helela, K. Juva, and T. Nikkari, *Scand. J. Clin. Lab. Invest.,* **23,** 85 (1969).

RIBOSOMES AND SUBUNITS

Birnie, G. D., Ed., *Subcellular Components: Preparation and Fractionation,* 2nd ed., Butterworth, London, and University Park Press, Baltimore, 1972.

Bonanou, S., R. A. Cox, B. Higginson, and K. Kanagalingam, *Biochem. J.,* **110,** 87 (1968).

Cline, G. B., G. L. Whitson, and B. H. Levedahl, *Biochem. Biophys. Res. Commun.,* **30,** 534 (1968).

Cartledge, T. G., L. Howells, and D. Lloyd, *Biochem. J.,* **116,** 40p (1970).

Fogel, S., and P. S. Sypherd, *J. Bacteriol.,* **96,** 358 (1968).

Ghosh, H. P., and H. G. Khorana, *Proc. Natl. Acad. Sci.,* **58,** 2455 (1967).

Goodenough, D. A., *J. Cell Biol.,* **61,** 557 (1974).

Hamilton, M. G., and M. E. Ruth, *Biochemistry,* **8,** 851 (1969).

Hill, W. E., G. P. Rossetti, and K. E. Van Holde, *J. Mol. Biol.,* **44,** 263 (1969).

Hill, W. E., J. D. Thompson, and J. W. Anderegg, *J. Mol. Biol.,* **44,** 89 (1969).

Kornguth, S. E., A. Flangas, J. Perrin, R. Geison, and G. Scott, *Prep. Biochem.,* **2,** 167 (1972).

Mendiola, L. R., A. P. Hirvonen, A. Kovacs, and C. A. Price, *Fed. Proc.*, **28**, 879 (1969).

Moore, P. B., R. R. Traut, H. Noller, P. Pearson, and H. Delius, *J. Mol. Biol.*, **31**, 441 (1968).

Pollack, M. S., and C. A. Price, *Anal. Biochem.*, **42**, 38 (1971).

Reboud, A. M., M. G. Hamilton, and M. L. Petermann, *Biochemistry*, **8**, 843 (1969).

Thomas, M. L., J. W. Clark, Jr., G. B. Cline, N. G. Anderson and H. Russell, *Appl. Microbiol.*, **23**, 714 (1972).

Whittaker, V. P., W. B. Essman, and G. H. C. Dowe, *Biochem. J.*, **128**, 833 (1972).

Wikman, J., E. Howard, and H. Busch, *J. Biol. Chem.*, **244**, 5471 (1969).

NUCLEI

Elrod, L. H., and N. G. Anderson, *J. Tenn. Acad. Sci.*, **42**, 52 (1967) (Abstract).

Elrod, L. H., N. G. Anderson, L. C. Patrick, and J. C. Shinpaugh, *J. Tenn. Acad. Sci.*, **43**, 48 (1967).

Elrod, L. H., L. C. Patrick, and N. G. Anderson, *Anal. Biochem.*, **30**, 230 (1969).

Enea, V., and N. D. Zinder, *Science*, **190**, 584 (1975).

Johnston, I. R., A. P. Mathias, F. Pennington, and D. Ridge, *Nature*, **220**, 668 (1968).

Soeiro, R., "Purification of Nucleoli and Ribonucleoprotein Particles by Isopycnic Banding in Renografin," in *Techniques of Biochemical and Biophysical Morphology*, Vol. 3, D. Glick, and R. M. Rosenbaum, Eds., Wiley, New York, 1977, pp. 45-57.

WHOLE CELLS

Baggiolini, M., J. G. Hirsch, and C. de Duve, *J. Cell. Biol.*, **40**, 529 (1969).

Binkley, F., and N. L. King, *Biochem. Biophys. Res. Commun.*, **33**, 99 (1968).

Binkley, F., N. King, E. Milikin, R. K. Wright, C. H. O'Neal, and I. J. Wundram, *Science*, **162**, 1009 (1968).

Flangas, A. L., *Prep. Biochem.*, **4**, 165 (1974).

Fisher, W. D., H. I. Adler, F. W. Shull, Jr., and A. Cohen, *J. Bacteriol.*, **97**, 500 (1969).

Grabske, R. J., S. Lake, B. L. Gledhill, and M. L. Meistrich, *J. Cell Biol.*, **63**, 119a (1974).

Griffith, O. M., and H. Wright, *Anal. Biochem.*, **47**, 575 (1972).

Griffith, O. M., *Anal. Biochem.*, **87**, 97 (1978).

Mahaley, M. S., Jr., E. D. Day, N. G. Anderson, R. F. Wilfong, and C. Brater, *Cancer Res.*, **28**, 1783 (1968).

McEwen, C. R., E. Th. Juhos, R. W. Stallard, J. V. Schnell, W. A. Siddiqui, and Q. M. Geiman, *J. Parasitol.*, **57**, 887 (1971).

Pine, L., R. G. Falcone, and C. J. Boone, *Mycopathol. Mycol. Appl.*, **37**, 1 (1969).

Sanderson, R. J., N. F. Palmer, and K. E. Bird, *Biophys. J.*, **15**, 321a (1975).

Vasconcelos, A., M. S. Pollack, L. R. Mendiola, H. P. Hoffmann, D. H. Brown, and C. A. Price, *Plant Physiol.*, **47**, 217 (1971).

MITOCHONDRIA, LYSOSOMES, AND PEROXISOMES

Barnett, W. E., and D. H. Brown, *Proc. Natl. Acad. Sci.*, **57**, 452 (1967).

Brightwell, R., D. Lloyd, G. Turner, and S. E. Venables, *Biochem. J.*, **109**, 42p (1968).

Brown, D. H., *Biochim. Biophys. Acta*, **162**, 152 (1968).

Canonico, P. G., and J. W. C. Bird, *J. Cell Biol.*, **43**, 367 (1969).

Cannock, M. J., P. R. Kirk, and A. P. Sturdee, *J. Cell Biol.*, **61**, 123 (1974).

Corbett, J. R., *Biochem. J.*, **102**, 43 (1967).

Müller, M., J. F. Hogg, and C. de Duve, *J. Biol. Chem.*, **243**, 5385 (1968).

Poole, B., T. Higashi, and C. deDuve, *J. Cell Biol.*, **45**, 408 (1970).

Windhorst, D. B., J. G. White, A. S. Zelickson, C. C. Clawson, P. B. Dent, B. Pollara, and R. A. Good, *Ann. N.Y. Acad. Sci.*, **155**, 818 (1968).

Woodward, D. O., *Fed. Proc.*, **27** (2), 1167 (1968).

MISCELLANEOUS

Barber, M. L. S., and R. E. Canning, *Nat. Cancer Inst. Monogr.*, **21**, 345 (1966).

Colten, H. R., H. E. Bond, T. Borsos, and H. J. Rapp, *J. Immunol.*, **103**, 862 (1969).

Colten, H. R., T. Borsos, and H. J. Rapp, *J. Immunol.*, **100**, 799 (1968).

Gerin, J. L., P. V. Holland, and R. H. Purcell, *J. Virol.*, **7**, 569 (1971).

Hinton, R. H., E. Klucis, A. A. El-Aaser, J. T. R. Fitzsimons, P. Alexander, and E. Reid, *Biochem. J.*, **105**, 14P (1967).

Jorgensen, P. L., J. C. Skou, and L. P. Solomonson, *Biochim. Biophys. Acta*, **233**, 381 (1971).

Kaplan, N. O., "Beef Heart TPNH-DPN Pyridine Nucleotide Transhydrogenases," in *Methods in Enzymology*, Vol. X, S. P. Colowick and N. O. Kaplan, Eds., Academic, New York, 1967, p. 317.

Lee, T. C., D. C. Swartzendruber, F. Snyder, *Biochem. Biophys. Res. Commun.*, **36** (5), 748 (1969).

Murdock, D. D., E. Katona, and M. A. Moscarello, *Can. J. Biochem.*, **47**, 818 (1969).

Pollara, B., A. Suran, J. Finstad, and R. A. Good, *Proc. Natl. Acad. Sci.,* **59,** 1307 (1968).

Popp, R. A., D. M. Popp, N. G. Anderson, and L. H. Elrod, *Biochim. Biophys. Acta,* **184,** 625 (1969).

Rankin, C. T., N. G. Anderson, and W. L. Rasmussen, *Oak Ridge National Laboratory Report No. 4171-Special,* 1967, pp. 65–72.

Rickwood, D., Ed., "Biological Separations in Iodinated Density-Gradient Media," *Colloquium on Use of Iodinated Density-Gradient Media for Biological Separation, Univ. of Glassgow,* Information Retrieval Ltd., London, 1976.

Tessier, H. M. Yaguchi, and D. Rose, *J. Dairy Sci.,* **52** (3), 139 (1969).

Wolf, A., *Transplantation,* **7,** 49 (1969).

AUTHOR INDEX

Parker, H. M., 87, 90, 110
Patterson, J. W., 304(132), 314
Pattyn, S. R., 340(50), 378
Paul, J., 353(58), 378
Pederson, K. O., 23, 134(11), 143(11), 182(11), 216, 217, 239, 244(43), 268, 312
Penman, S., 353(61), 378
Pennington, F., 363(82), 369(82), 379
Penton, J. R., 301(117), 314
Persign, J. P., 297(108), 313
Peters, D. G., 279(79), 282(79), 312
Pfander, W. H., 303(121), 314
Pham, H. D., 193, 218
Philips, R. W., 296(105), 297(105), 313
Pickels, E. G., 209(63), 218
Pigford, T. H., 50, 51
Pivnick, H., 303(122), 314
Plotz, C. M., 257(63, 66), 312
Pontryagin, L. S., 152(29), 217
Powell, M. E. A., 359(74), 379
Prader, E., 257(63), 312
Pressman, B. C., 357(65), 379
Pretlow, T. G., 146, 338(31), 377
Price, C. A., 159(38), 217
Pringle, C. R., 340(49), 378
Prigogine, I., 32(12), 35(12), 48
Proudman, J., 103, 110
Putnam, H. D., 304(132), 314

Rankin, C. T., 119, 176, 209(66), 210(66), 347(53), 378
Raynaud, A., 338(26), 377
Redgrave, T. G., 371(87), 379
Rees, A., 340(46), 378
Reid, E., 337(21), 358(21), 377
Reimer, C. B., 209, 340(51), 379
Reithorst, A., 297(108), 313
Rickwood, D., 339(37, 41, 42, 43, 45), 378
Ridge, E., 363(82), 369(82), 379
Robbins, J. B., 293(93), 313
Roberts, D. C. K., 371(87), 379
Rolsten, C., 296(104), 297(104), 313
Rooydn, D. B., 358(67), 379
Roseland, P. A., 305(136, 137), 314
Rosenblum, J., 146, 214

St. John, P. A., 286(81), 312
Sakamoto, T., 337(25), 377
Samorajski, T., 296(104), 297(104), 313
Sanchez, R., 333, 377
Santiago, R., 333, 377

Sarquis, J. L., 235(31), 254(51), 311
Sartory, W. K., 134(12), 146, 150, 154, 216
Savory, J., 293(89, 90, 91, 92), 313
Scatchard, G., 239(40, 41), 311
Schachman, H. K., 250, 311
Scheffler, I. E., 339(38), 378
Scheinberg, J. H., 293(93), 313
Schmidt, F. H., 296(103), 297(103), 313
Schneider, W. C., 357(64), 379
Schnittger, J. R., 409, 431
Schrager, R. I., 146(18), 217
Schuel, H., 358(69, 70), 379
Schuknecht, B., 303(119), 314
Schumaker, V. N., 137(22), 146(21), 146, 176, 192(22), 217, 250(46), 340(46), 378, 311
Schwartz, M. K., 295(102), 297(102, 109), 313
Schweder, J., 146(21), 217
Scott, C. D., 219(10, 12, 13, 14), 274, 275(71), 278(9, 12, 78), 287(12), 293(12), 299(116), 301(78), 303(116), 310, 312
Sellinger, O. Z., 338(35), 378
Senn, D. R., 303(125), 314
Serwer, P., 339(40), 378
Shah, V. S., 159(45), 217
Shaw, P. D., 303(124), 314
Shiga, S., 397(9), 404(9), 431
Shoji, H., 397(9), 404(9), 431
Shorten, P. E., 159(32), 217
Simmons, J. F., 159(32), 217
Sinclair, R., 337(23), 338(23), 377
Singer, J. M., 257, 261(65), 312
Sinsheimer, R. L., 333(11, 12), 377
Skarstrom, C., 65, 110
Skeggs, L. T., 359(72), 379
Slimming, T. K., 353(58), 378
Smith, J. C., 389(3), 431
Smith, L., 359(78), 379
Smith, M. J. H., 359(74), 379
Smyth, H. D., 1(2), 21, 51
Snyder, C. F., 170(51, 52), 172(51), 218
Spencer, R. D., 294(97), 313
Spooner, R. J., 304(137), 314
Spragg, S. P., 119, 216
Stahl, F. W., 255, 334, 312, 377
Stanley, W. M., 209(64), 218
Stede, L., 295(99), 313
Steenbeck, M., 50(5), 87, 109
Stevens, R. H., 219(8), 310
Striebich, M. J., 357(64), 379
Svedberg, T., 2, 23, 134, 143(10, 11), 147, 182, 210, 220, 244

SUBJECT INDEX